Jens Wittenburg

Dynamics of Multibody Systems

Jens Wittenburg

Dynamics of Multibody Systems

Second Edition

Springer

Professor Dr.-Ing. Jens Wittenburg
University of Karlsruhe (TH)
Institute of Engineering Mechanics
Kaiserstrasse 12
76128 Karlsruhe, Germany
Email: wittenburg@itm.uni-karlsruhe.de

Originally published under: Dynamics of Systems of Rigid Bodies, in the LAMM series, Teubner 1977

Library of Congress Control Number: 2007932332

ISBN 978-3-540-73913-5 Springer Berlin Heidelberg New York

Springer is a part of Springer Science+Business Media

springer.com

Typesetting and production: LE-TEX Jelonek, Schmidt & Vöckler GbR, Leipzig
Cover design: eStudio Calamar S.L., F. Steinen-Broo, Girona, Spain

SPIN 11680666 07/3180/YL – 5 4 3 2 1 0 Printed on acid-free paper

to my parents

Preface

Preface to the Second Edition

The first edition of this book published thirty years ago by Teubner had the title *Dynamics of Systems of Rigid Bodies* [97]. Soon after publication the term *multibody system* became the name of this new and rapidly developing branch of engineering mechanics. For this reason, the second edition published by Springer appears under the title *Dynamics of Multibody Systems*. Because of the success of the first edition (translations into Russian (1980), Chinese (1986) and Vietnamese (2000); use as textbook in advanced courses in Germany and abroad) little material has been added in the new edition. In Chaps. 1–4 nothing has changed except for the incorporation of short sections on quaternions and on raccording axodes. Chapters 5 and 6 have been rewritten in a new form. Both chapters are still devoted to multibody systems composed of rigid bodies with frictionless joints. Many years of teaching have led to simpler mathematical formulations in various places. Also, the order of topics has changed. Multibody systems with spherical joints and with equations of motion allowing purely analytical investigations are no longer treated first but last. The emphasis is placed on a general formalism for multibody systems with arbitrary joints and with arbitrary system structure. This formalism has found important engineering applications in many branches of industry. The first software tool based on the formalism was a FORTRAN program written by the author in 1975 for Daimler-Benz AG for simulating the dynamics of a human dummy in car accidents (passenger inside the car or pedestrian outside). Wolz [106] created the software tool *MESA VERDE* (**ME**chanism, **SA**tellite, **VE**hicle, **R**obot **D**ynamics **E**quations). Its characteristic feature is the generation of kinematics and dynamics equations in symbolic form. Using the same formalism Salecker [71], Wei [91], Weber [89], Bührle [11] and Reif [62] developed equations of motion as well as software tools for multibody systems composed of flexible bodies and for systems with electrical and hydraulic components. As a result of collaboration with *IPG Automotive, Karlsruhe* MESA VERDE-generated kinematics and dynamics equations for vehicles became the backbone of IPG's *CarMaker®* product range, which has become a powerful tool for vehicle dynamics analysis and for Hardware-in-the-Loop testing of vehicle electronic control systems. CarMaker is the basis of *AVL InMotion®* which is used for Hardware-in-the Loop development and testing of engines and entire powertrains. MESA VERDE-generated equations are used in the software tool *FADYNA* developed by IPG for Daimler-Chrysler. MESA VERDE is also used by Renault, PSA Peugeot Citroen and Opel.

It is a pleasure to thank Prof. Lothar Gaul for encouraging Springer as well as the author to publish this second edition. The author is indebted to Günther Stelzner and to Christian Simonides for their frequent advice in using TEX and to Marc Hiller for producing the data of all figures. Finally, I would like to thank the publisher for their technical advice and for their patience in waiting for the completion of the manuscript.

Karlsruhe,
June 2007 *Jens Wittenburg*

Preface to the First Edition

A system of rigid bodies in the sense of this book may be any finite number of rigid bodies interconnected in some arbitrary fashion by joints with ideal holonomic, nonholonomic, scleronomic and/or rheonomic constraints. Typical examples are the solar system, mechanisms in machines and living mechanisms such as the human body provided its individual members can be considered as rigid. Investigations into the dynamics of any such system require the formulation of nonlinear equations of motion, of energy expressions, kinematic relationships and other quantities. It is common practice to develop these for each system separately and to consider the labor necessary for deriving, for example, equations of motion from Lagrange's equation, as inevitable. It is the main purpose of this book to describe in detail a formalism which substantially simplifies this task. The formalism is general in that it provides mathematical expressions and equations which are valid for any system of rigid bodies. It is flexible in that it leaves the choice of generalized coordinates to the user. At the same time it is so explicit that its application to any particular system requires only little more than a specification of the system geometry. The book is addressed to advanced graduate students and to research workers. It tries to attract the interest of the theoretician as well as of the practitioner.

The first four out of six chapters are concerned with basic principles and with classical material. In Chap. 1 the reader is made familiar with symbolic vector and tensor notation which is used throughout this book for its compact form. In order to facilitate the transition from symbolically written equations to scalar coordinate equations matrices of vector and tensor coordinates are introduced. Transformation rules for such matrices are discussed, and methods are developed for translating compound vector-tensor expressions from symbolic into scalar coordinate form. For the purpose of compact formulations of systems of symbolically written equations matrices are introduced whose elements are vectors or tensors. Generalized multiplication rules for such matrices are defined.

In Chap. 2 on rigid body kinematics direction cosines, Euler angles, Bryan angles and Euler parameters are discussed. The notion of angular velocity is introduced, and kinematic differential equations are developed which relate the angular velocity to the rate of change of generalized coordinates. In Chap. 3 basic principles of rigid body dynamics are discussed. The definitions of both kinetic energy and angular momentum leads to the introduction of the inertia tensor. Formulations of the law of angular momentum for a rigid body are derived from Euler's axiom and also from d'Alembert's principle. Because of severe limitations on the length of the manuscript only those subjects are covered which are necessary for the later chapters. Other important topics such as cyclic variables or quasicoordinates, for example, had to be left out. In Chap. 4 some classical problems of rigid body mechanics are treated for which closed-form solutions exist. Chapter 5 which makes up one half of the book is devoted to the presentation of a general formalism for the dynamics of systems of rigid bodies. Kinematic relationships, nonlinear equations of motion, energy expressions and other quantities are developed which are suitable for both numerical and nonnumerical investigations. The unform description valid for any system of rigid bodies rests primarily on the application of concepts of graph theory (the first application to mechanics at the time of [66]). This mathematical tool in combination with matrix and symbolic vector and tensor notation leads to expressions which can easily be interpreted in physical terms. The usefulness of the formalism is demonstrated by means of some illustrative examples of nontrivial nature. Chapter 6 deals with phenomena which occur when a multibody system is subject to a collision either with another system or between two of its own bodies. Instantaneous changes of velocities and internal impulses in joints between bodies caused by such collisions are determined. The investigation reveals an interesting analogy to the law of Maxwell and Betti in elastostatics.

The material presented in subsections 1, 2, 4, 6, 8 and 9 of Sect. 5.2 was developed in close cooperation with Prof. R.E. Roberson (Univ. of Calif. at San Diego) with whom the author has a continuous exchange of ideas and results since 1965. Numerous mathematical relationships resulted from long discussions so that authorship is not claimed by any one person. It is a pleasant opportunity to express my gratitude for this fruitful cooperation. I also thank Dr. L. Lilov (Bulgarian Academy of Sciences) with whom I enjoyed close cooperation on the subject. He had a leading role in applying methods of analytical mechanics (subject of Sect. 5.2.8) and he contributed important ideas to Sect. 5.2.5. Finally, I thank the publishers for their kind patience in waiting for the completion of the manuscript.

Hannover,
February 1977 *Jens Wittenburg*

Contents

1

Mathematical Notation

In rigid body mechanics, vectors, tensors and matrices play an important role. Vectors are characterized by bold letters. In a right-handed cartesian reference base with unit base vectors $\mathbf{e}_1, \mathbf{e}_2$ and $\mathbf{e}_3$ a vector $\mathbf{v}$ is decomposed in the form

$$\mathbf{v} = v_1\mathbf{e}_1 + v_2\mathbf{e}_2 + v_3\mathbf{e}_3 \ . \tag{1.1}$$

The scalar quantities v_1, v_2 and v_3 are the coordinates of $\mathbf{v}$. Note that the term vector is used only for the quantity $\mathbf{v}$ and not as an abbreviation for the coordinate triple $[v_1,\ v_2,\ v_3]$ as is usually done in tensor calculus[1]. The unit base vectors satisfy the orthonormality conditions

$$\mathbf{e}_i \cdot \mathbf{e}_j = \delta_{ij} \qquad (i, j = 1, 2, 3) \tag{1.2}$$

and the right-handedness condition

$$\mathbf{e}_1 \cdot \mathbf{e}_2 \times \mathbf{e}_3 = +1 \ . \tag{1.3}$$

In rigid body mechanics it is necessary to work with more than one vector base. Throughout this book only right-handed cartesian bases are used. Let $\mathbf{e}_i^1$ $(i = 1, 2, 3)$ be the base vectors of one base and let $\mathbf{e}_i^2$ $(i = 1, 2, 3)$ be the base vectors of another base[2]. The bases themselves will be referred to as base $\underline{\mathbf{e}}^1$ and base $\underline{\mathbf{e}}^2$. The base vector $\mathbf{e}_i^2$ $(i = 1, 2, 3)$ of $\underline{\mathbf{e}}^2$ can be decomposed

[1] For different interpretations of the term vector see [41]. In some books on vector algebra the coordinates v_1, v_2 and v_3 are referred to as components. In the present book a component is understood to be itself a vector. Thus, $v_1\mathbf{e}_1$ in (1.1) is a component of $\mathbf{v}$.

[2] In equations such as (1.4) the superscript 2 will not be misunderstood as exponent 2. In the entire book there are only very few places where the superscript 2 and the exponent 2 occur together in a mathematical expression. In such places the superscript is placed in parentheses. Example: The moment of inertia $J_{11}^{(2)} = mr^2$ in base $\underline{\mathbf{e}}^2$.

in base $\underline{\mathbf{e}}^1$:

$$\mathbf{e}_i^2 = \sum_{j=1}^{3} a_{ij}^{21} \mathbf{e}_j^1 \qquad (i = 1, 2, 3) \, . \tag{1.4}$$

The altogether nine scalars a_{ij}^{21} $(i, j = 1, 2, 3)$ are the coordinates of the three base vectors. Each coordinate is the cosine of the angle between two base vectors:

$$a_{ij}^{21} = \mathbf{e}_i^2 \cdot \mathbf{e}_j^1 = \cos \sphericalangle \left(\mathbf{e}_i^2, \mathbf{e}_j^1 \right) \qquad (i, j = 1, 2, 3) \, . \tag{1.5}$$

For this reason the coordinates are called direction cosines. The three equations (1.4) are combined in the single matrix equation

$$\underline{\mathbf{e}}^2 = \underline{A}^{21} \underline{\mathbf{e}}^1 \, . \tag{1.6}$$

Here and throughout this book matrices are characterized by underlined letters. The (3×3)-matrix $\underline{A}^{21}$ is called *direction cosine matrix*. Note the mnemonic position of the superscripts 2 and 1. The symbol $\underline{\mathbf{e}}^2$, until now simply the name of the base, denotes the column matrix of the unit base vectors: $\underline{\mathbf{e}}^2 = [\, \mathbf{e}_1^2 \; \mathbf{e}_2^2 \; \mathbf{e}_3^2 \,]^T$. The exponent ^{T} denotes transposition. The use of bold letters indicates that the elements of $\underline{\mathbf{e}}^2$ are vectors. Equation (1.4) shows that the matrix product $\underline{A}^{21} \underline{\mathbf{e}}^1$ is evaluated following the rule of ordinary matrix algebra, although one of the matrices has vectors as elements and the other scalars. With two matrices each having vectors as elements one can form the inner product (dot product) as well as the outer product (cross product). Example: $\underline{\mathbf{e}}^1 \cdot \underline{\mathbf{e}}^{1^T} = \underline{I}$ (unit matrix). Scalar multiplication of (1.6) from the right by $\underline{\mathbf{e}}^{1^T}$ produces for the direction cosine matrix the explicit expression

$$\underline{A}^{21} = \underline{\mathbf{e}}^2 \cdot \underline{\mathbf{e}}^{1^T} \, . \tag{1.7}$$

This equation represents the matrix form of the nine Eqs. (1.5). In what follows properties of the direction cosine matrix are discussed. Each row contains the coordinates of one of the unit base vectors of $\underline{\mathbf{e}}^2$. From this it follows that the determinant of the matrix is the mixed product $\mathbf{e}_1^2 \cdot \mathbf{e}_2^2 \times \mathbf{e}_3^2$. According to (1.3) this equals $+1$. Hence,

$$\det \underline{A}^{21} = +1 \, . \tag{1.8}$$

From the orthonormality conditions (1.2) it follows that the scalar product of any two rows i and j of $\underline{A}^{21}$ equals the Kronecker delta:

$$\sum_{k=1}^{3} a_{ik}^{21} a_{jk}^{21} = \delta_{ij} \qquad (i, j = 1, 2, 3) \, . \tag{1.9}$$

A matrix having these properties is called orthogonal matrix. Because of the orthogonality the product $\underline{A}^{21} \underline{A}^{21^T}$ equals the unit matrix. Thus, the matrix has the important property that its inverse equals its transpose:

$$(\underline{A}^{21})^{-1} = \underline{A}^{21^T} . \qquad (1.10)$$

From this it follows that the inverse of (1.6) reads

$$\underline{\mathbf{e}}^1 = \underline{A}^{12}\underline{\mathbf{e}}^2 = \underline{A}^{21^T}\underline{\mathbf{e}}^2 . \qquad (1.11)$$

The identity (1.10) can also be explained as follows. In (1.4) the unit base vectors $\mathbf{e}_i^2$ $(i = 1,2,3)$ are decomposed in base $\underline{\mathbf{e}}^1$. If, instead, the unit base vectors $\mathbf{e}_i^1$ $(i = 1,2,3)$ are decomposed in base $\underline{\mathbf{e}}^2$ then the coordinates are the same direction cosines (1.5), but with indices interchanged. Equation (1.6) is replaced by the equation $\underline{\mathbf{e}}^1 = \underline{A}^{12}\underline{\mathbf{e}}^2$ with $\underline{A}^{12} = \underline{A}^{21^T}$. But the same original Eq. (1.6) yields also $\underline{A}^{12} = (\underline{A}^{21})^{-1}$. From this follows again the identity of the inverse matrix with its transpose. Furthermore, since each column of $\underline{A}^{21}$ contains the three coordinates of a unit base vector of $\underline{\mathbf{e}}^1$, the scalar product of any two columns i and j of $\underline{A}^{21}$ equals the Kronecker delta (cf. (1.9)):

$$\sum_{k=1}^{3} a_{ki}^{21} a_{kj}^{21} = \delta_{ij} \qquad (i,j = 1,2,3) . \qquad (1.12)$$

Consider, again, the vector $\mathbf{v}$ in (1.1). The right-hand side is given the form of a matrix product. For this purpose the column matrix $\underline{v} = [v_1\ v_2\ v_3]^T$ of the coordinates of $\mathbf{v}$ is introduced (a shorter name for $\underline{v}$ is coordinate matrix of $\mathbf{v}$ in base $\underline{\mathbf{e}}$). Then, (1.1) can be written in the two alternative forms

$$\mathbf{v} = \underline{\mathbf{e}}^T\underline{v} , \qquad \mathbf{v} = \underline{v}^T\underline{\mathbf{e}} . \qquad (1.13)$$

In two different bases $\underline{\mathbf{e}}^2$ and $\underline{\mathbf{e}}^1$ the vector $\mathbf{v}$ has different coordinate matrices. They are denoted $\underline{v}^2$ and $\underline{v}^1$, respectively. Thus,

$$\mathbf{v} = \underline{\mathbf{e}}^{2^T}\underline{v}^2 = \underline{\mathbf{e}}^{1^T}\underline{v}^1 . \qquad (1.14)$$

On the right-hand side (1.11) is substituted for $\underline{\mathbf{e}}^1$. This yields $\underline{\mathbf{e}}^{2^T}\underline{v}^2 = \underline{\mathbf{e}}^{2^T}\underline{A}^{21}\underline{v}^1$ and, consequently,

$$\underline{v}^2 = \underline{A}^{21}\underline{v}^1 . \qquad (1.15)$$

This equation represents the transformation rule for vector coordinates. It states that the direction cosine matrix is also the coordinate transformation matrix. Note the mnemonic position of the superscripts 2 and 1.

The scalar product of two vectors $\mathbf{a}$ and $\mathbf{b}$ can be written as a matrix product. Let $\underline{a}^1$ and $\underline{b}^1$ be the coordinate matrices of $\mathbf{a}$ and $\mathbf{b}$, respectively, in some vector base $\underline{\mathbf{e}}^1$. Then, $\mathbf{a}\cdot\mathbf{b} = \underline{a}^{1^T}\underline{b}^1 = \underline{b}^{1^T}\underline{a}^1$. Often the coordinate matrices of two vectors $\mathbf{a}$ and $\mathbf{b}$ are known in two different bases, say $\underline{a}^1$ in $\underline{\mathbf{e}}^1$ and $\underline{b}^2$ in $\underline{\mathbf{e}}^2$. Then, $\mathbf{a}\cdot\mathbf{b} = \underline{a}^{1^T}\underline{A}^{12}\underline{b}^2$.

Besides vectors second-order tensors play an important role in rigid body dynamics. Tensors are characterized by sans-serif upright letters. In its most general form a tensor D is a sum of so-called dyadic products of two vectors each:

$$\mathsf{D} = \mathbf{a}_1\mathbf{b}_1 + \mathbf{a}_2\mathbf{b}_2 + \mathbf{a}_3\mathbf{b}_3 + \dots\ . \tag{1.16}$$

A tensor is an operator. Its scalar product from the right with a vector $\mathbf{v}$ is defined as the vector

$$\begin{aligned}\mathsf{D}\cdot\mathbf{v} &= (\mathbf{a}_1\mathbf{b}_1 + \mathbf{a}_2\mathbf{b}_2 + \mathbf{a}_3\mathbf{b}_3 + \dots)\cdot\mathbf{v}\\ &= \mathbf{a}_1\mathbf{b}_1\cdot\mathbf{v} + \mathbf{a}_2\mathbf{b}_2\cdot\mathbf{v} + \mathbf{a}_3\mathbf{b}_3\cdot\mathbf{v} + \dots\ .\end{aligned} \tag{1.17}$$

No parentheses around the scalar products $\mathbf{b}_1\cdot\mathbf{v}$ etc. are necessary. Similarly, the scalar product of D from the left with $\mathbf{v}$ is defined as

$$\mathbf{v}\cdot\mathsf{D} = \mathbf{v}\cdot\mathbf{a}_1\mathbf{b}_1 + \mathbf{v}\cdot\mathbf{a}_2\mathbf{b}_2 + \mathbf{v}\cdot\mathbf{a}_3\mathbf{b}_3 + \dots\ . \tag{1.18}$$

If in all dyadic products of D the order of the factors is reversed a new tensor is obtained. It is called the conjugate of D and it is denoted by the symbol $\bar{\mathsf{D}}$:

$$\left.\begin{aligned}\mathsf{D} &= \mathbf{a}_1\mathbf{b}_1 + \mathbf{a}_2\mathbf{b}_2 + \mathbf{a}_3\mathbf{b}_3 + \dots\ ,\\ \bar{\mathsf{D}} &= \mathbf{b}_1\mathbf{a}_1 + \mathbf{b}_2\mathbf{a}_2 + \mathbf{b}_3\mathbf{a}_3 + \dots\ .\end{aligned}\right\} \tag{1.19}$$

In vector algebra the distributive law is valid:

$$\mathbf{a}\mathbf{b}_1\cdot\mathbf{v}+\mathbf{a}\mathbf{b}_2\cdot\mathbf{v} = \mathbf{a}(\mathbf{b}_1+\mathbf{b}_2)\cdot\mathbf{v}\ , \qquad \mathbf{a}_1\mathbf{b}\cdot\mathbf{v}+\mathbf{a}_2\mathbf{b}\cdot\mathbf{v} = (\mathbf{a}_1+\mathbf{a}_2)\mathbf{b}\cdot\mathbf{v}\ . \tag{1.20}$$

Hence, the dyadic products of a tensor are also distributive:

$$\mathbf{a}\mathbf{b}_1 + \mathbf{a}\mathbf{b}_2 = \mathbf{a}(\mathbf{b}_1 + \mathbf{b}_2)\ , \qquad \mathbf{a}_1\mathbf{b} + \mathbf{a}_2\mathbf{b} = (\mathbf{a}_1 + \mathbf{a}_2)\mathbf{b}\ . \tag{1.21}$$

It is, therefore, possible to resolve all vectors on the right-hand side of (1.16) in some vector base $\underline{\mathbf{e}}$ and to regroup the resulting expression in the form

$$\mathsf{D} = \sum_{i=1}^{3}\sum_{j=1}^{3} D_{ij}\mathbf{e}_i\mathbf{e}_j\ . \tag{1.22}$$

The nine scalars D_{ij} are the coordinates of D in base $\underline{\mathbf{e}}$ (note that not this set of coordinates but only the quantity D is referred to as a tensor). They are combined in the (3×3) coordinate matrix $\underline{D}$. With this matrix the tensor becomes

$$\mathsf{D} = \underline{\mathbf{e}}^T\underline{D}\,\underline{\mathbf{e}}\ . \tag{1.23}$$

It is a straightforward procedure to construct the matrix $\underline{D}$ from the coordinate matrices of the vectors $\mathbf{a}_1$, $\mathbf{b}_1$, $\mathbf{a}_2$, $\mathbf{b}_2$ etc. Let these latter matrices be $\underline{a}_1$, $\underline{a}_2$, $\underline{b}_1$, $\underline{b}_2$ etc. With the notation of (1.13) (1.16) becomes

$$\begin{aligned}\mathsf{D} &= \underline{\mathbf{e}}^T\underline{a}_1\,\underline{b}_1^T\underline{\mathbf{e}} + \underline{\mathbf{e}}^T\underline{a}_2\,\underline{b}_2^T\underline{\mathbf{e}} + \underline{\mathbf{e}}^T\underline{a}_3\,\underline{b}_3^T\underline{\mathbf{e}} + \cdots\\ &= \underline{\mathbf{e}}^T\left(\underline{a}_1\,\underline{b}_1^T + \underline{a}_2\,\underline{b}_2^T + \underline{a}_3\,\underline{b}_3^T + \cdots\right)\underline{\mathbf{e}}\ .\end{aligned} \tag{1.24}$$

Comparison with (1.23) shows that

$$\underline{D} = \underline{a}_1\,\underline{b}_1^T + \underline{a}_2\,\underline{b}_2^T + \underline{a}_3\,\underline{b}_3^T + \cdots \,. \tag{1.25}$$

From this and from (1.22) it follows that the coordinate matrix of the conjugate of D is the transpose of the coordinate matrix of D. With (1.22) and (1.1) the vector $\mathsf{D}\cdot\mathbf{v}$ is

$$\mathsf{D}\cdot\mathbf{v} = \sum_{i=1}^{3}\sum_{j=1}^{3} D_{ij}\mathbf{e}_i\mathbf{e}_j\cdot\mathbf{v} = \sum_{i=1}^{3}\sum_{j=1}^{3} D_{ij}v_j\mathbf{e}_i \,. \tag{1.26}$$

Its coordinate matrix in base $\underline{\mathbf{e}}$ is, therefore, the product $\underline{D}\,\underline{v}$ of the coordinate matrices of D and $\mathbf{v}$ in $\underline{\mathbf{e}}$. The same result is obtained in a more formal way when (1.23) and the first Eq. (1.13) are substituted for D and $\mathbf{v}$, respectively:

$$\mathsf{D}\cdot\mathbf{v} = \underline{\mathbf{e}}^T\underline{D}\,\underline{\mathbf{e}}\cdot\underline{\mathbf{e}}^T\underline{v} = \underline{\mathbf{e}}^T\underline{D}\,\underline{v} \,. \tag{1.27}$$

Of particular interest is the tensor

$$\mathsf{I} = \mathbf{e}_1\mathbf{e}_1 + \mathbf{e}_2\mathbf{e}_2 + \mathbf{e}_3\mathbf{e}_3 = \underline{\mathbf{e}}^T\underline{\mathbf{e}} \tag{1.28}$$

whose coordinate matrix is the unit matrix. When this tensor is scalar multiplied with an arbitrary vector $\mathbf{v}$ the result is $\mathbf{v}$ itself: $\mathsf{I}\cdot\mathbf{v}\equiv\mathbf{v}$ and $\mathbf{v}\cdot\mathsf{I}\equiv\mathbf{v}$. For this reason I is called unit tensor.

With the help of (1.11) it is a simple matter to establish the law by which the coordinate matrix of a tensor is transformed when instead of a base $\underline{\mathbf{e}}^1$ another base $\underline{\mathbf{e}}^2$ is used for decomposition. Let $\underline{D}^1$ and $\underline{D}^2$ be the coordinate matrices of D in the two bases, respectively, so that by (1.23) the identity

$$\underline{\mathbf{e}}^{2T}\,\underline{D}^2\,\underline{\mathbf{e}}^2 = \underline{\mathbf{e}}^{1T}\,\underline{D}^1\,\underline{\mathbf{e}}^1 \tag{1.29}$$

holds. On the right-hand side (1.11) is substituted for $\underline{\mathbf{e}}^1$. This yields

$$\underline{\mathbf{e}}^{2T}\,\underline{D}^2\,\underline{\mathbf{e}}^2 = \underline{\mathbf{e}}^{2T}\,\underline{A}^{21}\underline{D}^1\,\underline{A}^{12}\underline{\mathbf{e}}^2 \tag{1.30}$$

whence follows

$$\underline{D}^2 = \underline{A}^{21}\underline{D}^1\,\underline{A}^{12} \,. \tag{1.31}$$

Note, here too, the mnemonic position of the superscripts. This transformation is referred to as similarity transformation.

In rigid body mechanics, tensors with symmetric and with skew-symmetric coordinate matrices are met. The inertia tensor which will be defined in Sect. 3.1 and the unit tensor I have symmetric coordinate matrices. Tensors with skew-symmetric coordinate matrices are found in connection with vector cross products. Consider, first, the double cross product $(\mathbf{a}\times\mathbf{b})\times\mathbf{v}$. It can be written in the form

$$(\mathbf{a}\times\mathbf{b})\times\mathbf{v} = \mathbf{ba}\cdot\mathbf{v} - \mathbf{ab}\cdot\mathbf{v} = (\mathbf{ba}-\mathbf{ab})\cdot\mathbf{v} \tag{1.32}$$

as scalar product of the tensor (**ba** − **ab**) with **v**. If $\underline{a}$ and $\underline{b}$ are the coordinate matrices of **a** and **b**, respectively, in some vector base then the coordinate matrix of the tensor in this base is the skew-symmetric matrix

$$\underline{b}\,\underline{a}^T - \underline{a}\,\underline{b}^T = \begin{bmatrix} 0 & b_1a_2 - b_2a_1 & b_1a_3 - a_3b_1 \\ & 0 & b_2a_3 - a_3b_2 \\ \text{skew} - \text{symm.} & & 0 \end{bmatrix} . \tag{1.33}$$

Also the single vector cross product **c** × **v** can be expressed as a scalar product of a tensor with **v**. For this purpose two vectors **a** and **b** are constructed which satisfy the equation **a** × **b** = **c**. The tensor is then (**ba** − **ab**) as before and its coordinate matrix is given by (1.33). This matrix is seen to be identical with

$$\underline{\tilde{c}} = \begin{bmatrix} 0 & -c_3 & c_2 \\ c_3 & 0 & -c_1 \\ -c_2 & c_1 & 0 \end{bmatrix} \tag{1.34}$$

where c_1, c_2 and c_3 are the coordinates of **c** in the same base in which **a** and **b** are measured. With the newly defined symbol $\underline{\tilde{c}}$ (pronounced *c* tilde) for this matrix the vector **c** × **v** has the coordinate matrix $\underline{\tilde{c}}\,\underline{v}$. This notation simplifies the transition from symbolic vector equations to scalar coordinate equations[3]. For making this transition also the following rules are needed. If k is a scalar then

$$\widetilde{(k\,\underline{a})} = k\underline{\tilde{a}} \, . \tag{1.35}$$

Furthermore,

$$\widetilde{(\underline{a} + \underline{b})} = \underline{\tilde{a}} + \underline{\tilde{b}} \, , \tag{1.36}$$

$$\text{if} \quad \underline{\tilde{a}} = \underline{\tilde{b}} \quad \text{then} \quad \underline{a} = \underline{b} \, . \tag{1.37}$$

The identity **a** × **b** = −**b** × **a** yields

$$\underline{\tilde{a}}\,\underline{b} = -\underline{\tilde{b}}\,\underline{a} \tag{1.38}$$

and for the special case **a** = **b**

$$\underline{\tilde{a}}\,\underline{a} = \underline{0} \, . \tag{1.39}$$

With the help of the unit tensor **I** the double vector cross product **a** × (**b** × **v**) can be written in the form

$$\mathbf{a} \times (\mathbf{b} \times \mathbf{v}) = \mathbf{ba} \cdot \mathbf{v} - \mathbf{a} \cdot \mathbf{bv} = (\mathbf{ba} - \mathbf{a} \cdot \mathbf{b}\,\mathsf{I}) \cdot \mathbf{v} \, . \tag{1.40}$$

The corresponding coordinate equation reads $\underline{\tilde{a}}\,\underline{\tilde{b}}\,\underline{v} = (\underline{b}\,\underline{a}^T - \underline{a}^T\underline{b}\,\underline{I})\underline{v}$ with the unit matrix $\underline{I}$. Since this equation holds for every **v** the identity

$$\underline{\tilde{a}}\,\underline{\tilde{b}} = \underline{b}\,\underline{a}^T - \underline{a}^T\underline{b}\,\underline{I} \tag{1.41}$$

[3] The notation $\underline{\tilde{c}}\,\underline{v}$ for the coordinates of **c** × **v** is equivalent to the notation $\epsilon_{ijk}c_jv_k$ $(i = 1, 2, 3)$ which is commonly used in tensor algebra.

is valid. According to (1.32) the coordinate matrix of $(\mathbf{a}\times\mathbf{b})\times\mathbf{v}$ is $(\underline{b}\,\underline{a}^T - \underline{a}\,\underline{b}^T)\underline{v}$. It can also be written in the form $(\widetilde{\tilde{\underline{a}}\,\underline{b}})\underline{v}$. Since both forms are identical for every $\mathbf{v}$ the identity

$$(\widetilde{\tilde{\underline{a}}\,\underline{b}}) = \underline{b}\,\underline{a}^T - \underline{a}\,\underline{b}^T \tag{1.42}$$

holds. Finally, the transformation rule (1.31) for tensor coordinates states that

$$\tilde{\underline{a}}^2 = (\widetilde{\underline{A}^{21}\underline{a}^1}) = \underline{A}^{21}\tilde{\underline{a}}^1\underline{A}^{12}\ . \tag{1.43}$$

Systems of linear vector equations can be written in a compact form if, in addition to matrices with vectorial elements, matrices with tensors as elements are used. Such matrices are characterized by underlined sans-serif upright letters. They have the general form

$$\underline{\mathsf{D}} = \begin{bmatrix} \mathsf{D}_{11} & \dots\ \mathsf{D}_{1r} \\ \vdots & \\ \mathsf{D}_{m1} & \dots\ \mathsf{D}_{mr} \end{bmatrix} \tag{1.44}$$

with arbitrary numbers of rows and columns. The scalar product $\underline{\mathsf{D}}\cdot\underline{\mathbf{b}}$ of the $(m\times r)$-matrix $\underline{\mathsf{D}}$ from the right with an $(r\times n)$-matrix $\underline{\mathbf{b}}$ with vectors $\mathbf{b}_{ij}$ is defined as an $(m\times n)$-matrix with the elements

$$\sum_{k=1}^{r}\mathsf{D}_{ik}\cdot\mathbf{b}_{kj}\qquad(i=1,\dots,m;\ j=1,\dots,n)\ . \tag{1.45}$$

A similar definition holds for the scalar product of $\underline{\mathsf{D}}$ from the left with an $(n\times m)$-matrix $\underline{\mathbf{b}}$ with vectors $\mathbf{b}_{ij}$. The following example illustrates the practical use of these notations. Suppose it is desired to write the scalar

$$c=\sum_{i=1}^{n}\sum_{j=1}^{n}\mathbf{a}_i\cdot\mathsf{D}_{ij}\cdot\mathbf{b}_j \tag{1.46}$$

as a matrix product. This can be done in symbolic form, $c=\underline{\mathbf{a}}^T\cdot\underline{\mathsf{D}}\cdot\underline{\mathbf{b}}$ with the factors

$$\underline{\mathbf{a}}=\begin{bmatrix}\mathbf{a}_1\\ \vdots\\ \mathbf{a}_n\end{bmatrix},\qquad \underline{\mathsf{D}}=\begin{bmatrix}\mathsf{D}_{11} & \dots\ \mathsf{D}_{1n}\\ \vdots & \\ \mathsf{D}_{n1} & \dots\ \mathsf{D}_{nn}\end{bmatrix},\qquad \underline{\mathbf{b}}=\begin{bmatrix}\mathbf{b}_1\\ \vdots\\ \mathbf{b}_n\end{bmatrix}. \tag{1.47}$$

When it is desired to calculate c numerically the following expression in terms of coordinate matrices is more convenient. Let $\underline{a}_i$, $\underline{b}_i$ and $\underline{D}_{ij}$ be the coordinate matrices of $\mathbf{a}_i$, $\mathbf{b}_i$ and D_{ij} $(i,j=1,\dots,n)$, respectively, in some common vector base. Then,

$$c=\sum_{i=1}^{n}\sum_{j=1}^{n}\underline{a}_i^T\underline{D}_{ij}\underline{b}_j\ . \tag{1.48}$$

This can, in turn, be written as the matrix product $c = \underline{a}^T \underline{D}\,\underline{b}$ with

$$\underline{a} = \begin{bmatrix} \underline{a}_1 \\ \vdots \\ \underline{a}_n \end{bmatrix}, \qquad \underline{D} = \begin{bmatrix} \underline{D}_{11} & \cdots & \underline{D}_{1n} \\ \vdots & & \\ \underline{D}_{n1} & \cdots & \underline{D}_{nn} \end{bmatrix}, \qquad \underline{b} = \begin{bmatrix} \underline{b}_1 \\ \vdots \\ \underline{b}_n \end{bmatrix}. \tag{1.49}$$

Problem 1.1. Given is the direction cosine matrix $\underline{A}^{21}$ relating the vector bases $\underline{\mathbf{e}}^1$ and $\underline{\mathbf{e}}^2$. Express the matrix products $\underline{\mathbf{e}}^1 \cdot \underline{\mathbf{e}}^{1T}$, $\underline{\mathbf{e}}^{1T} \cdot \underline{\mathbf{e}}^1$, $\underline{\mathbf{e}}^1 \times \underline{\mathbf{e}}^{1T}$, $\underline{\mathbf{e}}^{1T} \times \underline{\mathbf{e}}^1$, $\underline{\mathbf{e}}^2 \cdot \underline{\mathbf{e}}^{1T}$, $\underline{\mathbf{e}}^1 \cdot \underline{\mathbf{e}}^{2T}$ and $\underline{\mathbf{e}}^{2T} \cdot \underline{\mathbf{e}}^1$ in terms of $\underline{A}^{21}$ or of elements of $\underline{A}^{21}$.

Problem 1.2. Let $\underline{\mathbf{a}}$ and $\underline{\mathbf{b}}$ be vectorial matrices and let $\underline{c}$ be a scalar matrix of such dimensions that the products $\underline{\mathbf{a}} \cdot \underline{c}\,\underline{\mathbf{b}}$ and $\underline{\mathbf{a}} \times \underline{c}\,\underline{\mathbf{b}}$ exist. Show that the former product is identical with $\underline{\mathbf{a}}\,\underline{c} \cdot \underline{\mathbf{b}}$ and the latter with $\underline{\mathbf{a}}\,\underline{c} \times \underline{\mathbf{b}}$.

Problem 1.3. $\underline{\mathbf{e}}^1$ and $\underline{\mathbf{e}}^2 = \underline{A}^{21}\underline{\mathbf{e}}^1$ are two vector bases, and $\mathbf{a}$, $\mathbf{b}$ and $\mathbf{c}$ are vectors whose coordinate matrices $\underline{a}^1$ and $\underline{b}^1$ in $\underline{\mathbf{e}}^1$ and $\underline{c}^2$ in $\underline{\mathbf{e}}^2$, respectively, are given. Furthermore, D is a tensor with the coordinate matrix $\underline{D}^2$ in $\underline{\mathbf{e}}^2$. Formulate in terms of $\underline{A}^{21}$ and of the given coordinate matrices the scalars 1. $\mathbf{a} \cdot \mathbf{b} \times \mathbf{c}$, 2. $\mathbf{a} \times \mathbf{b} \cdot \mathbf{b} \times \mathbf{c}$, 3. $\mathbf{c} \cdot \mathsf{D} \cdot \mathbf{a}$ and 4. $\mathbf{c} \cdot \mathbf{b} \times \mathsf{D} \cdot \mathbf{c}$ as well as the coordinate matrices in $\underline{\mathbf{e}}^1$ of the vectors 5. $\mathbf{a} \times \mathbf{b}$, 6. $\mathbf{a} \times \mathbf{c}$, 7. $\mathbf{a} \times (\mathbf{c} \times \mathbf{b})$, 8. $\mathbf{c} \times \mathsf{D} \cdot \mathbf{a}$ and 9. $\mathbf{a} \times [(\mathsf{D} \cdot \mathbf{b}) \times \mathbf{c}]$.

Problem 1.4. Rewrite the vector equations

$$\begin{aligned} \mathbf{a}_1 &= \mathbf{b} \times (\mathbf{v}_1 \times \mathbf{b} + \mathbf{v}_2 \times \mathbf{c}) + \mathbf{d} \times \mathbf{v}_2\,, \\ \mathbf{a}_2 &= \mathbf{c} \times (\mathbf{v}_1 \times \mathbf{b} + \mathbf{v}_2 \times \mathbf{c}) - \mathbf{d} \times \mathbf{v}_1 \end{aligned}$$

in the form

$$\begin{bmatrix} \mathbf{a}_1 \\ \mathbf{a}_2 \end{bmatrix} = \begin{bmatrix} \mathsf{D}_{11} & \mathsf{D}_{12} \\ \mathsf{D}_{21} & \mathsf{D}_{22} \end{bmatrix} \cdot \begin{bmatrix} \mathbf{v}_1 \\ \mathbf{v}_2 \end{bmatrix}$$

with explicit expressions for the tensors D_{ij} $(i, j = 1, 2)$. How are D_{12} and D_{21} related to one another? In some vector base the vectors in the original equations have the coordinate matrices $\underline{a}_1$, $\underline{a}_2$, $\underline{v}_1$, $\underline{v}_2$, $\underline{b}$, $\underline{c}$ and $\underline{d}$, respectively. Write down the coordinate matrix equation

$$\begin{bmatrix} \underline{a}_1 \\ \underline{a}_2 \end{bmatrix} = \begin{bmatrix} \underline{D}_{11} & \underline{D}_{12} \\ \underline{D}_{21} & \underline{D}_{22} \end{bmatrix} \cdot \begin{bmatrix} \underline{v}_1 \\ \underline{v}_2 \end{bmatrix}$$

giving explicit expressions for the (3×3) submatrices $\underline{D}_{ij}$ $(i, j = 1, 2)$. What can be said about the (6×6) matrix on the right-hand side?

2

Rigid Body Kinematics

In rigid body kinematics purely geometrical aspects of individual positions and of continuous motions of rigid bodies are studied. Forces and torques which are the cause of motions are not considered. In this chapter only some basic material is presented.

2.1 Generalized Coordinates of Angular Orientation

In order to specify the angular orientation of a rigid body in a vector base $\underline{\mathbf{e}}^1$ it is sufficient to specify the angular orientation of a vector base $\underline{\mathbf{e}}^2$ which is rigidly attached to the body. This can be done, for instance, by means of the direction cosine matrix (see (1.6)):

$$\underline{\mathbf{e}}^2 = \underline{A}^{21}\underline{\mathbf{e}}^1 \,. \tag{2.1}$$

The nine elements of this matrix are generalized coordinates which describe the angular orientation of the body in base $\underline{\mathbf{e}}^1$. Between these coordinates there exist the six constraint Eqs. (1.9):

$$\sum_{k=1}^{3} a_{ik}^{21} a_{jk}^{21} = \delta_{ij} \qquad (i,j = 1,2,3) \,. \tag{2.2}$$

It is often inconvenient to work with nine coordinates and six constraint equations. There are several useful systems of three coordinates without constraint equations and of four coordinates with one constraint equation which can be used as alternatives to direction cosines. In the following subsections generalized coordinates known as Euler angles, Bryan angles, rotation parameters and Euler–Rodrigues parameters will be discussed.

2.1.1 Euler Angles

The angular orientation of the body-fixed base $\underline{\mathbf{e}}^2$ is thought to be the result of three successive rotations. Prior to the first rotation the base $\underline{\mathbf{e}}^2$ coincides

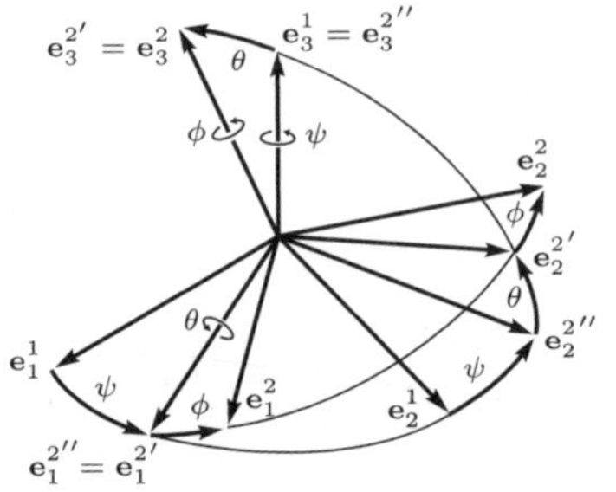

Fig. 2.1. Euler angles ψ, θ, ϕ

with the base $\underline{\mathbf{e}}^1$. The first rotation is carried out about the axis $\mathbf{e}_3^1$ through an angle ψ. It carries the base from its original orientation to an intermediate orientation denoted $\underline{\mathbf{e}}^{2''}$ (Fig. 2.1). The second rotation through the angle θ about the axis $\mathbf{e}_1^{2''}$ results in another intermediate orientation denoted $\underline{\mathbf{e}}^{2'}$. The third rotation through the angle ϕ about the axis $\mathbf{e}_3^{2'}$ produces the final orientation of the base. It is denoted $\underline{\mathbf{e}}^2$ in Fig. 2.1. A characteristic property of Euler angles is that each rotation is carried out about a base vector of the body-fixed base in a position which is the result of all previous rotations. A further characteristic is the sequence (3, 1, 3) of indices of rotation axes. The desired presentation of the transformation matrix $\underline{A}^{21}$ in terms of ψ, θ and ϕ is found from the transformation equations for the individual rotations which are according to Fig. 2.1.

$$\underline{\mathbf{e}}^2 = \underline{A}_\phi \underline{\mathbf{e}}^{2'} , \qquad \underline{\mathbf{e}}^{2'} = \underline{A}_\theta \underline{\mathbf{e}}^{2''} , \qquad \underline{\mathbf{e}}^{2''} = \underline{A}_\psi \underline{\mathbf{e}}^1 \tag{2.3}$$

with

$$\underline{A}_\phi = \begin{bmatrix} \cos\phi & \sin\phi & 0 \\ -\sin\phi & \cos\phi & 0 \\ 0 & 0 & 1 \end{bmatrix} , \quad \underline{A}_\theta = \begin{bmatrix} 1 & 0 & 0 \\ 0 & \cos\theta & \sin\theta \\ 0 & -\sin\theta & \cos\theta \end{bmatrix} ,$$

$$\underline{A}_\psi = \begin{bmatrix} \cos\psi & \sin\psi & 0 \\ -\sin\psi & \cos\psi & 0 \\ 0 & 0 & 1 \end{bmatrix} . \tag{2.4}$$

From (2.3) it follows that $\underline{A}^{21} = \underline{A}_\phi \underline{A}_\theta \underline{A}_\psi$. Multiplying out and using the abbreviations c_ψ, c_θ, c_ϕ for $\cos\psi$, $\cos\theta$, $\cos\phi$ and s_ψ, s_θ, s_ϕ for $\sin\psi$, $\sin\theta$, $\sin\phi$, respectively, one obtains the final result

$$\underline{A}^{21} = \begin{bmatrix} c_\psi c_\phi - s_\psi c_\theta s_\phi & s_\psi c_\phi + c_\psi c_\theta s_\phi & s_\theta s_\phi \\ -c_\psi s_\phi - s_\psi c_\theta c_\phi & -s_\psi s_\phi + c_\psi c_\theta c_\phi & s_\theta c_\phi \\ s_\psi s_\theta & -c_\psi s_\theta & c_\theta \end{bmatrix} . \tag{2.5}$$

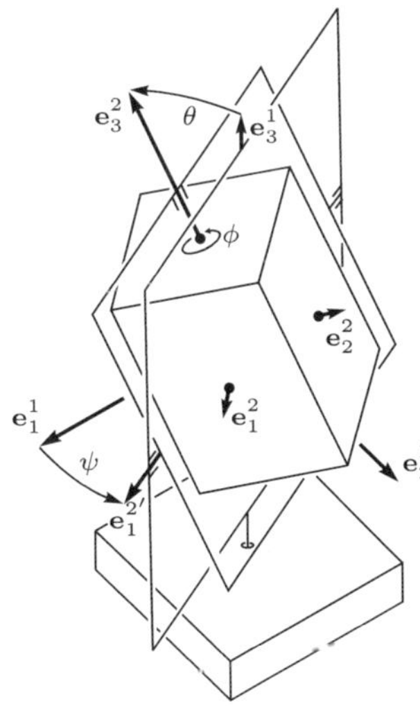

Fig. 2.2. Euler angles in a two-gimbal suspension

The advantage of having only three coordinates and no constraint equation is paid for by the disadvantage that the direction cosines are complicated functions of the three coordinates. There is still another problem. Figure 2.1 shows that in the case $\theta = n\pi$ $(n = 0, \pm 1, \ldots)$ the axis of the third rotation coincides with the axis of the first rotation. This has the consequence that ψ and ϕ cannot be distinguished.

Euler angles can be illustrated by means of a rigid body in a two-gimbal suspension system (Fig. 2.2). The bases $\underline{\mathbf{e}}^1$ and $\underline{\mathbf{e}}^2$ are attached to the material base and to the suspended body, respectively. The angles ψ, θ and ϕ are, in this order, the rotation angle of the outer gimbal relative to the material base, of the inner gimbal relative to the outer gimbal and of the body relative to the inner gimbal. With this device all three angles can be adjusted independently since the intermediate bases $\underline{\mathbf{e}}^{2''}$ and $\underline{\mathbf{e}}^{2'}$ are materially realized by the gimbals. For $\theta = n\pi$ $(n = 0, 1, \ldots)$ the planes of the two gimbals coincide (gimbal lock).

Euler angles are ideally suited as position variables for the study of motions in which $\theta(t)$ is either exactly or approximately constant, whereas ψ and ϕ are (exactly or approximately) proportional to time, i.e. $\dot{\psi} \approx$ const and $\dot{\phi} \approx$ const. Euler angles are advantageous also whenever there exist two physically significant directions, one fixed in the reference base $\underline{\mathbf{e}}^1$ and the other fixed in the body-fixed base $\underline{\mathbf{e}}^2$. In such cases the base vectors $\mathbf{e}_3^1$ and $\mathbf{e}_3^2$ are given these directions so that θ is the angle between the two. For examples see Sects. 4.1.4, 4.2 and 4.4. The use of Euler angles is, however, not restricted to such special cases.

It is often necessary to calculate the Euler angles which correspond to a numerically given matrix $\underline{A}^{21}$. For this purpose the following formulas are deduced from (2.5):

$$\left.\begin{aligned} &\cos\theta = a_{33}^{21}\,, && \sin\theta = \sigma\sqrt{1-\cos^2\theta} \quad (\sigma=+1 \ \text{ or } \ -1)\,,\\ &\cos\psi = -a_{32}^{21}/\sin\theta\,, && \sin\psi = a_{31}^{21}/\sin\theta\,,\\ &\cos\phi = a_{23}^{21}/\sin\theta\,, && \sin\phi = a_{13}^{21}/\sin\theta\,. \end{aligned}\right\} \tag{2.6}$$

If (ψ,θ,ϕ) are the angles associated with $\sigma=+1$ then the angles associated with $\sigma=-1$ are $(\pi+\psi,-\theta,\pi+\phi)$. Both triples produce one and the same final position of the base $\underline{\mathbf{e}}^2$. Numerical difficulties arise when θ is close to one of the critical values $n\pi$ $(n=0,1,\ldots)$.

2.1.2 Bryan Angles

These angles are also referred to as Cardan angles. The angular orientation of the body-fixed base $\underline{\mathbf{e}}^2$ is, again, represented as the result of a sequence of three rotations at the beginning of which the base $\underline{\mathbf{e}}^2$ coincides with the reference base $\underline{\mathbf{e}}^1$. The first rotation through an angle ϕ_1 is carried out about the axis $\mathbf{e}_1^1$. It results in the intermediate base $\underline{\mathbf{e}}^{2''}$ (Fig. 2.3). The second rotation through an angle ϕ_2 about the axis $\mathbf{e}_2^{2''}$ produces another intermediate base $\underline{\mathbf{e}}^{2'}$. The third rotation through an angle ϕ_3 about the axis $\mathbf{e}_3^{2'}$ gives the body-fixed base its final orientation denoted $\underline{\mathbf{e}}^2$ in Fig. 2.3. The transformation equations for the individual rotations are

$$\underline{\mathbf{e}}^2 = \underline{A}_3\underline{\mathbf{e}}^{2'}\,, \qquad \underline{\mathbf{e}}^{2'} = \underline{A}_2\underline{\mathbf{e}}^{2''}\,, \qquad \underline{\mathbf{e}}^{2''} = \underline{A}_1\underline{\mathbf{e}}^1 \tag{2.7}$$

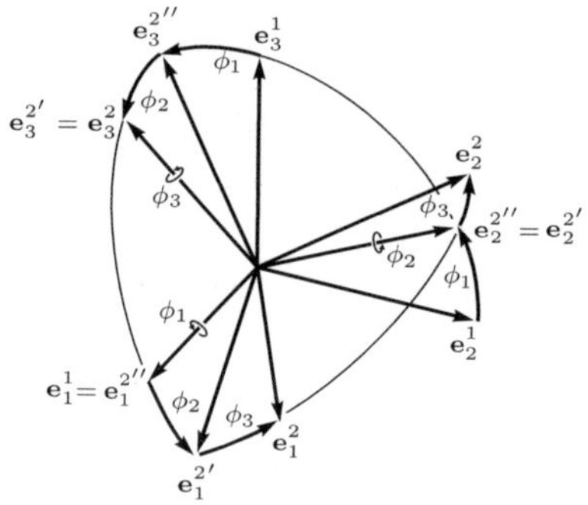

Fig. 2.3. Bryan angles ϕ_1, ϕ_2, ϕ_3

with

$$\underline{A}_3 = \begin{bmatrix} \cos\phi_3 & \sin\phi_3 & 0 \\ -\sin\phi_3 & \cos\phi_3 & 0 \\ 0 & 0 & 1 \end{bmatrix}, \quad \underline{A}_2 = \begin{bmatrix} \cos\phi_2 & 0 & -\sin\phi_2 \\ 0 & 1 & 0 \\ \sin\phi_2 & 0 & \cos\phi_2 \end{bmatrix},$$

$$\underline{A}_1 = \begin{bmatrix} 1 & 0 & 0 \\ 0 & \cos\phi_1 & \sin\phi_1 \\ 0 & -\sin\phi_1 & \cos\phi_1 \end{bmatrix}. \tag{2.8}$$

The desired direction cosine matrix relating the bases $\underline{\mathbf{e}}^1$ and $\underline{\mathbf{e}}^2$ is $\underline{A}^{21} = \underline{A}_3\underline{A}_2\underline{A}_1$. Multiplying out and using the abbreviations $c_i = \cos\phi_i$, $s_i = \sin\phi_i$ $(i = 1, 2, 3)$ one obtains the final result

$$\underline{A}^{21} = \begin{bmatrix} c_2c_3 & c_1s_3 + s_1s_2c_3 & s_1s_3 - c_1s_2c_3 \\ -c_2s_3 & c_1c_3 - s_1s_2s_3 & s_1c_3 + c_1s_2s_3 \\ s_2 & -s_1c_2 & c_1c_2 \end{bmatrix}. \tag{2.9}$$

The only significant difference as compared with Euler angles is the sequence (1, 2, 3) of indices of rotation axes. Bryan angles, too, can be illustrated by means of a rigid body in a two-gimbal suspension system. The arrangement is shown in Fig. 2.4. The bases $\underline{\mathbf{e}}^1$ and $\underline{\mathbf{e}}^2$ are attached to the material base and to the suspended body, respectively. The angles ϕ_1, ϕ_2 and ϕ_3 are, in this order, the rotation angle of the outer gimbal relative to the material base, of the inner gimbal relative to the outer gimbal and of the body relative to the inner gimbal. The three angles can be adjusted independently since the intermediate bases $\underline{\mathbf{e}}^{2''}$ and $\underline{\mathbf{e}}^{2'}$ are materially realized by the gimbals. For $\phi_2 = 0$ the three rotation axes are mutually orthogonal. As with Euler angles there exists a critical case in which the axes of the first and of the third rotation coincide. This occurs if $\phi_2 = \pi/2 + n\pi$ $(n = 0, 1, \ldots)$.

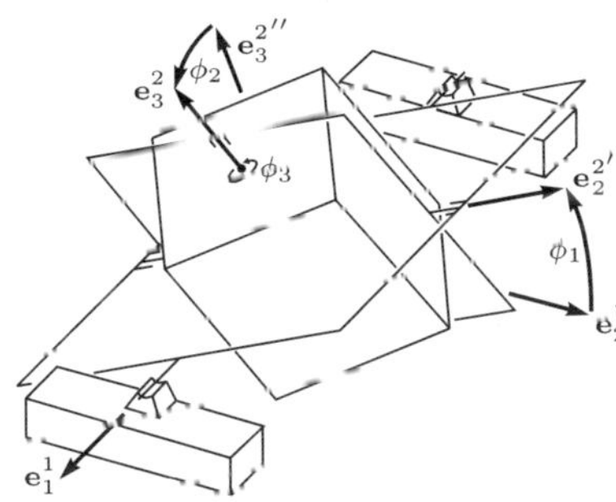

Fig. 2.4. Bryan angles in a two-gimbal suspension

In contrast to Euler angles linearization for small angles is possible. In the case $|\phi_i| \ll 1$ $(i = 1,2,3)$ one has $\sin\phi_i \approx \phi_i$, $\cos\phi_i \approx 1$ and with this

$$\underline{A}^{21} \approx \begin{bmatrix} 1 & \phi_3 & -\phi_2 \\ -\phi_3 & 1 & \phi_1 \\ \phi_2 & -\phi_1 & 1 \end{bmatrix} = \underline{I} - \underline{\tilde{\phi}}\,. \tag{2.10}$$

For the skew-symmetric matrix $\underline{\tilde{\phi}}$ and its role in vector cross-products see (1.34). The occurrence of this matrix suggests to interpret ϕ_1, ϕ_2, ϕ_3 as coordinates of a *rotation vector* $\boldsymbol{\phi}$. Note that in the linear approximation the coordinates are the same in both bases. Indeed, transformation, i.e. multiplication of the coordinate matrix $\underline{\phi} = [\phi_1\ \phi_2\ \phi_3]^T$ with $\underline{A}^{21}$, causes no change (see (1.39)): $\underline{A}^{21}\underline{\phi} \approx (\underline{I} - \underline{\tilde{\phi}})\underline{\phi} = \underline{\phi}$.

The rotation vector $\boldsymbol{\phi}$ can be used as follows. Let $\mathbf{r}$ and $\mathbf{r}^*$ be the positions of an arbitrary body-fixed vector before and after the small rotation. If, as usual, $\underline{r}^{*1}$ and $\underline{r}^{*2}$ denote the coordinate matrices of $\mathbf{r}^*$ in $\underline{\mathbf{e}}^1$ and in $\underline{\mathbf{e}}^2$, respectively, then $\underline{r}^{*1} = \underline{A}^{12}\underline{r}^{*2}$. The coordinate matrix $\underline{r}^{*2}$ is identical with the coordinate matrix $\underline{r}^1$ of $\mathbf{r}$ in base $\underline{\mathbf{e}}^1$ since before the rotation $\underline{\mathbf{e}}^2$ coincides with $\underline{\mathbf{e}}^1$ and the body-fixed vector coincides with $\mathbf{r}$. Therefore,

$$\underline{r}^{*1} = \underline{A}^{12}\underline{r}^1\,. \tag{2.11}$$

With (2.10) this is the equation $\underline{r}^{*1} \approx (\underline{I} + \underline{\tilde{\phi}})\underline{r}^1$. It is the coordinate form of the vector equation

$$\mathbf{r}^* \approx \mathbf{r} + \boldsymbol{\phi} \times \mathbf{r} \qquad \text{(valid for small rotations only)}\,. \tag{2.12}$$

The conclusion of these considerations is that within linear approximations small rotation angles can be added like vectors.

What follows is not restricted to small angles. If the matrix $\underline{A}^{21}$ in (2.9) is given then the associated Bryan angles are calculated from the equations

$$\left.\begin{aligned} &\sin\phi_2 = a_{31}^{21}\,, && \cos\phi_2 = \sigma\sqrt{1-\sin^2\phi_2} \quad (\sigma = +1 \text{ or } -1)\,, \\ &\sin\phi_1 = -a_{32}^{21}/\cos\phi_2\,, && \cos\phi_1 = a_{33}^{21}/\cos\phi_2\,, \\ &\sin\phi_3 = -a_{21}^{21}/\cos\phi_2\,, && \cos\phi_3 = a_{11}^{21}/\cos\phi_2\,. \end{aligned}\right\} \tag{2.13}$$

If (ϕ_1, ϕ_2, ϕ_3) are the angles associated with $\sigma = +1$ then the angles associated with $\sigma = -1$ are $(\pi + \phi_1, \pi - \phi_2, \pi + \phi_3)$. Both triples produce one and the same final position of the base $\underline{\mathbf{e}}^2$. Numerical difficulties arise when ϕ_2 is close to one of the critical values $\pi/2 + n\pi$ $(n = 0, 1, \ldots)$.

2.1.3 Rotation Tensor

In Chap. 1 it has been shown that a direction cosine matrix $\underline{A}^{21}$ relating two bases $\underline{\mathbf{e}}^1$ and $\underline{\mathbf{e}}^2$ has the determinant $+1$ and that its inverse equals its transpose (see (1.8) and (1.10)). Now, another property of fundamental importance is revealed. The eigenvalue problem $\underline{A}^{21}\underline{v} = \lambda\underline{v}$ or

$$(\underline{A}^{21} - \lambda\underline{I})\underline{v} = \underline{0} \tag{2.14}$$

is investigated. This is the transformation rule $\underline{A}^{21}\underline{v}^1 = \underline{v}^2$ in the special case $\underline{v}^2 = \lambda\underline{v}^1$. Since the absolute value of a vector does not change under a transformation it can be predicted that all three eigenvalues have the absolute value one. The eigenvalues are the roots of the characteristic equation $\det(\underline{A}^{21} - \lambda\underline{I}) = 0$. Omitting the superscript this is the cubic equation

$$\begin{aligned}&-\lambda^3 + \lambda^2 \operatorname{tr}\underline{A}\\ &-\lambda[(a_{11}a_{22} - a_{12}a_{21}) + (a_{22}a_{33} - a_{23}a_{32}) + (a_{33}a_{11} - a_{31}a_{13})]\\ &+\det\underline{A} = 0\,.\end{aligned} \tag{2.15}$$

The determinant is +1. In the coefficient of λ every expression in parentheses represents the so-called co-factor of one diagonal element of $\underline{A}^{21}$. Since the determinant is +1 the co-factor is identical with the diagonal element. Consequently, the coefficient of λ represents the trace of the matrix. Thus, the equation reads

$$-\lambda^3 + \lambda^2 \operatorname{tr}\underline{A}^{21} - \lambda \operatorname{tr}\underline{A}^{21} + 1 = 0\,. \tag{2.16}$$

It shows that every direction cosine matrix $\underline{A}^{21}$ has the eigenvalue $\lambda = +1$. Division by $(\lambda - 1)$ produces for the other eigenvalues the quadratic equation $\lambda^2 - (\operatorname{tr}\underline{A}^{21} - 1)\lambda + 1 = 0$. It has the roots

$$\begin{aligned}\lambda_{2,3} &= \frac{\operatorname{tr}\underline{A}^{21} - 1}{2} \pm \mathrm{i}\sqrt{1 - \left(\frac{\operatorname{tr}\underline{A}^{21} - 1}{2}\right)^2}\\ &= \cos\varphi \pm \mathrm{i}\sin\varphi = \mathrm{e}^{\pm\mathrm{i}\varphi}\end{aligned} \tag{2.17}$$

with

$$\cos\varphi = \frac{\operatorname{tr}\underline{A}^{21} - 1}{2}\,. \tag{2.18}$$

If $\underline{A}^{21}$ is the unit matrix, then it has the triple eigenvalue +1. In the case $\operatorname{tr}\underline{A}^{21} = -1$ it has the double eigenvalue $\lambda_{2,3} = -1$.

Let $\underline{n}$ be the normalized eigenvector associated with the eigenvalue $\lambda = +1$. It is calculated from the equations

$$(\underline{A}^{21} - \underline{I})\,\underline{n} = \underline{0}\,, \qquad n_1^2 + n_2^2 + n_3^2 = 1\,. \tag{2.19}$$

This eigenvector $\underline{n}$ represents the coordinate matrix of a unit vector $\mathbf{n}$ which has identical coordinate matrices in the two bases $\underline{\mathbf{e}}^1$ and $\underline{\mathbf{e}}^2$ related by $\underline{A}^{21}$. Imagine that, starting from the initial position, the body-fixed base $\underline{\mathbf{e}}^2$ is rotated about this vector $\mathbf{n}$. Every value of the rotation angle is associated with a final position. Independent of the angle the vector $\mathbf{n}$ has identical coordinate matrices in $\underline{\mathbf{e}}^1$ and in $\underline{\mathbf{e}}^2$. The existence of the eigenvector guarantees the existence of an angle which carries the base from its initial position to the final position which is given by the matrix $\underline{A}^{21}$. Hence, the

Theorem 2.1 (Euler). *The displacement of a body-fixed base from an initial position* $\underline{\mathbf{e}}^1$ *to an arbitrary final position* $\underline{\mathbf{e}}^2$ *is achieved by a rotation through a certain angle about an axis which is fixed in both bases. The axis has the direction of the eigenvector associated with the eigenvalue* $\lambda = +1$ *of the direction cosine matrix* $\underline{A}^{21}$.

Proposition: The rotation angle in Theorem 2.1 is the angle φ in the eigenvalues λ_2 and λ_3 in (2.17). The proof is achieved by solving the inverse problem. Given is a rotation through the axial unit vector $\mathbf{n}$ and the angle of rotation φ (both arbitrary). The rotation is called rotation $(\mathbf{n}, \varphi)$. Note: The rotations $(\mathbf{n}, \varphi)$, $(-\mathbf{n}, -\varphi)$ and $(\mathbf{n}, \varphi + 2\pi)$ produce the same final position. For this reason, they are called equal. In what follows an expression for the direction cosine matrix $\underline{A}^{21}$ in terms of $\mathbf{n}$ and φ is developed. The matrix is found from a comparison of the coordinate matrices in the bases $\underline{\mathbf{e}}^1$ and $\underline{\mathbf{e}}^2$ of a body-fixed vector. In Fig. 2.5 this vector is shown in its positions $\mathbf{r}$ and $\mathbf{r}^*$ before and after the rotation, respectively. In both positions the vector lies on a circular cone whose axis is defined by the unit vector $\mathbf{n}$. Let $\underline{r}^{*1}$ and $\underline{r}^1$ be the coordinate matrices of $\mathbf{r}^*$ and of $\mathbf{r}$, respectively, in $\underline{\mathbf{e}}^1$. For reasons explained earlier these two matrices are related through (2.11):

$$\underline{r}^{*1} = \underline{A}^{12}\underline{r}^1 \,. \tag{2.20}$$

According to Fig. 2.5 the vectors $\mathbf{r}^*$ and $\mathbf{r}$ are related through the equation $\mathbf{r}^* = \mathbf{r} + (1 - \cos\varphi)\mathbf{b} + \sin\varphi\,\mathbf{a}$ or, recognizing that $\mathbf{a} = \mathbf{n} \times \mathbf{r}$ and $\mathbf{b} = \mathbf{n} \times \mathbf{a}$,

$$\mathbf{r}^* = \mathbf{r} + (1 - \cos\varphi)\mathbf{n} \times (\mathbf{n} \times \mathbf{r}) + \sin\varphi\,\mathbf{n} \times \mathbf{r} \tag{2.21}$$

$$= \cos\varphi\,\mathbf{r} + (1 - \cos\varphi)\mathbf{n}\mathbf{n} \cdot \mathbf{r} + \sin\varphi\,\mathbf{n} \times \mathbf{r} \,. \tag{2.22}$$

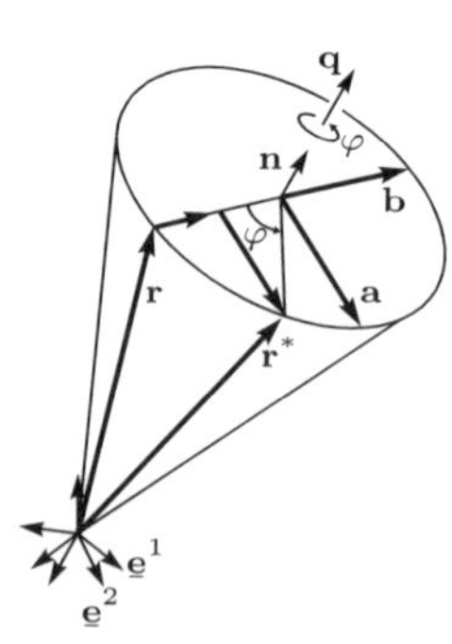

Fig. 2.5. Rotation of a body-fixed vector $\mathbf{r}$ about an axial unit vector $\mathbf{n}$

Decomposition in base $\underline{\mathbf{e}}^1$ results in the coordinate equation

$$\underline{r}^{*1} = [\cos\varphi\,\underline{I} + (1-\cos\varphi)\underline{n}\,\underline{n}^T + \sin\varphi\,\underline{\tilde{n}}\,]\,\underline{r}^1\,. \tag{2.23}$$

Equation (2.20) shows that the expression in brackets is the matrix $\underline{A}^{12}$. Hence, the transpose $\underline{A}^{21}$ is, with the abbreviations $c = \cos\varphi$ and $s = \sin\varphi$,

$$\underline{A}^{21} = \begin{bmatrix} n_1^2 + (1-n_1^2)c & n_1n_2(1-c) + n_3s & n_1n_3(1-c) - n_2s \\ n_1n_2(1-c) - n_3s & n_2^2 + (1-n_2^2)c & n_2n_3(1-c) + n_1s \\ n_1n_3(1-c) + n_2s & n_2n_3(1-c) - n_1s & n_3^2 + (1-n_3^2)c \end{bmatrix}. \tag{2.24}$$

The diagonal elements can be given other forms if use is made of the constraint equation $n_1^2 + n_2^2 + n_3^2 = 1$. The diagonal elements yield the first equation below and the off-diagonal elements yield the second:

$$\cos\varphi = \frac{\operatorname{tr}\underline{A}^{21} - 1}{2}\,, \qquad n_i\sin\varphi = \frac{a_{jk}^{21} - a_{kj}^{21}}{2} \qquad (i,j,k = 1,2,3 \text{ cyclic})\,. \tag{2.25}$$

The first equation is identical with (2.18). This proves that the rotation angle φ is the angle in the eigenvalues $\lambda_{2,3} = \mathrm{e}^{\pm i\varphi}$ of the direction cosine matrix.

When the matrix $\underline{A}^{21}$ is given numerically, the rotation parameters φ and n_1, n_2, n_3 can be determined in two ways. Either $\underline{n}$ is calculated from (2.19): $(\underline{A}^{21} - \underline{I})\underline{n} = \underline{0}$. Then, the two Eqs. (2.25) together determine φ uniquely. The alternative way is to take one of the two solutions φ satisfying the first Eq. (2.25) and to calculate n_1, n_2, n_3 from the second equation. This second equation for n_1, n_2, n_3 fails in the particularly simple case $\varphi = \pi$. In this case $\underline{A}^{21}$ is the symmetric matrix

$$\underline{A}^{21} = 2\underline{n}\,\underline{n}^T - \underline{I}\,. \tag{2.26}$$

The trace is -1, whence follows that the matrix has the real double eigenvalue $\lambda_{2,3} = -1$. A symmetric matrix has mutually orthogonal eigenvectors. Let $\underline{z}$ be the eigenvectors associated with the double eigenvalue $\lambda = -1$. Equation (2.14) for $\underline{z}$ has the form $2\underline{n}\underline{n}^T\underline{z} = \underline{0}$ whence follows $\underline{n}^T\underline{z} = 0$. This means that any vector perpendicular to the rotation axis $\mathbf{n}$ is an eigenvector. The kinematical interpretation of these eigenvectors is the following. By (2.22) any body-fixed vector $\mathbf{r}$ perpendicular to $\mathbf{n}$ is rotated into the position $\mathbf{r}^* = -\mathbf{r}$. This is the characteristic of eigenvectors associated with the eigenvalue -1.

The general formulation (2.24) of the matrix $\underline{A}^{21}$ in terms of the four rotation parameters n_1, n_2, n_3 and φ is particularly useful if a body is rotating about a fixed axis which is not aligned with one of the base vectors. Then, n_1, n_2, n_3 are constants and only φ is a variable. For small rotations with $|\varphi| \ll 1$ the Taylor expansion of (2.23) up to second-order terms yields the approximation

$$\underline{A}^{21} \approx \underline{I} - \varphi\,\underline{\tilde{n}} + \frac{1}{2}\varphi^2(\underline{n}\,\underline{n}^T - \underline{I})\,. \tag{2.27}$$

The linear approximation $\underline{A}^{21} \approx \underline{I} - \varphi\,\underline{\tilde{n}}$ is known from (2.10).

2.1.4 Euler–Rodrigues Parameters

In what follows (2.21) is considered again. It is given a new form. By means of the relations

$$1 - \cos\varphi = 2\sin^2\frac{\varphi}{2}\,, \qquad \sin\varphi = 2\sin\frac{\varphi}{2}\cos\frac{\varphi}{2} \tag{2.28}$$

the transition to the half-angle is made. New quantities are defined as follows:

$$q_0 = \cos\frac{\varphi}{2}\,, \qquad \mathbf{q} = \mathbf{n}\sin\frac{\varphi}{2}\,. \tag{2.29}$$

The vector $\mathbf{q}$ lies in the rotation axis. Therefore, it has identical coordinates in the bases $\underline{\mathbf{e}}^1$ and $\underline{\mathbf{e}}^2$. These coordinates are denoted q_1, q_2, q_3 and the coordinate matrix is called $\underline{q}$. The four quantities $q_0, \ldots, q_3$ are called *Euler–Rodrigues parameters.* They satisfy the constraint equation (here and in other places the exponent 2 of Euler–Rodrigues parameters will not be misunderstood as superscript 2 refering to a reference base)

$$q_0^2 + \mathbf{q}^2 = \sum_{i=0}^{3} q_i^2 = 1\,. \tag{2.30}$$

This can be written in the forms

$$1 - 2\mathbf{q}^2 = q_0^2 - \mathbf{q}^2 = 2q_0^2 - 1\,. \tag{2.31}$$

From (2.29) it is seen: A change of the signs of all four parameters means that either $\cos\frac{\varphi}{2}$ and $\sin\frac{\varphi}{2}$ change signs or $\cos\frac{\varphi}{2}$ and $\mathbf{n}$ change signs. The former has the effect that $(\mathbf{n}, \varphi)$ is replaced by $(\mathbf{n}, \varphi + 2\pi)$. The latter means that $(\mathbf{n}, \varphi)$ is replaced by $(-\mathbf{n}, -\varphi)$. Neither one of these changes has an effect on the rotation.

With (2.28), (2.29) and (2.31) one gets for (2.21) the new form

$$\mathbf{r}^* = \mathbf{r} + 2\mathbf{q}\times(\mathbf{q}\times\mathbf{r}) + 2q_0\mathbf{q}\times\mathbf{r} \tag{2.32}$$

$$= (q_0^2 - \mathbf{q}^2)\mathbf{r} + 2(\mathbf{q}\mathbf{q}\cdot\mathbf{r} + q_0\mathbf{q}\times\mathbf{r})\,. \tag{2.33}$$

Decomposition in base $\underline{\mathbf{e}}^1$ results in the coordinate equation

$$\underline{r}^{*1} = \left[(q_0^2 - \mathbf{q}^2)\underline{I} + 2(\underline{q}\,\underline{q}^T + q_0\underline{\tilde{q}})\right]\underline{r}^1\,. \tag{2.34}$$

This is the new form of (2.23). With the same arguments as before the expression in brackets is the matrix $\underline{A}^{12}$. Thus, for the transpose $\underline{A}^{21}$ the new form is obtained:

$$\underline{A}^{21} = \begin{bmatrix} q_0^2 + q_1^2 - q_2^2 - q_3^2 & 2(q_1q_2 + q_0q_3) & 2(q_1q_3 - q_0q_2) \\ 2(q_1q_2 - q_0q_3) & q_0^2 + q_2^2 - q_3^2 - q_1^2 & 2(q_2q_3 + q_0q_1) \\ 2(q_1q_3 + q_0q_2) & 2(q_2q_3 - q_0q_1) & q_0^2 + q_3^2 - q_1^2 - q_2^2 \end{bmatrix}\,. \tag{2.35}$$

With (2.31) the diagonal elements can be given the alternative form $a_{ii}^{21} = 2(q_0^2 + q_i^2) - 1$ $(i = 1, 2, 3)$.

In what follows the inverse problem is considered, again. Given is the matrix $\underline{A}^{21}$. To be determined are the Euler–Rodrigues parameters q_0, q_1, q_2, q_3. The diagonal elements yield the first formula below and the off-diagonal elements yield the second:

$$q_0 = \frac{1}{2}\sqrt{1 + \operatorname{tr}\underline{A}^{21}}\,, \qquad q_i = \frac{a_{jk}^{21} - a_{kj}^{21}}{4q_0}\,, \qquad (i, j, k = 1, 2, 3 \text{ cyclic})\,. \tag{2.36}$$

The formulas for q_1, q_2, q_3 fail in the case q_0. This is the case $\varphi = \pi$, again. By definition, in this case $q_i = n_i$ $(i = 1, 2, 3)$. The eigenvector $\underline{n}$ is determined from (2.19).

In contrast to Euler angles and to Bryan angles (and to any other set of three generalized coordinates) there is no critical case in which either the four parameters n_1, n_2, n_3, φ or the four Euler–Rodrigues parameters are indeterminate[1].

2.1.5 Euler–Rodrigues Parameters in Terms of Euler Angles

Expressions for Euler–Rodrigues parameters in terms of Euler angles are obtained when in (2.36) for $\underline{A}^{21}$ (2.5) is substituted. First, one calculates

$$\begin{aligned} 1 + \operatorname{tr}\underline{A}^{21} &= 1 + c_\theta + c_\theta(c_\psi c_\phi - s_\psi s_\phi) + (c_\psi c_\phi - s_\psi s_\phi) \\ &= (1 + c_\theta)[1 + (c_\psi c_\phi - s_\psi s_\phi)] \\ &= (1 + \cos\theta)[1 + \cos(\psi + \phi)] = 4\cos^2\frac{\theta}{2}\,\cos^2\frac{\psi + \phi}{2}\,. \end{aligned} \tag{2.37}$$

This yields $q_0 = \cos\frac{\theta}{2}\,\cos\frac{\psi+\phi}{2}$ (sign arbitrary). From (2.36) first q_1 is calculated and for this purpose

$$a_{23}^{21} - a_{32}^{21} = s_\theta(c_\psi + c_\phi) = 4\sin\frac{\theta}{2}\,\cos\frac{\theta}{2}\,\cos\frac{\psi + \phi}{2}\,\cos\frac{\psi - \phi}{2}\,. \tag{2.38}$$

From this and from the result for q_0 one obtains $q_1 = \sin\frac{\theta}{2}\,\cos\frac{\psi-\phi}{2}$. In a similar manner also q_2 and q_3 are calculated. All four formulas together read

$$\left.\begin{aligned} q_0 &= \cos\frac{\theta}{2}\,\cos\frac{\psi + \phi}{2}\,, \qquad & q_2 &= \sin\frac{\theta}{2}\,\sin\frac{\psi - \phi}{2}\,, \\ q_1 &= \sin\frac{\theta}{2}\,\cos\frac{\psi - \phi}{2}\,, \qquad & q_3 &= \cos\frac{\theta}{2}\,\sin\frac{\psi + \phi}{2}\,. \end{aligned}\right\} \tag{2.39}$$

---

[1] Hopf [29] was the first to prove that no representation of finite rotations by three parameters is possible without singular points. For a simpler proof see Stuelpnagel [78].

From these equations follow the relationships

$$\cos^2\frac{\theta}{2} = q_0^2 + q_3^2\,, \qquad \sin^2\frac{\theta}{2} = q_1^2 + q_2^2\,, \tag{2.40}$$

$$\tan\frac{\psi+\phi}{2} = \frac{q_3}{q_0}\,, \qquad \tan\frac{\psi-\phi}{2} = \frac{q_2}{q_1}\,. \tag{2.41}$$

2.1.6 Quaternions

A quaternion, abbreviated Q, is composed of a scalar u and a vector $\mathbf{v}$, i.e. of four quantities altogether. Therefore, the name quaternion. The quaternion is denoted $Q = (u, \mathbf{v})$. The product of a quaternion $(u, \mathbf{v})$ by a scalar λ is defined to be the quaternion $(\lambda u, \lambda\mathbf{v})$. The sum and the product of two quaternions Q_1 and Q_2 are defined as follows:

$$Q_1 + Q_2 = Q_2 + Q_1 = (u_1 + u_2, \mathbf{v}_1 + \mathbf{v}_2)\,, \tag{2.42}$$

$$Q_2Q_1 = (u_2, \mathbf{v}_2)(u_1, \mathbf{v}_1) = (u_2u_1 - \mathbf{v}_2\cdot\mathbf{v}_1\,,\; u_2\mathbf{v}_1 + u_1\mathbf{v}_2 + \mathbf{v}_2\times\mathbf{v}_1)\,. \tag{2.43}$$

Both the sum and the product are themselves quaternions. Because of the term $\mathbf{v}_2 \times \mathbf{v}_1$ multiplication is not commutative. It is associative, however, as can be verified by multiplying out: $Q_3Q_2Q_1 = Q_3(Q_2Q_1) = (Q_3Q_2)Q_1$.

The special quaternion $(1, \mathbf{0})$ is called unit quaternion because multiplication with an arbitrary quaternion Q yields Q:

$$(1, \mathbf{0})Q = Q(1, \mathbf{0}) \equiv Q\,. \tag{2.44}$$

The *conjugate* of $Q = (u, \mathbf{v})$ is defined to be the quaternion $\tilde{Q} = (u, -\mathbf{v})$. The product of a quaternion with its own conjugate is, according to the rule (2.43)

$$Q\tilde{Q} = (u, \mathbf{v})(u, -\mathbf{v}) = (u^2 + \mathbf{v}^2, \mathbf{0}) = (u^2 + \mathbf{v}^2)(1, \mathbf{0})\,. \tag{2.45}$$

Thus, it is a non-negative scalar multiple of the unit quaternion. The square root of this scalar is called the *norm* of Q, abbreviated

$$\|Q\| = \sqrt{u^2 + \mathbf{v}^2}\,. \tag{2.46}$$

An arbitrary quaternion Q with the norm $\|Q\| \neq 0$ satisfies the equation $(\tilde{Q}/\|Q\|^2)Q = (1, \mathbf{0})$. Because of this property $\tilde{Q}/\|Q\|^2$ is called the inverse of Q.

Let Q_1 and Q_2 be the quaternions from (2.43), again. Then, the conjugate of the product is (the vector part is multiplied by -1)

$$\widetilde{Q_2Q_1} = (u_2u_1 - \mathbf{v}_2\cdot\mathbf{v}_1,\; -u_2\mathbf{v}_1 - u_1\mathbf{v}_2 - \mathbf{v}_2\times\mathbf{v}_1)\,. \tag{2.47}$$

Following the rule (2.43) also the product is calculated:

$$\tilde{Q}_1\tilde{Q}_2 = (u_1, -\mathbf{v}_1)(u_2, -\mathbf{v}_2) = (u_1u_2 - \mathbf{v}_1 \cdot \mathbf{v}_2,\ -u_1\mathbf{v}_2 - u_2\mathbf{v}_1 + \mathbf{v}_1 \times \mathbf{v}_2)\ . \tag{2.48}$$

Comparison reveals the formula

$$\tilde{Q}_1\tilde{Q}_2 = \widetilde{Q_2Q_1}\ . \tag{2.49}$$

It is natural to define a quaternion which is composed of the Euler–Rodrigues parameters q_0 and $\mathbf{q}$ of a rotation $(\mathbf{n}, \varphi)$. It is denoted D:

$$D = (q_0, \mathbf{q}) = \Big(\cos\frac{\varphi}{2},\ \mathbf{n}\sin\frac{\varphi}{2}\Big)\ . \tag{2.50}$$

It has the norm $\|D\| = \sqrt{q_0^2 + \mathbf{q}^2} = 1$. Thus, its inverse equals its conjugate:

$$\tilde{D} = (q_0,\ -\mathbf{q})\ , \qquad \tilde{D}D = (1,\ \mathbf{0})\ . \tag{2.51}$$

The conjugate is the quaternion of the inverse rotation $(\mathbf{n}, -\varphi)$. The quaternion of the null rotation $(\varphi = 0)$ is the unit quaternion $(1, \mathbf{0})$.

With an arbitrary vector $\mathbf{r}$ the special quaternion $(0, \mathbf{r})$ can be constructed. With the vector $\mathbf{r}$ shown in Fig. 2.5 and with the quaternion D of the rotation in this figure the product is calculated:

$$D(0,\ \mathbf{r})\tilde{D} = (q_0, \mathbf{q})(0,\ \mathbf{r})(q_0, -\mathbf{q})\ . \tag{2.52}$$

In a first step one calculates

$$(0,\ \mathbf{r})(q_0, -\mathbf{q}) = (\mathbf{r} \cdot \mathbf{q},\ q_0\mathbf{r} - \mathbf{r} \times \mathbf{q})\ . \tag{2.53}$$

With this expression the scalar part of $D(0, \mathbf{r})\tilde{D}$ is

$$q_0\mathbf{r} \cdot \mathbf{q} - \mathbf{q} \cdot (q_0\mathbf{r} - \mathbf{r} \times \mathbf{q}) = 0\ . \tag{2.54}$$

The vector part is

$$\begin{aligned} &q_0(q_0\mathbf{r} - \mathbf{r} \times \mathbf{q}) + (\mathbf{r} \cdot \mathbf{q})\mathbf{q} + \mathbf{q} \times (q_0\mathbf{r} - \mathbf{r} \times \mathbf{q}) \\ &= q_0^2\mathbf{r} + q_0\mathbf{q} \times \mathbf{r} + \mathbf{q}(\mathbf{q} \cdot \mathbf{r}) + q_0\mathbf{q} \times \mathbf{r} + \mathbf{q} \times (\mathbf{q} \times \mathbf{r})\ . \end{aligned} \tag{2.55}$$

Reformulation of the double cross product yields the expression

$$(q_0^2 - \mathbf{q}^2)\mathbf{r} + 2(\mathbf{q}\mathbf{q} \cdot \mathbf{r} + q_0\mathbf{q} \times \mathbf{r})\ . \tag{2.56}$$

Comparison with (2.33) shows that this is the vector $\mathbf{r}^*$ of Fig. 2.5. Thus, one has the equation

$$(0,\ \mathbf{r}^*) = D(0,\ \mathbf{r})\tilde{D}\ . \tag{2.57}$$

Next, two consecutive rotations with quaternions D_1 (first rotation) and D_2 are carried out. The result of the first rotation is $(0, \mathbf{r}^*) = D_1(0, \mathbf{r})\tilde{D}_1$.

The second rotation carries the vector $\mathbf{r}^*$ into the new position $\mathbf{r}^{**}$ given by the equation $(0, \mathbf{r}^{**}) = D_2(0, \mathbf{r}^*)\tilde{D}_2$. For $\mathbf{r}^*$ the expression from the previous equation is substituted. This yields the relationship $(0, \mathbf{r}^{**}) = D_2 D_1(0, \mathbf{r})\tilde{D}_1\tilde{D}_2$ or, because of (2.49),

$$(0,\ \mathbf{r}^{**}) = (D_2 D_1)(0,\ \mathbf{r})\widetilde{(D_2 D_1)}\ . \tag{2.58}$$

This has the form (2.57). Thus, one has the[2]

Theorem 2.2. *The quaternion D_{res} of the resultant of two consecutive rotations with quaternions D_1 (first rotation) and D_2 is the product*

$$\begin{aligned} D_{\mathrm{res}} &= D_2 D_1 = (q_{02}, \mathbf{q}_2)(q_{01}, \mathbf{q}_1) \\ &= \Big(\cos\frac{\varphi_2}{2},\ \mathbf{n}_2\sin\frac{\varphi_2}{2}\Big)\Big(\cos\frac{\varphi_1}{2},\ \mathbf{n}_1\sin\frac{\varphi_1}{2}\Big)\ . \end{aligned} \tag{2.59}$$

The multiplication rule (2.43) yields the formulas

$$q_{0_{\mathrm{res}}} = q_{02}q_{01} - \mathbf{q}_2\cdot\mathbf{q}_1\ , \qquad \mathbf{q}_{\mathrm{res}} = q_{02}\mathbf{q}_1 + q_{01}\mathbf{q}_2 + \mathbf{q}_2\times\mathbf{q}_1\ . \tag{2.60}$$

More explicitly, these are formulas for the rotation angle φ_{res} and for the unit vector $\mathbf{n}_{\mathrm{res}}$ of the resultant rotation:

$$\cos\frac{\varphi_{\mathrm{res}}}{2} = \cos\frac{\varphi_2}{2}\cos\frac{\varphi_1}{2} - \mathbf{n}_2\cdot\mathbf{n}_1\sin\frac{\varphi_2}{2}\sin\frac{\varphi_1}{2}\ , \tag{2.61}$$

$$\begin{aligned} \mathbf{n}_{\mathrm{res}}\sin\frac{\varphi_{\mathrm{res}}}{2} &= \mathbf{n}_1\cos\frac{\varphi_2}{2}\sin\frac{\varphi_1}{2} + \mathbf{n}_2\cos\frac{\varphi_1}{2}\sin\frac{\varphi_2}{2} \\ &\quad + \mathbf{n}_2\times\mathbf{n}_1\sin\frac{\varphi_2}{2}\sin\frac{\varphi_1}{2}\ . \end{aligned} \tag{2.62}$$

Because of the term $\mathbf{n}_2 \times \mathbf{n}_1$ the axis of the resultant rotation is not located in the plane of the axes of the other two rotations. Its location depends upon the order in which the two rotations are carried out. In contrast, the angle φ_{res} is independent of the order.

Linearization in the case of small angles φ_1 and φ_2 yields the approximations

$$\cos\frac{\varphi_{\mathrm{res}}}{2} \approx 1\ , \qquad \mathbf{n}_{\mathrm{res}}\varphi_{\mathrm{res}} \approx \mathbf{n}_1\varphi_1 + \mathbf{n}_2\varphi_2\ . \tag{2.63}$$

From the first equation it follows that also φ_{res} is small. The second equation proves that in this approximation it is possible to define rotation vectors $\boldsymbol{\varphi}_{\mathrm{res}} = \mathbf{n}_{\mathrm{res}}\varphi_{\mathrm{res}}$, $\boldsymbol{\varphi}_i = \mathbf{n}_i\varphi_i$ and to calculate the resultant vector by the parallelogram rule

$$\boldsymbol{\varphi}_{\mathrm{res}} \approx \boldsymbol{\varphi}_1 + \boldsymbol{\varphi}_2\ . \tag{2.64}$$

The vector of a small rotation was first introduced in the context of (2.10).

---

[2] The invention of quaternions is attributed to Hamilton with the date Oct. 16, 1843. However, Theorem 2.2 was published by Rodrigues [67] in 1840 already. Euler [34] communicated quaternions in a letter to Goldbach on May 4, 1748.

Problem 2.1. Determine the eigenvalues and eigenvectors of the matrix

$$\begin{bmatrix} 1 & 0 & 0 \\ 0 & \cos\varphi & \sin\varphi \\ 0 & -\sin\varphi & \cos\varphi \end{bmatrix}$$

Problem 2.2. Given are the direction cosine matrices

$$\underline{A}_1 = \begin{bmatrix} \frac{2}{3} & \frac{1}{3} & \frac{2}{3} \\ \frac{1}{3} & \frac{2}{3} & \frac{-2}{3} \\ \frac{-2}{3} & \frac{2}{3} & \frac{1}{3} \end{bmatrix}, \quad \underline{A}_2 = \begin{bmatrix} \frac{2}{3} & \frac{1}{3} & \frac{2}{3} \\ \frac{-11}{15} & \frac{2}{15} & \frac{2}{3} \\ \frac{2}{15} & \frac{-14}{15} & \frac{1}{3} \end{bmatrix},$$

$$\underline{A}_3 = \frac{1}{9}\begin{bmatrix} -7 & 4 & 4 \\ 4 & -1 & 8 \\ 4 & 8 & -1 \end{bmatrix}.$$

Determine the corresponding Euler angles and Euler–Rodrigues parameters.

Problem 2.3. Determine all rotations $(\mathbf{n}, \varphi)$ which result in positions that can be produced by Bryan angles in the critical case $\cos\phi_2 = 0$.

Problem 2.4. Determine the resultant of two successive 180°-rotations about intersecting axes $\mathbf{n}_1$ and $\mathbf{n}_2$ which enclose the angle α.

Problem 2.5. A body-fixed base $\underline{\mathbf{e}}^2$ which is initially coincident with a reference base $\underline{\mathbf{e}}^1$ is subjected to three successive rotations. The first rotation is carried out about the axis $\mathbf{e}_1^1$ through the angle ϕ_1, the second about $\mathbf{e}_2^1$ through ϕ_2 and the third about $\mathbf{e}_3^1$ through ϕ_3. Note that in contrast to Bryan angles all three rotations are carried out about base vectors of the reference base $\underline{\mathbf{e}}^1$. Express the direction cosine matrix $\underline{A}^{21}$ relating the final orientation of $\underline{\mathbf{e}}^2$ to $\underline{\mathbf{e}}^1$ as product of three matrices, each representing one of the three rotations. Express the quaternion D_{res} of the resultant rotation as product of three quaternions, each representing one of the three rotations. Evaluate $\underline{A}^{21}$ and D_{res} for the three sets of angles $(\phi_1, \phi_2, \phi_3) = (\pi/2, \pi/2, \pi/2)$, $(0, \pi/2, 0)$ and (π, π, π). Check the results experimentally.

2.2 Kinematics of Continuous Motion

2.2.1 Angular Velocity. Angular Acceleration

Let $\underline{\mathbf{e}}^1$ be some arbitrarily moving base. Relative to this base a rigid body is in arbitrary motion. Fixed on this body there is a base $\mathbf{e}^2$ with origin A. Furthermore, a point P is considered which is moving relative to the body. With the notations of Fig. 2.6

$$\mathbf{r} = \mathbf{r}_{\text{A}} + \boldsymbol{\varrho}\,. \tag{2.65}$$

The goal of the present investigation is an expression for the velocity $\mathbf{v}$ of P relative to the base $\underline{\mathbf{e}}^1$ in terms of the velocity $\mathbf{v}_{\text{rel}}$ of P relative to the body, of the velocity $\mathbf{v}_{\text{A}}$ of A relative to $\underline{\mathbf{e}}^1$ and of some as yet unknown

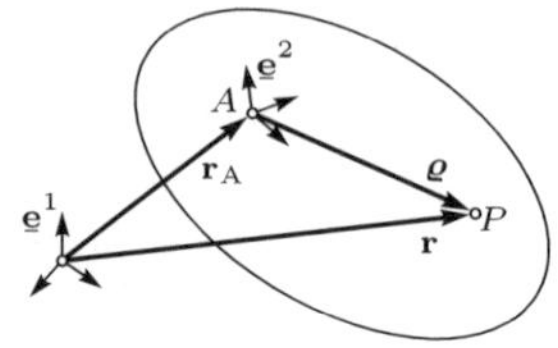

Fig. 2.6. Radius vectors of two points A (body-fixed) and P (moving relative to the body)

quantity which accounts for changes of the body angular orientation in base $\underline{\mathbf{e}}^1$. A velocity relative to a specified base is calculated as time derivative of the radius vector in this base. More precisely, the velocity coordinates are the time derivatives of the coordinates of the radius vector. In what follows the base in which a vector is differentiated is denoted by an upper index (1) or (2) placed in front. With this notation the two velocities of P are

$$\mathbf{v} = \frac{^{(1)}\mathrm{d}\mathbf{r}}{\mathrm{d}t}\,, \qquad \mathbf{v}_{\mathrm{rel}} = \frac{^{(2)}\mathrm{d}\boldsymbol{\varrho}}{\mathrm{d}t}\,. \tag{2.66}$$

In contrast to vectors scalar quantities have identical time derivatives in different bases. These derivatives are denoted by a dot as usual.

The relationship between time derivatives of a vector in two bases $\underline{\mathbf{e}}^1$ and $\underline{\mathbf{e}}^2$ is met frequently and not only with position vectors. For this reason it is formulated first for a vector $\mathbf{p}$ of arbitrary physical nature. Let p_i $(i = 1, 2, 3)$ be the coordinates of $\mathbf{p}$ in $\underline{\mathbf{e}}^2$. Then

$$\frac{^{(1)}\mathrm{d}\mathbf{p}}{\mathrm{d}t} = \frac{^{(1)}\mathrm{d}}{\mathrm{d}t}\sum_{i=1}^{3} p_i\,\mathbf{e}_i^2 = \sum_{i=1}^{3} \dot{p}_i\,\mathbf{e}_i^2 + \sum_{i=1}^{3} p_i\,\frac{^{(1)}\mathrm{d}}{\mathrm{d}t}\,\mathbf{e}_i^2\,. \tag{2.67}$$

By definition, the first sum represents the time derivative of $\mathbf{p}$ in $\underline{\mathbf{e}}^2$. Hence,

$$\frac{^{(1)}\mathrm{d}\mathbf{p}}{\mathrm{d}t} = \frac{^{(2)}\mathrm{d}\mathbf{p}}{\mathrm{d}t} + \sum_{i=1}^{3} p_i\,\frac{^{(1)}\mathrm{d}}{\mathrm{d}t}\,\mathbf{e}_i^2\,. \tag{2.68}$$

The derivative of $\mathbf{e}_i^2$ is a vector. Let a_{ij} $(j = 1, 2, 3)$ be its unknown coordinates in base $\underline{\mathbf{e}}^2$:

$$\frac{^{(1)}\mathrm{d}}{\mathrm{d}t}\,\mathbf{e}_i^2 = \sum_{j=1}^{3} a_{ij}\,\mathbf{e}_j^2 \qquad (i = 1, 2, 3)\,. \tag{2.69}$$

The base vectors satisfy the orthonormality conditions $\mathbf{e}_i^2 \cdot \mathbf{e}_k^2 = \delta_{ik}$ $(i, k = 1, 2, 3)$. Differentiation of this equation in base $\underline{\mathbf{e}}^1$ yields

$$\left(\frac{^{(1)}\mathrm{d}}{\mathrm{d}t}\,\mathbf{e}_i^2\right)\cdot\mathbf{e}_k^2 + \mathbf{e}_i^2\cdot\left(\frac{^{(1)}\mathrm{d}}{\mathrm{d}t}\,\mathbf{e}_k^2\right) = 0 \qquad (i, k = 1, 2, 3)\,. \tag{2.70}$$

The two derivatives in parentheses are expressed by means of (2.69):

$$\left(\sum_{j=1}^{3} a_{ij}\,\mathbf{e}_j^2\right)\cdot\mathbf{e}_k^2 \;+\; \mathbf{e}_i^2\cdot\left(\sum_{j=1}^{3} a_{kj}\,\mathbf{e}_j^2\right) = 0 \qquad (i,k=1,2,3)\,. \qquad (2.71)$$

Because of the orthonormality conditions this reduces to $a_{ik}+a_{ki}=0$ $(i,k=1,2,3)$. It follows that $a_{11}=a_{22}=a_{33}=0$. Instead of nine unknowns a_{ij} there are only three. The non zero ones are given the new names $a_{ij}=-a_{ji}=\omega_k$ $(i,j,k=1,2,3$ cyclic). Then, (2.69) gets the form

$$\frac{{}^{(1)}\mathrm{d}}{\mathrm{d}t}\,\mathbf{e}_i^2 = -\omega_j\,\mathbf{e}_k^2+\omega_k\,\mathbf{e}_j^2 \qquad (i,j,k=1,2,3 \text{ cyclic})\,. \qquad (2.72)$$

Now, the three quantities ω_i $(i=1,2,3)$ are interpreted as coordinates of a vector $\boldsymbol{\omega}_{21}$ in base $\underline{\mathbf{e}}^2$. Then, (2.72) is the equation

$$\frac{{}^{(1)}\mathrm{d}}{\mathrm{d}t}\,\mathbf{e}_i^2 = \boldsymbol{\omega}_{21}\times\mathbf{e}_i^2 \qquad (i=1,2,3)\,. \qquad (2.73)$$

The vector $\boldsymbol{\omega}_{21}$ is called *angular velocity* of the body, i.e. of the base $\underline{\mathbf{e}}^2$, relative to $\underline{\mathbf{e}}^1$. In (2.68) – (2.73) the origin A of the base $\underline{\mathbf{e}}^2$ does not play any role. Hence, the angular velocity is independent of the choice of A. In figures the angular velocity vector can be attached to any point of the body. It is not wrong but sometimes misleading to talk about the angular velocity *about* A. This is particularly true if the body has a fixed axis which does not pass through A.

With (2.73) one gets for (2.68) the form

$$\frac{{}^{(1)}\mathrm{d}\mathbf{p}}{\mathrm{d}t} = \frac{{}^{(2)}\mathrm{d}\mathbf{p}}{\mathrm{d}t} + \boldsymbol{\omega}_{21}\times\mathbf{p} \quad (\boldsymbol{\omega}_{21} = \text{angular velocity of } \underline{\mathbf{e}}^2 \text{ relative to } \underline{\mathbf{e}}^1)\,. \qquad (2.74)$$

This is the desired relationship between the time derivatives of an arbitrary vector $\mathbf{p}$ in two different bases. The application to $\boldsymbol{\omega}_{21}$ itself shows that the time derivatives in the two bases are identical. Therefore, without distinction one can simply write $\dot{\boldsymbol{\omega}}_{21}$:

$$\frac{{}^{(1)}\mathrm{d}\boldsymbol{\omega}_{21}}{\mathrm{d}t} = \frac{{}^{(2)}\mathrm{d}\boldsymbol{\omega}_{21}}{\mathrm{d}t} = \dot{\boldsymbol{\omega}}_{21}\,. \qquad (2.75)$$

This vector is called *angular acceleration* of the body relative to $\underline{\mathbf{e}}^1$.

After these preparations we return to (2.65): $\mathbf{r}=\mathbf{r}_\mathrm{A}+\boldsymbol{\varrho}$. The equation is differentiated with respect to time in base $\underline{\mathbf{e}}^1$. The derivative of $\mathbf{r}_\mathrm{A}$ is the velocity $\mathbf{v}_\mathrm{A}$ of the body-fixed point A relative to $\underline{\mathbf{e}}^1$. For differentiating $\boldsymbol{\varrho}$ (2.74) is used. Instead of $\boldsymbol{\omega}_{21}$ simply $\boldsymbol{\omega}$ is written. Using the notations (2.66) one obtains for the velocity $\mathbf{v}$ of P the expression

$$\mathbf{v} = \mathbf{v}_\mathrm{A}+\boldsymbol{\omega}\times\boldsymbol{\varrho}+\mathbf{v}_\mathrm{rel}\,. \qquad (2.76)$$

Let $\mathbf{a}$ be the acceleration of P relative to $\underline{\mathbf{e}}^1$. It is found by differentiating $\mathbf{v}$ one more time in $\underline{\mathbf{e}}^1$. In doing so (2.74) is applied to the second and to the third vector. Each of these vectors contributes the term $\boldsymbol{\omega} \times \mathbf{v}_{\text{rel}}$. Taking into account also (2.75) one obtains the expression

$$\mathbf{a} = \mathbf{a}_{\text{A}} + \dot{\boldsymbol{\omega}} \times \boldsymbol{\varrho} + \boldsymbol{\omega} \times (\boldsymbol{\omega} \times \boldsymbol{\varrho}) + 2\boldsymbol{\omega} \times \mathbf{v}_{\text{rel}} + \mathbf{a}_{\text{rel}} \,. \tag{2.77}$$

2.2.2 Inverse Motion

The motion of the base $\underline{\mathbf{e}}^1$ relative to $\underline{\mathbf{e}}^2$ is called the *inverse* of the motion of $\underline{\mathbf{e}}^2$ relative to $\underline{\mathbf{e}}^1$. Let $\boldsymbol{\omega}_{\text{rel}}$ and $\dot{\boldsymbol{\omega}}_{\text{rel}}$ be the angular velocity and the angular acceleration, respectively, of $\underline{\mathbf{e}}^1$ relative to $\underline{\mathbf{e}}^2$. Furthermore, let $\mathbf{v}_{\text{rel}}$ and $\mathbf{a}_{\text{rel}}$ be the velocity and the acceleration, respectively, relative to $\underline{\mathbf{e}}^2$, of a point fixed in $\underline{\mathbf{e}}^1$. These quantities are determined as follows. Equation (2.74) is solved for the right-hand side time derivative of $\mathbf{p}$. This is achieved by interchanging the indices 1 and 2. The result is: $\boldsymbol{\omega}_{12} = -\boldsymbol{\omega}_{21}$. By definition, $\boldsymbol{\omega}_{12}$ is the desired angular velocity $\boldsymbol{\omega}_{\text{rel}}$. This yields the first equation below. The second equation follows by the argument used for (2.75) (again $\boldsymbol{\omega}$ is written instead of $\boldsymbol{\omega}_{21}$):

$$\boldsymbol{\omega}_{\text{rel}} = -\boldsymbol{\omega} \,, \qquad \dot{\boldsymbol{\omega}}_{\text{rel}} = -\dot{\boldsymbol{\omega}} \,. \tag{2.78}$$

Next, (2.76) is applied twice, once to the point A fixed in $\underline{\mathbf{e}}^2$ and once to that point fixed in $\underline{\mathbf{e}}^1$ which instantaneously coincides with A. In the first case the equation is $\mathbf{v} = \mathbf{v}_{\text{A}}$. In the second case one has $\mathbf{v} = \mathbf{0}$, and $\mathbf{v}_{\text{rel}}$ is the desired velocity of the inverse motion. The equation reads: $\mathbf{0} = \mathbf{v}_{\text{A}} + \mathbf{v}_{\text{rel}}$. By combining both equations one obtains the result

$$\mathbf{v}_{\text{rel}} = -\mathbf{v} \,. \tag{2.79}$$

In the same way an expression for $\mathbf{a}_{\text{rel}}$ is obtained from (2.77). The two applications yield the equations $\mathbf{a} = \mathbf{a}_{\text{A}}$ and $\mathbf{0} = \mathbf{a}_{\text{A}} + 2\boldsymbol{\omega} \times \mathbf{v}_{\text{rel}} + \mathbf{a}_{\text{rel}}$. The combination of these two equations with (2.79) yields

$$\mathbf{a}_{\text{rel}} = -\mathbf{a} + 2\boldsymbol{\omega} \times \mathbf{v} \,. \tag{2.80}$$

Here, too, the quantities on the left-hand side belong to the inverse motion and the ones on the right-hand side to the motion. Equations (2.78)–(2.80) are summarized in the

Theorem 2.3. *The switch from motion to inverse motion has the consequence that angular velocity, angular acceleration and velocities of arbitrary points are multiplied by* (-1)*. For accelerations this rule is valid only for those points which satisfy the condition* $\boldsymbol{\omega} \times \mathbf{v} = \mathbf{0}$*. These are all points of the instantaneous screw axis.*

2.2.3 Instantaneous Screw Axis. Raccording Axodes

In the special case $\mathbf{v}_{\text{rel}} = \mathbf{0}$, $\mathbf{a}_{\text{rel}} = \mathbf{0}$ (2.76) and (2.77) express the velocity and the acceleration, respectively, of the body-fixed point P defined by the position vector $\boldsymbol{\varrho}$:

$$\mathbf{v} = \mathbf{v}_{\text{A}} + \boldsymbol{\omega} \times \boldsymbol{\varrho} \,, \tag{2.81}$$

$$\mathbf{a} = \mathbf{a}_{\text{A}} + \dot{\boldsymbol{\omega}} \times \boldsymbol{\varrho} + \boldsymbol{\omega} \times (\boldsymbol{\omega} \times \boldsymbol{\varrho}) \,. \tag{2.82}$$

The first equation describes the velocity distribution of a rigid body. All body-fixed points along an arbitrary line parallel to $\boldsymbol{\omega}$ have, instantaneously, identical velocities (different for different lines). There exists a single line parallel to $\boldsymbol{\omega}$ the points of which have a velocity in the direction of $\boldsymbol{\omega}$, i.e. the velocity $\mathbf{v} = p\boldsymbol{\omega}$ with a scalar p of dimension length. Hence, for all points on this line $p\boldsymbol{\omega} = \mathbf{v}_{\text{A}} + \boldsymbol{\omega} \times \boldsymbol{\varrho}$. Let $\mathbf{u}$ be the perpendicular from A onto this particular line. Then, also $p\boldsymbol{\omega} = \mathbf{v}_{\text{A}} + \boldsymbol{\omega} \times \mathbf{u}$. Cross- and dot-multiplying this equation by $\boldsymbol{\omega}$ and making use of the orthogonality $\boldsymbol{\omega} \cdot \mathbf{u} = 0$ one gets for $\mathbf{u}$ and for p the expressions

$$\mathbf{u} = \frac{\boldsymbol{\omega} \times \mathbf{v}_{\text{A}}}{\omega^2} \,, \qquad p = \frac{\boldsymbol{\omega} \cdot \mathbf{v}_{\text{A}}}{\omega^2} \,. \tag{2.83}$$

If in (2.81) as point A an arbitrary point on the line determined by $\mathbf{u}$ is chosen then the velocity distribution in the body is

$$\mathbf{v} = p\boldsymbol{\omega} + \boldsymbol{\omega} \times \boldsymbol{\varrho} \,, \tag{2.84}$$

i.e. the superposition of a rotation with angular velocity $\boldsymbol{\omega}$ about this particular line and of the translation with the velocity $p\boldsymbol{\omega}$ along the line. This is the velocity distribution of a screw motion. The particular line is the *instantaneous screw axis* (ISA) and p is the *pitch* of the screw. The velocity $\mathbf{v}$ of an arbitrary body-fixed point not located on the ISA has the direction of the helix through this point.

In general, the ISA is time-varying. It moves relative to the reference system and relative to the body. Thereby, it is the *generator* of a ruled surface fixed in the reference system and of another ruled surface fixed in the body and moving with the body. The two ruled surfaces are called *fixed axode* F_{f} (fixed in the reference system) and *moving axode* F_{m} (fixed in the body).

By definition, the ISA is common to both F_{f} and F_{m}. Proposition: At every point along the common ISA the tangent planes of both F_{f} and F_{m} coincide. Proof: Let k_{m} be an arbitrary curve fixed on F_{m}, i.e. fixed on the moving body, which intersects the generators of F_{m}. The generator is the ISA which, in the course of time, is sweeping out F_{m}. Let P be the point which at all times t is located on both k_{m} and the ISA(t), and let, furthermore, $\mathbf{v}_{\text{rel}}$ be the velocity of P along k_{m}. On F_{f} P is moving along a different trajectory k_{f}. Its velocity along k_{f} is, according to (2.76) and (2.84), $\mathbf{v} = p\boldsymbol{\omega} + \mathbf{v}_{\text{rel}}$. The plane spanned by $p\boldsymbol{\omega}$ and $\mathbf{v}$ coincides with the plane spanned by $p\boldsymbol{\omega}$

and $\mathbf{v}_{\rm rel}$. Since these planes represent the tangent planes of $\mathrm{F_f}$ and of $\mathrm{F_m}$, respectively, also the tangent planes coincide. End of proof. These statements are summarized in the

Theorem 2.4 (Painlevé). *Every continuous motion of a rigid body can be interpreted as raccording motion of the body-fixed axode* $\mathrm{F_m}$ *on the axode* $\mathrm{F_f}$ *fixed in the reference-base. The axodes are generated by the* ISA. *They are in tangential contact along the* ISA. *The raccording motion is the superposition of the translation with velocity* $p\boldsymbol{\omega}$ *along the* ISA *and the rotation (rolling motion) with angular velocity* $\boldsymbol{\omega}$ *about the* ISA.

In what follows the orientation of the tangent planes in all points along the ISA is investigated at a fixed time $t = \text{const}$ (arbitrary). With the abbreviations

$$\mathbf{r}(0) = \mathbf{r}_{\rm A} + \mathbf{u}, \qquad \mathbf{e} = \frac{\boldsymbol{\omega}}{|\boldsymbol{\omega}|} \tag{2.85}$$

and with a free parameter λ of dimension length the ISA has the parameter equation

$$\mathbf{r}(\lambda) = \mathbf{r}(0) + \lambda\mathbf{e}\,. \tag{2.86}$$

The plane tangent to the axodes at the point $\mathbf{r}(\lambda)$ is spanned by the vectors $\mathbf{e}$ and $\dot{\mathbf{r}} = \dot{\mathbf{r}}(0) + \lambda\dot{\mathbf{e}}$. These two vectors define the unit normal vector $\mathbf{n}(\lambda)$ of the tangent plane at this point:

$$\mathbf{n}(\lambda) = \frac{(\dot{\mathbf{r}}(0) + \lambda\dot{\mathbf{e}}) \times \mathbf{e}}{|(\dot{\mathbf{r}}(0) + \lambda\dot{\mathbf{e}}) \times \mathbf{e}\,|}\,. \tag{2.87}$$

In the infinitely distant points $\lambda \to -\infty$ and $\lambda \to +\infty$ the unit normal vectors of the tangent planes are the mutually opposite vectors

$$\mathbf{n}_{-\infty} = \frac{\mathbf{e} \times \dot{\mathbf{e}}}{|\mathbf{e} \times \dot{\mathbf{e}}\,|}\,, \qquad \mathbf{n}_{+\infty} = -\frac{\mathbf{e} \times \dot{\mathbf{e}}}{|\mathbf{e} \times \dot{\mathbf{e}}\,|}\,. \tag{2.88}$$

These formulas show: When traveling from the infinite point $\lambda \to -\infty$ to the infinite point $\lambda \to +\infty$ the tangent plane rotates through the angle π. The particular point on the ISA where one half of this rotation is executed is referred to as *striction point* S on the ISA. The unit vector $\mathbf{n}_{\rm S}$ normal the tangent plane at S is

$$\mathbf{n}_{\rm S} = \mathbf{n}_{-\infty} \times \mathbf{e} = \frac{(\mathbf{e} \times \dot{\mathbf{e}}) \times \mathbf{e}}{|\mathbf{e} \times \dot{\mathbf{e}}\,|}\,. \tag{2.89}$$

From $\boldsymbol{\omega} = \omega\mathbf{e}$ and $\dot{\boldsymbol{\omega}} = \dot{\omega}\mathbf{e} + \omega\dot{\mathbf{e}}$ it follows that $(\boldsymbol{\omega} \times \dot{\boldsymbol{\omega}}) \times \boldsymbol{\omega} = \omega^3(\mathbf{e} \times \dot{\mathbf{e}}) \times \mathbf{e}$. This yields the expression

$$\mathbf{n}_{\rm S} = \frac{(\boldsymbol{\omega} \times \dot{\boldsymbol{\omega}}) \times \boldsymbol{\omega}}{|\boldsymbol{\omega} \times \dot{\boldsymbol{\omega}}|\,|\boldsymbol{\omega}|}\,. \tag{2.90}$$

The cartesian base with the origin at the striction point S and with axes along $\boldsymbol{\omega}$, $\boldsymbol{\omega}\times\dot{\boldsymbol{\omega}}$ and $\mathbf{n}_S$ constitutes the so-called *instantaneous canonical reference frame* of the raccording axodes (of both axodes).

Let $\varphi(\lambda)$ be the angle through which the normal vector $\mathbf{n}(\lambda)$ of the tangent plane at the point $\mathbf{r}(\lambda)$ is rotated against $\mathbf{n}_S$ (see Fig. 2.7). The unknown function $\varphi(\lambda)$ as well as the unknown location of the striction point on the ISA are found from properties of the acceleration of body-fixed points which momentarily coincide with the ISA. The point coinciding with $\mathbf{r}(\lambda)$ has the acceleration (see (2.82))

$$\mathbf{a}(\lambda) = \mathbf{a}(0) - \frac{\lambda}{|\boldsymbol{\omega}|}\,\boldsymbol{\omega}\times\dot{\boldsymbol{\omega}} \tag{2.91}$$

where $\mathbf{a}(0)$ is the acceleration of the body-fixed point at the foot of the perpendicular $\mathbf{u}$ on the ISA:

$$\mathbf{a}(0) = \mathbf{a}_A + \dot{\boldsymbol{\omega}}\times\mathbf{u} - \omega^2\mathbf{u} = \mathbf{a}_A - \boldsymbol{\omega}\times\mathbf{v}_A + \frac{\dot{\boldsymbol{\omega}}\times(\boldsymbol{\omega}\times\mathbf{v}_A)}{\omega^2}\ . \tag{2.92}$$

Proposition: The acceleration $\mathbf{a}(\lambda)$ lies in the plane spanned by the ISA and by $\mathbf{n}(\lambda)$. Proof: The acceleration has a component along the ISA due to the translatory part of the raccording motion and a component $\mathbf{a}_r(\lambda)$ due to the rolling motion. Consider the rolling motion alone. The body-fixed point which is in rolling contact at $\mathbf{r}(\lambda)$ is passing through a cusp of its trajectory with the normal unit vector $\mathbf{n}(\lambda)$ being, in the limit, the tangent to the trajectory. From this it follows that $\mathbf{a}_r(\lambda)$ has the direction of $\mathbf{n}(\lambda)$. End of proof.

In particular, the acceleration $\mathbf{a}_S = \mathbf{a}(\lambda_S)$ of the body-fixed point coinciding with the striction point lies in the plane spanned by $\boldsymbol{\omega}$ and $\mathbf{n}_S$. From this together with (2.90) it follows that the striction point is characterized by coplanarity of the vectors $\mathbf{a}_S$, $\boldsymbol{\omega}$ and $\dot{\boldsymbol{\omega}}$:

$$\mathbf{a}_S\cdot\boldsymbol{\omega}\times\dot{\boldsymbol{\omega}} = 0\ . \tag{2.93}$$

Into this equation the expression (2.91) with $\lambda = \lambda_S$ is substituted. This equation yields

$$\frac{\lambda_S}{|\boldsymbol{\omega}|} = \frac{\mathbf{a}(0)\cdot\boldsymbol{\omega}\times\dot{\boldsymbol{\omega}}}{(\boldsymbol{\omega}\times\dot{\boldsymbol{\omega}})^2}\ . \tag{2.94}$$

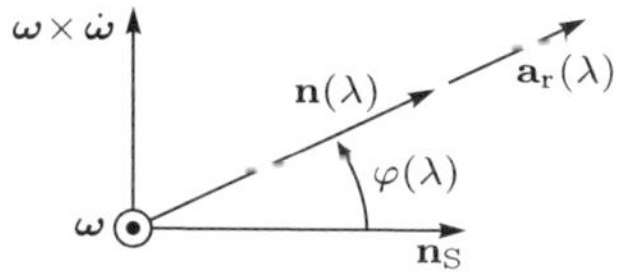

Fig. 2.7. Collinear vectors $\mathbf{n}(\lambda)$ and $\mathbf{a}_r(\lambda)$ and angle $\varphi(\lambda)$ in the plane of the mutually orthogonal vectors $\mathbf{n}_S$ and $\boldsymbol{\omega}\times\dot{\boldsymbol{\omega}}$ perpendicular to the ISA

Substitution into (2.86) yields the desired expression for the position vector $\mathbf{r}_S$ of the striction point:

$$\mathbf{r}_S = \mathbf{r}_A + \frac{\boldsymbol{\omega} \times \mathbf{v}_A}{\omega^2} + \frac{\mathbf{a}(0) \cdot \boldsymbol{\omega} \times \dot{\boldsymbol{\omega}}}{(\boldsymbol{\omega} \times \dot{\boldsymbol{\omega}})^2}\, \boldsymbol{\omega}\,. \tag{2.95}$$

The function $\varphi(\lambda)$ is found as follows. Equation (2.91) is written in the form

$$\mathbf{a}(\lambda) = \mathbf{a}_S + (\lambda_S - \lambda)\, \frac{\boldsymbol{\omega} \times \dot{\boldsymbol{\omega}}}{|\boldsymbol{\omega}|}\,. \tag{2.96}$$

In Fig. 2.7 the collinear vectors $\mathbf{n}(\lambda)$ and $\mathbf{a}_r(\lambda)$ are shown in the canonical reference frame. They are located in the plane spanned by $\mathbf{n}_S$ and $\boldsymbol{\omega} \times \dot{\boldsymbol{\omega}}$. The angle $\varphi(\lambda)$ is determined by

$$\cos\varphi(\lambda) = \frac{\mathbf{a}_r(\lambda) \cdot \mathbf{n}_S}{|\mathbf{a}_r(\lambda)|}\,, \qquad \sin\varphi(\lambda) = \frac{\mathbf{a}_r(\lambda) \cdot \boldsymbol{\omega} \times \dot{\boldsymbol{\omega}}}{|\mathbf{a}_r(\lambda)|\, |\boldsymbol{\omega} \times \dot{\boldsymbol{\omega}}|}\,. \tag{2.97}$$

The numerator expressions are with (2.90), (2.91), (2.93) and (2.96)

$$\mathbf{a}_r(\lambda) \cdot \mathbf{n}_S = \mathbf{a}(\lambda) \cdot \mathbf{n}_S = \mathbf{a}(0) \cdot \mathbf{n}_S = \frac{\mathbf{a}(0) \cdot [(\boldsymbol{\omega} \times \dot{\boldsymbol{\omega}}) \times \boldsymbol{\omega}]}{|\boldsymbol{\omega} \times \dot{\boldsymbol{\omega}}|\, |\boldsymbol{\omega}|}\,, \tag{2.98}$$

$$\mathbf{a}_r(\lambda) \cdot \boldsymbol{\omega} \times \dot{\boldsymbol{\omega}} = (\lambda_S - \lambda)\, \frac{(\boldsymbol{\omega} \times \dot{\boldsymbol{\omega}})^2}{|\boldsymbol{\omega}|}\,. \tag{2.99}$$

From these expressions it follows that

$$\tan\varphi(\lambda) = \frac{\lambda_S - \lambda}{\delta} \tag{2.100}$$

with

$$\delta = \frac{\mathbf{a}(0) \cdot [(\boldsymbol{\omega} \times \dot{\boldsymbol{\omega}}) \times \boldsymbol{\omega}]}{(\boldsymbol{\omega} \times \dot{\boldsymbol{\omega}})^2}\,. \tag{2.101}$$

Thus, one has the simple result: On the ISA at $t = \text{const}$ (arbitrary) $\tan\varphi(\lambda)$ is proportional to the distance from the striction point. The constant parameter δ is called *distribution parameter*. In (2.98) also $\mathbf{a}(\lambda) \cdot \mathbf{n}_S = \mathbf{a}_S \cdot \mathbf{n}_S$ is correct. This yields the more appealing formula

$$\delta = \frac{\mathbf{a}_S \cdot [(\boldsymbol{\omega} \times \dot{\boldsymbol{\omega}}) \times \boldsymbol{\omega}]}{(\boldsymbol{\omega} \times \dot{\boldsymbol{\omega}})^2}\,. \tag{2.102}$$

At the beginning of this section it has been shown that the two raccording axodes are in tangential contact everywhere along the ISA. For this to be the case it suffices that the two axodes have, at every instant of time, the same striction point and the same distribution parameter. The moving axode and the fixed axode exchange their roles when the transition from motion to inverse motion is made. According to Theorem 2.3 the quantities $\mathbf{v}_A$, $\mathbf{a}(0)$, $\mathbf{a}_S$, $\boldsymbol{\omega}$ and $\dot{\boldsymbol{\omega}}$ are multiplied by -1. Neither $\mathbf{r}_S$ nor δ is effected by this change.

Through (2.101), (2.95) and (2.90)[3] the distribution parameter δ, the striction point $\mathbf{r}_S$ and the normal $\mathbf{n}_S$ and, with it, the canonical reference frame are expressed in terms of the same five kinematical quantities $\mathbf{r}_A$, $\mathbf{v}_A$, $\mathbf{a}_A$, $\boldsymbol{\omega}$ and $\dot{\boldsymbol{\omega}}$ which determine the velocity screw and the raccording axodes ($\mathbf{u}$ and p in (2.83)). Imagine an arbitrary spatial 1-d.o.f. mechanism. As a result of a kinematics analysis the five kinematical quantities are explicitly available for each link of the mechanism as functions of a single input variable φ and of its derivatives $\dot{\varphi}$ and $\ddot{\varphi}$. With the said equations the ISA, the striction point, the canonical reference frame and the distribution parameter are known for every link. The spatial trajectories of the striction point S on the fixed axode and on the moving axode are called *striction lines.* The line on the fixed axode is determined by the coordinates of $\mathbf{r}_S(\varphi)$ in the reference base $\underline{\mathbf{e}}^1$. The line on the moving axode is determined by the coordinates of $\mathbf{r}_S(\varphi) - \mathbf{r}_A(\varphi)$ in base $\underline{\mathbf{e}}^2$. In general, the two striction lines are not tangent to one another at the striction point.

The shape of the two axodes and the raccording motion are particularly simple if the body under consideration is moving about a point A fixed in the reference base $\underline{\mathbf{e}}^1$. Equations (2.83) yield $\mathbf{u} = \mathbf{0}$ and $p = 0$. This means that the ISA is always passing through the fixed point A. Both axodes are, therefore, general cones, one fixed in the body and the other fixed in the reference base (Fig. 2.8). The raccording motion is pure rolling without slipping. That the two cones are in tangential contact along the ISA is a consequence of (2.75) which states that $\boldsymbol{\omega}(t)$ is sweeping out both cones with equal rates of change.

In the simplest possible case, the case of planar motion, the raccording axodes are general cylinders which are rolling one on the other without translation along the ISA. It suffices to visualize the intersection curves of the cylinders with the plane of motion. These curves are called centrodes and the point of contact, i.e. the intersection of the ISA with the plane of motion, is the instantaneous center of rotation.

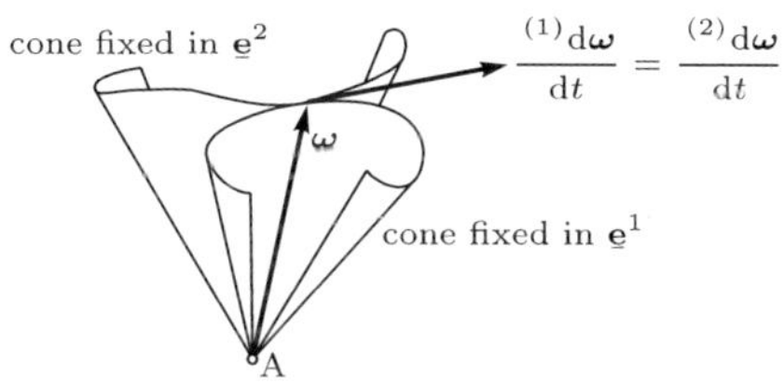

Fig. 2.8. The cones generated by $\boldsymbol{\omega}$ in two bases $\underline{\mathbf{e}}^1$ and $\underline{\mathbf{e}}^2$, $\boldsymbol{\omega}$ being the angular velocity of $\underline{\mathbf{e}}^2$ relative to $\underline{\mathbf{e}}^1$

[3] Reported in Wittenburg [101].

Problem 2.6. Derive from (2.82) an expression for the jerk $\mathbf{j}$ (the time derivative of the acceleration $\mathbf{a}$) of body-fixed points relative to $\underline{\mathbf{e}}^1$.

Problem 2.7. Given are in some reference base the instantaneous position vectors $\mathbf{r}_1$, $\mathbf{r}_2$, $\mathbf{r}_3$ of three noncollinear points P_1, P_2, P_3 of a rigid body as well as the instantaneous velocities $\mathbf{v}_1$, $\mathbf{v}_2$, $\mathbf{v}_3$ of these points. To be determined is the instantaneous angular velocity $\boldsymbol{\omega}$ of the body. Since none of the three points is dominant in any way it is possible to find an expression for $\boldsymbol{\omega}$ which is cyclically symmetric with respect to the indices 1, 2, 3.

2.3 Kinematic Differential Equations

The angular velocity of a body cannot, in general, be represented as time derivative of another vector. This is possible only in the trivial case when the direction of $\boldsymbol{\omega}$ is constant. The coordinates ω_1, ω_2 and ω_3 of $\boldsymbol{\omega}$ in a body-fixed vector base do not, therefore, represent generalized velocities in the sense of analytical mechanics. From this it follows that generalized coordinates for the angular orientation of a body cannot be determined from $\omega_i(t)$ $(i = 1, 2, 3)$ by simple integration. Instead, differential equations must be solved in which $\omega_i(t)$ $(i = 1, 2, 3)$ appear as variable coefficients. These equations will now be formulated for direction cosines, Euler angles, Bryan angles and Euler–Rodrigues parameters as generalized coordinates.

2.3.1 Direction Cosines

Let $\mathbf{r}$ be the position vector of an arbitrary body-fixed point. Its constant coordinate matrix $\underline{r}^2$ in $\underline{\mathbf{e}}^2$ and its time-varying coordinate matrix $\underline{r}^1(t)$ in $\underline{\mathbf{e}}^1$ are related through the equation $\underline{r}^1(t) = \underline{A}^{12}(t)\,\underline{r}^2$. Differentiation yields the velocity coordinates relative to $\underline{\mathbf{e}}^1$ and decomposed in $\underline{\mathbf{e}}^1$:

$$\dot{\underline{r}}^1 = \dot{\underline{A}}^{12}\underline{r}^2\,. \tag{2.103}$$

The velocity vector is also given in the form $\boldsymbol{\omega} \times \mathbf{r}$. Its coordinate matrix in $\underline{\mathbf{e}}^2$ is $\underline{\tilde{\omega}}\,\underline{r}^2$ and in $\underline{\mathbf{e}}^1$ it is $\underline{A}^{12}\underline{\tilde{\omega}}\,\underline{r}^2$. Comparison with (2.103) yields the equation $\dot{\underline{A}}^{12}\underline{r}^2 = \underline{A}^{12}\underline{\tilde{\omega}}\,\underline{r}^2$. This is valid for any arbitrary matrix $\underline{r}^2$. Hence, also the preceding factors on both sides are equal. Transposition yields

$$\dot{\underline{A}}^{21} = -\underline{\tilde{\omega}}\,\underline{A}^{21}\,. \tag{2.104}$$

These are, in matrix form, the desired differential equations for direction cosines. They are called *Poisson's equations.* They are linear equations with time-varying coefficients. Equations for individual direction cosines are found by multiplying out:

$$\dot{a}^{21}_{11} = \omega_3 a^{21}_{21} - \omega_2 a^{21}_{31} \qquad \text{etc.} \tag{2.105}$$

Because of the six constraint Eqs. (2.2) only three differential equations need be integrated.

The kinematic differential equations determine the direction cosines when $\omega_1(t), \omega_2(t), \omega_3(t)$ are known from an analytical or numerical integration of dynamics equations of motion. Frequently, dynamics equations of motion contain as unknowns not only $\omega_1, \omega_2, \omega_3$ and $\dot{\omega}_1, \dot{\omega}_2, \dot{\omega}_3$ but also the direction cosines themselves. In such cases the dynamics equations of motion and the kinematic differential equations must be integrated simultaneously.

Equally interesting is the inverse problem: Given is as function of time the direction cosine matrix. Then, (2.104) yields for the coordinates of $\boldsymbol{\omega}$ the equation

$$\underline{\tilde{\omega}} = -\underline{\dot{A}}^{21}(t)\underline{A}^{21^T}(t) . \tag{2.106}$$

2.3.2 Euler Angles

Figure 2.1 yields for the angular velocity $\boldsymbol{\omega}$ of the base $\underline{\mathbf{e}}^2$ the expression

$$\boldsymbol{\omega} = \dot{\psi}\mathbf{e}_3^1 + \dot{\theta}\mathbf{e}_1^{2'} + \dot{\phi}\mathbf{e}_3^2 . \tag{2.107}$$

The vectors are decomposed in base $\underline{\mathbf{e}}^2$. With the help of (2.3) this results in the coordinate equations

$$\begin{bmatrix} \omega_1 \\ \omega_2 \\ \omega_3 \end{bmatrix} = \begin{bmatrix} \sin\theta\sin\phi & \cos\phi & 0 \\ \sin\theta\cos\phi & -\sin\phi & 0 \\ \cos\theta & 0 & 1 \end{bmatrix} \begin{bmatrix} \dot{\psi} \\ \dot{\theta} \\ \dot{\phi} \end{bmatrix} . \tag{2.108}$$

Inversion yields the desired differential equations

$$\begin{bmatrix} \dot{\psi} \\ \dot{\theta} \\ \dot{\phi} \end{bmatrix} = \begin{bmatrix} \sin\phi/\sin\theta & \cos\phi/\sin\theta & 0 \\ \cos\phi & -\sin\phi & 0 \\ -\sin\phi\cot\theta & -\cos\phi\cot\theta & 1 \end{bmatrix} \begin{bmatrix} \omega_1 \\ \omega_2 \\ \omega_3 \end{bmatrix} . \tag{2.109}$$

These equations are nonlinear. Numerical problems arise when θ gets close to one of the critical values $n\pi$ $(n = 0, \pm 1, \ldots)$.

2.3.3 Bryan Angles

Figure 2.3 yields for the angular velocity $\boldsymbol{\omega}$ of the base $\underline{\mathbf{e}}^2$ the expression

$$\boldsymbol{\omega} = \dot{\phi}_1\mathbf{e}_1^1 + \dot{\phi}_2\mathbf{e}_2^{2'} + \dot{\phi}_3\mathbf{e}_3^2 . \tag{2.110}$$

The vectors are decomposed in base $\underline{\mathbf{e}}^2$. With the help of (2.7) this results in the coordinate equations

$$\begin{bmatrix} \omega_1 \\ \omega_2 \\ \omega_3 \end{bmatrix} = \begin{bmatrix} \cos\phi_2\cos\phi_3 & \sin\phi_3 & 0 \\ -\cos\phi_2\sin\phi_3 & \cos\phi_3 & 0 \\ \sin\phi_2 & 0 & 1 \end{bmatrix} \begin{bmatrix} \dot{\phi}_1 \\ \dot{\phi}_2 \\ \dot{\phi}_3 \end{bmatrix} . \tag{2.111}$$

Inversion yields the desired differential equations

$$\begin{bmatrix} \dot\phi_1 \\ \dot\phi_2 \\ \dot\phi_3 \end{bmatrix} = \begin{bmatrix} \cos\phi_3/\cos\phi_2 & -\sin\phi_3/\cos\phi_2 & 0 \\ \sin\phi_3 & \cos\phi_3 & 0 \\ -\cos\phi_3\tan\phi_2 & \sin\phi_3\tan\phi_2 & 1 \end{bmatrix} \begin{bmatrix} \omega_1 \\ \omega_2 \\ \omega_3 \end{bmatrix} . \tag{2.112}$$

These equations are nonlinear. Numerical problems arise when ϕ_2 gets close to one of the critical values $\pi/2 + n\pi$ $(n = 0, \pm 1, \ldots)$.

In Sect. 2.1.2 it has been shown that in the case of small angles ϕ_1, ϕ_2, ϕ_3 linearization of the direction cosine matrix is possible (see (2.10)). Linearization of (2.111) yields $\omega_i \approx \dot\phi_i$ $(i = 1, 2, 3)$. Thus, the angular orientation of the body is the result of simple integration:

$$\phi_i \approx \int \omega_i \,\mathrm{d}t \qquad (|\phi_i| \ll 1;\ i = 1, 2, 3) . \tag{2.113}$$

2.3.4 Euler–Rodrigues Parameters

Kinematic differential equations for Euler–Rodrigues parameters are established by two different methods. The first method starts out from (2.60) for the Euler–Rodrigues parameters of the resultant of two consecutive rotations:

$$q_{0_{\mathrm{res}}} = q_{02}q_{01} - \mathbf{q}_2 \cdot \mathbf{q}_1 , \qquad \mathbf{q}_{\mathrm{res}} = q_{02}\mathbf{q}_1 + q_{01}\mathbf{q}_2 + \mathbf{q}_2 \times \mathbf{q}_1 . \tag{2.114}$$

The parameters $(q_{01}, \mathbf{q}_1)$ are attributed to time t during a continuous motion and the parameters $(q_{0_{\mathrm{res}}}, \mathbf{q}_{\mathrm{res}})$ are attributed to time $t + \mathrm{d}t$. The quantities $(q_{02}, \mathbf{q}_2)$ represent the Euler–Rodrigues parameters of the differential rotation $\boldsymbol{\omega}\,\mathrm{d}t = \mathbf{e}_\omega \omega\,\mathrm{d}t$ during the time interval $\mathrm{d}t$ (unit vector $\mathbf{e}_\omega$). These parameters are

$$q_{02} = \cos\left(\frac{1}{2}\omega\,\mathrm{d}t\right) = 1 , \qquad \mathbf{q}_2 = \mathbf{e}_\omega \sin\left(\frac{1}{2}\omega\,\mathrm{d}t\right) = \frac{1}{2}\boldsymbol{\omega}\,\mathrm{d}t . \tag{2.115}$$

With these expressions Eqs. (2.114) take the forms

$$q_0(t+\mathrm{d}t) = q_0(t) - \frac{1}{2}\boldsymbol{\omega}\cdot\mathbf{q}(t)\,\mathrm{d}t , \quad \mathbf{q}(t+\mathrm{d}t) = \mathbf{q}(t) + \frac{1}{2}[q_0(t)\boldsymbol{\omega} + \boldsymbol{\omega}\times\mathbf{q}(t)]\,\mathrm{d}t . \tag{2.116}$$

Division by $\mathrm{d}t$ yields for $\dot q_0$ and for the derivative of $\mathbf{q}$ in the reference base $\underline{\mathbf{e}}^1$ the differential equations

$$\dot q_0 = -\frac{1}{2}\boldsymbol{\omega}\cdot\mathbf{q} , \qquad \frac{{}^{(1)}\mathrm{d}\mathbf{q}}{\mathrm{d}t} = \frac{1}{2}(q_0\boldsymbol{\omega} + \boldsymbol{\omega}\times\mathbf{q}) . \tag{2.117}$$

In kinematics as well as in dynamics the derivative of $\mathbf{q}$ in the body-fixed base $\underline{\mathbf{e}}^2$ is needed. By the general rule (2.74) this derivative is

$$\frac{{}^{(2)}\mathrm{d}\mathbf{q}}{\mathrm{d}t} = \frac{1}{2}(q_0\boldsymbol{\omega} - \boldsymbol{\omega}\times\mathbf{q}) . \tag{2.118}$$

Decomposition of this equation in $\underline{\mathbf{e}}^2$ yields the desired kinematic differential equations:

$$\begin{bmatrix} \dot{q}_0 \\ \dot{\underline{q}} \end{bmatrix} = \frac{1}{2} \begin{bmatrix} 0 & -\underline{\omega}^T \\ \underline{\omega} & -\tilde{\underline{\omega}} \end{bmatrix} \begin{bmatrix} q_0 \\ \underline{q} \end{bmatrix} . \tag{2.119}$$

The equations are linear with a time-varying skew-symmetric coefficient matrix.

The second method leading to the same equations uses relationships between Euler–Rodrigues parameters and Euler angles. The elements $(3,3)$ of the direction cosine matrices in (2.5) and in (2.35) are identical, whence follows

$$\cos\theta = q_0^2 - q_1^2 - q_2^2 + q_3^2 . \tag{2.120}$$

Furthermore, from (2.41) the relations are known

$$\tan\frac{\psi+\phi}{2} = \frac{q_3}{q_0} , \qquad \tan\frac{\psi-\phi}{2} = \frac{q_2}{q_1} . \tag{2.121}$$

This yields

$$\psi = \tan^{-1}\frac{q_3}{q_0} + \tan^{-1}\frac{q_2}{q_1} , \qquad \phi = \tan^{-1}\frac{q_3}{q_0} - \tan^{-1}\frac{q_2}{q_1} . \tag{2.122}$$

Differentiation with respect to time produces the equations

$$\dot{\psi} = \frac{q_0\dot{q}_3 - q_3\dot{q}_0}{q_0^2+q_3^2} + \frac{q_1\dot{q}_2 - q_2\dot{q}_1}{q_1^2+q_2^2} , \qquad \dot{\phi} = \frac{q_0\dot{q}_3 - q_3\dot{q}_0}{q_0^2+q_3^2} - \frac{q_1\dot{q}_2 - q_2\dot{q}_1}{q_1^2+q_2^2} . \tag{2.123}$$

These expressions and the expression for $\cos\theta$ from (2.120) are substituted into the third differential Eq. (2.108) for Euler angles: $\omega_3 = \dot{\psi}\cos\theta + \dot{\phi}$. This results in the equation

$$\begin{aligned} \omega_3 = {} & \frac{q_0\dot{q}_3 - q_3\dot{q}_0}{q_0^2+q_3^2} \underbrace{(q_0^2 - q_1^2 - q_2^2 + q_3^2 + 1)}_{2(q_0^2+q_3^2)} \\ & + \frac{q_1\dot{q}_2 - q_2\dot{q}_1}{q_1^2+q_2^2} \underbrace{(q_0^2 - q_1^2 - q_2^2 + q_3^2 - 1)}_{-2(q_1^2+q_2^2)} \\ = {} & 2(q_0\dot{q}_3 - q_3\dot{q}_0 - q_1\dot{q}_2 + q_2\dot{q}_1) . \end{aligned} \tag{2.124}$$

Equations for ω_2 and ω_1 are obtained by cyclic permutation of the indices 1, 2 and 3. The three equations constitute the rows 2, 3 and 4 of the matrix equation below. The first row represents the time derivative of the constraint equation $1 = q_0^2 + q_1^2 + q_2^2 + q_3^2$.

$$\begin{bmatrix} 0 \\ \omega_1 \\ \omega_2 \\ \omega_3 \end{bmatrix} = 2 \begin{bmatrix} q_0 & q_1 & q_2 & q_3 \\ -q_1 & q_0 & q_3 & -q_2 \\ -q_2 & -q_3 & q_0 & q_1 \\ -q_3 & q_2 & -q_1 & q_0 \end{bmatrix} \begin{bmatrix} \dot{q}_0 \\ \dot{q}_1 \\ \dot{q}_2 \\ \dot{q}_3 \end{bmatrix} . \tag{2.125}$$

The coefficient matrix is orthogonal (the scalar product of any two rows or columns i and j equals δ_{ij}). Consequently, the inverse equals the transpose. Thus, an explicit expression is obtained for the column matrix $[\dot{q}_0\ \dot{q}_1\ \dot{q}_2\ \dot{q}_3]^T$. Rearranging this expression results again in (2.119).

In the course of numerical integration of the equations inevitable numerical errors have the effect that computed quantities $q_i(t)$ $(i = 0, 1, 2, 3)$ do not strictly satisfy the constraint equation $q_0^2 + q_1^2 + q_2^2 + q_3^2 = 1$. Such faulty quantities must not be used for calculating from (2.35) a direction cosine matrix for the transformation of vector coordinates. Before doing so the quantities must be replaced by corrected quantities which satisfy the constraint equation. Corrected quantities q_i^* $(i = 0, \ldots, 3)$ are calculated such that the sum of squares of the corrections, i.e. $\sum_{i=0}^{3}(q_i^* - q_i)^2$, is minimal. This criterion yields the formulas

$$q_i^* = \frac{q_i}{\sqrt{\sum_{j=0}^{3} q_j^2}} \qquad (i = 0, \ldots, 3)\,. \tag{2.126}$$

Problem 2.8. A rigid body is suspended in two gimbals as is shown in Fig. 2.9. In the outer gimbal the two axes are offset from 90° by an angle α and in the inner gimbal by an angle β. Let ϕ_1, ϕ_2 and ϕ_3 be defined like Bryan angles, i.e. as rotation angles of the outer gimbal about $\mathbf{e}_1^1$, of the inner gimbal relative to the outer gimbal and of the body relative to the inner gimbal, respectively. For $\phi_1 = \phi_2 = \phi_3 = 0$ the planes of the gimbals are perpendicular to one another and, furthermore, the base vectors $\mathbf{e}_1^1$ and $\mathbf{e}_2^1$ of the reference base as well as the body-fixed vector $\mathbf{e}_1^2$ lie in the plane of the outer gimbal. Develop (i) an expression for the direction cosine matrix $\underline{A}^{21}$ and (ii) kinematic differential equations similar to (2.112).

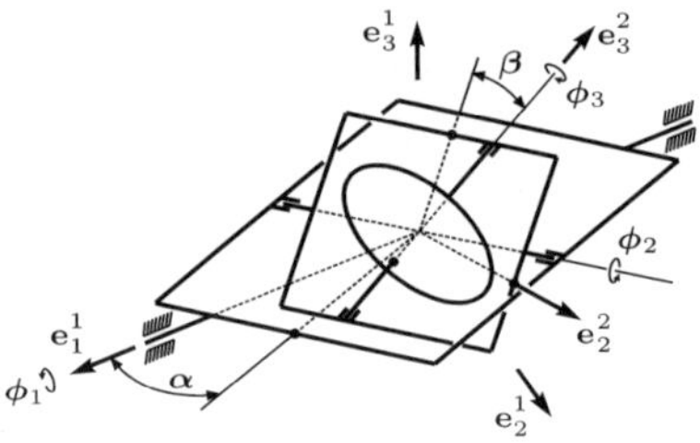

Fig. 2.9. Two-gimbal suspension with nonorthogonal gimbal axes

3

Basic Principles of Rigid Body Dynamics

The two most important physical quantities in rigid body dynamics are kinetic energy and angular momentum. Both lead directly to the definition of the inertia tensor.

3.1 Kinetic Energy

The kinetic energy is a scalar quantity. For a point mass m it is defined as $T = m\dot{\mathbf{r}}^2/2$ where $\dot{\mathbf{r}}$ is the absolute velocity of m, i.e. its velocity relative to an inertial reference base. Throughout this chapter a dot over a vector designates differentiation with respect to time in an inertial base. For a rigid body as for any extended body the kinetic energy is the integral

$$T = \frac{1}{2}\int_m \dot{\mathbf{r}}^2 \,\mathrm{d}m \,. \tag{3.1}$$

Let A be an arbitrary point fixed on the body (Fig. 3.1). The absolute velocity $\dot{\mathbf{r}}$ of a mass particle $\mathrm{d}m$ is according to (2.81) $\dot{\mathbf{r}} = \dot{\mathbf{r}}_\mathrm{A} + \boldsymbol{\omega} \times \boldsymbol{\varrho}$ where $\dot{\mathbf{r}}_\mathrm{A}$ is the absolute velocity of the reference point A, $\boldsymbol{\omega}$ the absolute angular velocity of the body and $\boldsymbol{\varrho}$ the radius vector from A to the mass particle. The

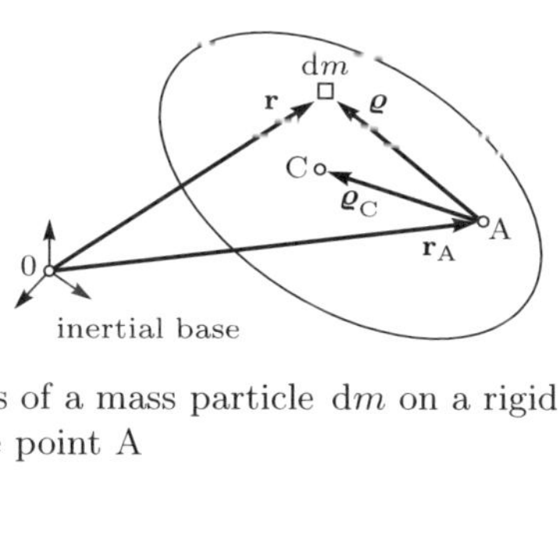

Fig. 3.1. Radius vectors of a mass particle $\mathrm{d}m$ on a rigid body. Center of mass C and body-fixed reference point A

point C in Fig. 3.1 at the radius vector $\boldsymbol{\varrho}_{\mathrm{C}} = \overrightarrow{\mathrm{AC}}$ indicates the body center of mass. Evaluation of the integral yields

$$T = \frac{1}{2}\dot{\mathbf{r}}_{\mathrm{A}}^2 m + \dot{\mathbf{r}}_{\mathrm{A}} \cdot (\boldsymbol{\omega} \times \boldsymbol{\varrho}_{\mathrm{C}}) m + \frac{1}{2}\int_m (\boldsymbol{\omega} \times \boldsymbol{\varrho})^2 \,\mathrm{d}m \,. \tag{3.2}$$

This expression is particularly simple if either the body-fixed point A is also fixed in inertial space or the center of mass C is used as reference point A. In the former case $\dot{\mathbf{r}}_{\mathrm{A}} = \mathbf{0}$ so that the first two terms equal zero. In the latter case $\boldsymbol{\varrho}_{\mathrm{C}} = \mathbf{0}$ so that the central term vanishes. The first term is then called kinetic energy of translation T_{trans} and the third term kinetic energy of rotation T_{rot}. For the integrand in the third term the identity holds[1]

$$(\boldsymbol{\omega} \times \boldsymbol{\varrho})^2 = \boldsymbol{\omega} \cdot [\boldsymbol{\varrho} \times (\boldsymbol{\omega} \times \boldsymbol{\varrho})] \,. \tag{3.3}$$

In this expression tensor notation is introduced (cf. (1.40)):

$$\boldsymbol{\varrho} \times (\boldsymbol{\omega} \times \boldsymbol{\varrho}) = \left(\varrho^2 \mathsf{I} - \boldsymbol{\varrho}\boldsymbol{\varrho}\right) \cdot \boldsymbol{\omega} \,. \tag{3.4}$$

Together with the previous equation this yields

$$\int_m (\boldsymbol{\omega} \times \boldsymbol{\varrho})^2 \,\mathrm{d}m = \boldsymbol{\omega} \cdot \mathsf{J}^{\mathrm{A}} \cdot \boldsymbol{\omega} \tag{3.5}$$

where J^{A} is the tensor

$$\mathsf{J}^{\mathrm{A}} = \int_m \left(\varrho^2 \mathsf{I} - \boldsymbol{\varrho}\boldsymbol{\varrho}\right) \,\mathrm{d}m \,. \tag{3.6}$$

It is called inertia tensor of the body with respect to A. In a body-fixed base $\underline{\mathbf{e}}$ in which $\boldsymbol{\varrho}$ has the coordinate matrix $\underline{\varrho}$ J^{A} has the coordinate matrix

$$\underline{J}^{\mathrm{A}} = \int_m \left(\underline{\varrho}^T \underline{\varrho}\, \underline{I} - \underline{\varrho}\, \underline{\varrho}^T\right) \,\mathrm{d}m \tag{3.7}$$

or explicitly

$$\underline{J}^{\mathrm{A}} = \begin{bmatrix} \int_m \left(\varrho_2^2 + \varrho_3^2\right) \,\mathrm{d}m & -\int_m \varrho_1\varrho_2 \,\mathrm{d}m & -\int_m \varrho_1\varrho_3 \,\mathrm{d}m \\ & \int_m \left(\varrho_3^2 + \varrho_1^2\right) \,\mathrm{d}m & -\int_m \varrho_2\varrho_3 \,\mathrm{d}m \\ \text{symmetric} & & \int_m \left(\varrho_1^2 + \varrho_2^2\right) \,\mathrm{d}m \end{bmatrix} . \tag{3.8}$$

This symmetric matrix is called inertia matrix of the body with respect to A and to the chosen body-fixed base $\underline{\mathbf{e}}$. It is a geometric quantity which is determined by the mass distribution of the body. The integrals along the diagonal are called moments of inertia J_{11}, J_{22} and J_{33}, and the off-diagonal elements including the minus signs are called products of inertia[2] J_{12}, J_{13}

[1] In mixed products the symbols of dot and cross multiplication can be interchanged so that $\boldsymbol{\omega} \times \boldsymbol{\varrho} \cdot \mathbf{c} = \boldsymbol{\omega} \cdot \boldsymbol{\varrho} \times \mathbf{c}$. Here, $\mathbf{c}$ equals $\boldsymbol{\omega} \times \boldsymbol{\varrho}$.

[2] In some books the integrals without minus signs are referred to as products of inertia.

and J_{23}:

$$J_{ii} = \int_m \left(\varrho_j^2 + \varrho_k^2\right)\,\mathrm{d}m \qquad (i, j, k \text{ different})\,, \tag{3.9}$$

$$J_{ij} = -\int_m \varrho_i \varrho_j\,\mathrm{d}m \qquad (i \neq j)\,. \tag{3.10}$$

Moments of inertia are nonnegative whereas products of inertia can be positive, negative or zero.

With (3.5) the kinetic energy expression in (3.2) becomes

$$T = \frac{1}{2}\dot{\mathbf{r}}_{\mathrm{A}}^2 m + \dot{\mathbf{r}}_{\mathrm{A}} \cdot (\boldsymbol{\omega} \times \boldsymbol{\varrho}_{\mathrm{C}})m + \frac{1}{2}\boldsymbol{\omega} \cdot \mathsf{J}^{\mathrm{A}} \cdot \boldsymbol{\omega}\,. \tag{3.11}$$

Problem 3.1. Let the input shaft (rotation angle q_1) and the output shaft (rotation angle q_2) of a device be coupled by a mechanism which produces a nonuniform transmission ratio $\dot{q}_1/\dot{q}_2 = i(q_1) \neq$ const. A typical example is a Hooke's joint. A rotor with moment of inertia J_1 is mounted on the input shaft and another rotor with moment of inertia J_2 on the output shaft. Compared with these rotors the masses and moments of inertia of the coupling mechanism are assumed to be negligible. Furthermore, it is assumed that no torques are applied to the two shafts and that no springs and dampers are present in the entire system. Under these conditions the system is idling with time-varying angular velocities. Determine, for given initial conditions, the function $\dot{q}_1 = f(q_1)$ defining the phase portrait.

3.2 Angular Momentum

The absolute angular momentum – also referred to as absolute moment of momentum – of a point mass m having an absolute velocity $\dot{\mathbf{r}}$ is a vector. For its definition the specification of a reference point is required. Let 0 be a point fixed in inertial space. The absolute angular momentum with respect to 0 is $\mathbf{L}^0 = \mathbf{r} \times \dot{\mathbf{r}}m$ where $\mathbf{r}$ is the radius vector from 0 to the point mass The expression explains the name moment of momentum since $\dot{\mathbf{r}}m$ is the linear momentum of the point mass. For a rigid body as for any extended body the absolute angular momentum with respect to 0 is the integral

$$\mathbf{L}^0 = \int_m \mathbf{r} \times \dot{\mathbf{r}}\,\mathrm{d}m\,. \tag{3.12}$$

The vectors under the integral are those of Fig. 3.1. With $\mathbf{r} = \mathbf{r}_{\mathrm{A}} + \boldsymbol{\varrho}$ and $\dot{\mathbf{r}} = \dot{\mathbf{r}}_{\mathrm{A}} + \boldsymbol{\omega} \times \boldsymbol{\varrho}$ the integral becomes

$$\begin{aligned}\mathbf{L}^0 &= \int_m (\mathbf{r}_{\mathrm{A}} + \boldsymbol{\varrho}) \times (\dot{\mathbf{r}}_{\mathrm{A}} + \boldsymbol{\omega} \times \boldsymbol{\varrho})\,\mathrm{d}m \\ &= \mathbf{r}_{\mathrm{A}} \times (\dot{\mathbf{r}}_{\mathrm{A}} + \boldsymbol{\omega} \times \boldsymbol{\varrho}_{\mathrm{C}})m + \boldsymbol{\varrho}_{\mathrm{C}} \times \dot{\mathbf{r}}_{\mathrm{A}} m + \int_m \boldsymbol{\varrho} \times (\boldsymbol{\omega} \times \boldsymbol{\varrho})\,\mathrm{d}m\,.\end{aligned} \tag{3.13}$$

The expression $(\dot{\mathbf{r}}_\mathrm{A} + \boldsymbol{\omega} \times \boldsymbol{\varrho}_\mathrm{C})$ is the absolute velocity $\dot{\mathbf{r}}_\mathrm{C}$ of the center of mass C so that $(\dot{\mathbf{r}}_\mathrm{A} + \boldsymbol{\omega} \times \boldsymbol{\varrho}_\mathrm{C})m$ represents the absolute linear momentum of the body. The expression for $\mathbf{L}^0$ is particularly simple if either the body-fixed point A is also fixed in inertial space or the center of mass C is used as reference point A. In the former case $\dot{\mathbf{r}}_\mathrm{A} = \mathbf{0}$ and in the latter $\boldsymbol{\varrho}_\mathrm{C} = \mathbf{0}$. In both cases the central term in (3.13) vanishes. The first term then represents the angular momentum with respect to 0 due to translation of the body center of mass and the last term represents the angular momentum caused by rotation of the body. Using tensor notation this last term becomes

$$\int_m \boldsymbol{\varrho} \times (\boldsymbol{\omega} \times \boldsymbol{\varrho})\,\mathrm{d}m = \int_m (\varrho^2 \mathsf{I} - \boldsymbol{\varrho}\boldsymbol{\varrho})\,\mathrm{d}m \cdot \boldsymbol{\omega} = \mathsf{J}^\mathrm{A} \cdot \boldsymbol{\omega}\,. \tag{3.14}$$

This introduces, again, the inertia tensor of the body with respect to A. With this expression the angular momentum $\mathbf{L}^0$ is

$$\mathbf{L}^0 = \mathbf{r}_\mathrm{A} \times (\dot{\mathbf{r}}_\mathrm{A} + \boldsymbol{\omega} \times \boldsymbol{\varrho}_\mathrm{C})m + \boldsymbol{\varrho}_\mathrm{C} \times \dot{\mathbf{r}}_\mathrm{A} m + \mathsf{J}^\mathrm{A} \cdot \boldsymbol{\omega}\,. \tag{3.15}$$

3.3 Properties of Moments and of Products of Inertia

Moments and products of inertia are defined with respect to a body-fixed reference point and with respect to a body-fixed base. Therefore, it is necessary to investigate how moments and products of inertia change when the reference point and/or the base are changed.

3.3.1 Change of Reference Point. Reference Base Unchanged

Given the inertia tensor J^A of a body with respect to a point A what is the inertia tensor J^P with respect to another point P? For solving this problem it is sufficient to establish the relationship between J^A and the central inertia tensor J^C with respect to the body center of mass C. The same relationship with P instead of A then exists between J^P and J^C so that the desired relationship between J^P and J^A is obtained by eliminating J^C.

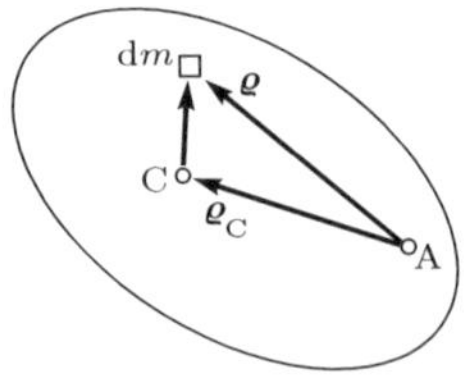

Fig. 3.2. Radius vectors of a mass particle $\mathrm{d}m$. Center of mass C and body-fixed reference point A

The points A and C and the vectors shown in Fig. 3.2 are the same as in Fig. 3.1. According to (3.6) the inertia tensor with respect to A is

$$\mathsf{J}^{\mathrm{A}} = \int_m \left(\boldsymbol{\varrho}^2 \mathsf{I} - \boldsymbol{\varrho}\boldsymbol{\varrho}\right)\,\mathrm{d}m\,. \tag{3.16}$$

The same formula with $\boldsymbol{\varrho} - \boldsymbol{\varrho}_{\mathrm{C}}$ instead of $\boldsymbol{\varrho}$ yields the inertia tensor J^{C}. Taking into account that $\int_m \boldsymbol{\varrho}\,\mathrm{d}m = \boldsymbol{\varrho}_{\mathrm{C}} m$ one calculates

$$\begin{aligned}\mathsf{J}^{\mathrm{C}} &= \int_m \left[(\boldsymbol{\varrho} - \boldsymbol{\varrho}_{\mathrm{C}})^2 \mathsf{I} - (\boldsymbol{\varrho} - \boldsymbol{\varrho}_{\mathrm{C}})(\boldsymbol{\varrho} - \boldsymbol{\varrho}_{\mathrm{C}})\right]\,\mathrm{d}m \\ &= \int_m \left[\left(\boldsymbol{\varrho}^2 - \boldsymbol{\varrho}_{\mathrm{C}}^2\right)\mathsf{I} - \left(\boldsymbol{\varrho}\boldsymbol{\varrho} - \boldsymbol{\varrho}_{\mathrm{C}}\boldsymbol{\varrho}_{\mathrm{C}}\right)\right]\,\mathrm{d}m \\ &= \mathsf{J}^{\mathrm{A}} - \left(\boldsymbol{\varrho}_{\mathrm{C}}^2 \mathsf{I} - \boldsymbol{\varrho}_{\mathrm{C}}\boldsymbol{\varrho}_{\mathrm{C}}\right) m\,.\end{aligned} \tag{3.17}$$

This is the desired relationship between J^{A} and J^{C}.

The tensors are now resolved in one and the same arbitrarily chosen body-fixed reference base. This yields the coordinate equation in matrix form

$$\underline{J}^{\mathrm{A}} = \underline{J}^{\mathrm{C}} + \left(\underline{\varrho}_{\mathrm{C}}^T \underline{\varrho}_{\mathrm{C}}\,\underline{I} - \underline{\varrho}_{\mathrm{C}}\,\underline{\varrho}_{\mathrm{C}}^T\right) m \tag{3.18}$$

and for the single moments and products of inertia

$$J_{ii}^{\mathrm{A}} = J_{ii}^{\mathrm{C}} + \left(\varrho_{\mathrm{C}j}^2 + \varrho_{\mathrm{C}k}^2\right) m \qquad (i, j, k \text{ different}) \tag{3.19}$$

$$J_{ij}^{\mathrm{A}} = J_{ij}^{\mathrm{C}} - \varrho_{\mathrm{C}i}\varrho_{\mathrm{C}j} m \qquad (i \neq j)\,. \tag{3.20}$$

These formulas are known as Huygens–Steiner formulas. It is seen that the moment of inertia J_{ii}^{C} about an axis $\mathbf{e}_i$ through the center of mass C is smaller than the moment of inertia J_{ii}^{A} about any parallel axis not passing through C. For products of inertia no such statement can be made because the term $\varrho_{\mathrm{C}i}\varrho_{\mathrm{C}j}$ can be positive or zero or negative.

3.3.2 Change of Reference Base. Reference Point Unchanged

Let $\underline{\mathbf{e}}^1$ and $\underline{\mathbf{e}}^2 = \underline{A}^{21}\underline{\mathbf{e}}^1$ be two vector bases fixed on one and the same body and let, furthermore, A be a point on this body. Given the inertia matrix $\underline{J}^1$ of the body with respect to A in base $\underline{\mathbf{e}}^1$ what is the inertia matrix $\underline{J}^2$ in base $\underline{\mathbf{e}}^2$ with respect to the same point A (for simplicity, the superscript A is omitted)? The answer is provided by the transformation equation (1.31) for tensor coordinates (similarity transformation):

$$\underline{J}^2 = \underline{A}^{21}\underline{J}^1\underline{A}^{12}\,. \tag{3.21}$$

From this matrix equation transformation formulas for individual moments and products of inertia are found by multiplying out the product on the right-hand side. These formulas are, however, rather complicated so that it is preferable to memorize only the matrix equation as a whole. Simple formulas result only in the case when $\underline{J}^1$ is a diagonal matrix with elements J_1, J_2, J_3 along the diagonal. These are the principal moments of inertia explained further below. With a_{ij} being the elements of $\underline{A}^{12}$ the formulas then read (in $J_{ij}^{(2)}$ the superscript (2) refers to base $\underline{\mathbf{e}}^2$)

$$J_{ij}^{(2)} = \sum_{k=1}^{3} J_k a_{ki} a_{kj} \qquad (i, j = 1, 2, 3) \, . \tag{3.22}$$

The similarity transformation and the Huygens–Steiner formulas (3.14) are commutative. This means: If moments and products of inertia are to be transformed from a reference point A in a base $\underline{\mathbf{e}}^1$ to another reference point P in another base $\underline{\mathbf{e}}^2$ then it is immaterial whether the transition from A to P in base $\underline{\mathbf{e}}^1$ is followed by the similarity transformation from $\underline{\mathbf{e}}^1$ to $\underline{\mathbf{e}}^2$ or whether the similarity transformation precedes the transition from A to P.

3.3.3 Principal Axes. Principal Moments of Inertia

Suppose that the inertia matrix $\underline{J}^1$ is known for a body with respect to a certain reference point A and for a certain body-fixed base $\underline{\mathbf{e}}^1$ and that it is not a diagonal matrix (the superscript A is omitted, again). Does another body-fixed base $\underline{\mathbf{e}}^2$ exist for which the inertia matrix $\underline{J}^2$ with respect to the same point A is diagonal? If so, how are the diagonal elements of $\underline{J}^2$ and the transformation matrix $\underline{A}^{12}$ relating $\underline{\mathbf{e}}^1$ and $\underline{\mathbf{e}}^2$ determined from $\underline{J}^1$? The answer to these questions is found as follows. The unknowns $\underline{J}^2$ and $\underline{A}^{12}$ are related to $\underline{J}^1$ by the similarity transformation (3.21). This equation is written in the form $\underline{J}^1 \underline{A}^{12} = \underline{A}^{12} \underline{J}^2$. Let J_1, J_2 and J_3 be the unknown diagonal elements of $\underline{J}^2$ and let $\underline{A}_i$ $(i = 1, 2, 3)$ be the ith column of $\underline{A}^{12}$, i.e. the coordinate matrix of the unknown base vector $\mathbf{e}_i^2$ in base $\underline{\mathbf{e}}^1$. The transformation equation is then equivalent to the set of equations $\underline{J}^1 \underline{A}_i = J_i \underline{A}_i$ $(i = 1, 2, 3)$. Each of these three equations represents the same eigenvalue problem

$$\left(\underline{J}^1 - J_i \underline{I}\right) \underline{A}_i = 0 \, . \tag{3.23}$$

The unknowns J_1, J_2 and J_3 are the eigenvalues. They are the solutions of the cubic equation

$$\det \left(\underline{J}^1 - J_i \underline{I}\right) = 0 \, . \tag{3.24}$$

The unknown column matrices $\underline{A}_i$ $(i = 1, 2, 3)$ are the corresponding eigenvectors. These results not only answer the question how $\underline{J}^2$ and $\underline{A}^{12}$ are determined from $\underline{J}^1$. They also show that for any inertia matrix $\underline{J}^1$ there exists a real base $\underline{\mathbf{e}}^2$ in which the inertia matrix $\underline{J}^2$ is diagonal and real. This follows from the fact that a symmetric matrix has real eigenvalues and eigenvectors and that, in addition, the eigenvectors are mutually orthogonal (see Gantmacher [19]).

The eigenvalues J_1, J_2 and J_3 are called principal moments of inertia (with respect to A), and the base vectors $\mathbf{e}_i^2$ $(i = 1,2,3)$ determine the directions of what is called principal axes of inertia (with respect to A). In determining these principal axes it must be distinguished whether all three eigenvalues are different from one another or whether (3.24) has a double or a triple root. In the case of three different eigenvalues each of the three coefficient matrices $(\underline{J}^1 - J_i\underline{I})$ in (3.23) has defect one. Each equation then determines the associated eigenvector $\underline{A}_i$, i.e. the coordinates of the principal axis base vector $\mathbf{e}_i^2$ (the sum of squares of the coordinates equals one). In the case of a double eigenvalue $J_1 = J_2$ the principal axis which corresponds to J_3 is determined uniquely as before. For the eigenvalue J_1 however, the coefficient matrix $(\underline{J}^1 - J_1\underline{I})$ in (3.23) has defect two so that the equation defines only a plane. This is the plane spanned by the two principal axes which correspond to J_1 and J_2. Any two mutually perpendicular axes in this plane (and passing through A) can serve as principal axes of inertia since for any such axis the moment of inertia has magnitude J_1. In the case of a triple eigenvalue $J_1 = J_2 = J_3$ the original matrix $\underline{J}^1$ is already diagonal. All axes passing through A are then principal axes of inertia. Examples of such bodies with A being the center of mass are homogeneous spheres, cubes, regular tetrahedra and cylinders with a ratio height/radius $= \sqrt{3}$.

3.3.4 Invariants. Inequalities

In connection with stability investigations and with other problems it is sometimes necessary to determine the signs of expressions which are composed of moments and products of inertia. In such cases the knowledge of invariants of the inertia matrix and of inequalities involving moments and products of inertia is helpful. Equation (3.24) represents a cubic equation for the principal moments of inertia. Since these moments are independent of the orientation of the vector base in which the inertia matrix $\underline{J}^1$ is measured, the coefficients of the cubic must also be independent. Omitting the superscript 1 this yields the invariants (J_{12}^2 is the square of J_{12})

$$\left.\begin{array}{l} \operatorname{tr}\underline{J} = J_{11} + J_{22} + J_{33} = J_1 + J_2 + J_3\,, \\ \left(J_{11}J_{22} - J_{12}^2\right) + \left(J_{22}J_{33} - J_{23}^2\right) + \left(J_{33}J_{11} - J_{31}^2\right) = J_1J_2 + J_2J_3 + J_3J_1\,, \\ \det\underline{J} = J_{11}J_{22}J_{33} - J_{11}J_{23}^2 - J_{22}J_{31}^2 - J_{33}J_{12}^2 + 2J_{12}J_{23}J_{31} = J_1J_2J_3\,. \end{array}\right\} \tag{3.25}$$

All three invariants are positive.

From the definitions of moments of inertia in (3.9) it follows that for i, j and k being any permutation of 1, 2 and 3

$$J_{ii} + J_{jj} = \int_m \left(\varrho_i^2 + \varrho_j^2 + 2\varrho_k^2\right)\,\mathrm{d}m = J_{kk} + 2\int_m \varrho_k^2\,\mathrm{d}m\,. \tag{3.26}$$

From this it follows that moments of inertia satisfy the so-called triangle inequalities

$$J_{ii} + J_{jj} \geq J_{kk} \qquad (i, j, k \text{ different}) \,. \tag{3.27}$$

Inequalities involving both moments and products of inertia are the following (J_{ij}^2 is the square of J_{ij}):

$$J_{ii} \geq 2|J_{jk}| \qquad (i, j, k \text{ different}) \,, \tag{3.28}$$

$$J_{ii}J_{jj} \geq J_{ij}^2 \qquad (i \neq j) \,. \tag{3.29}$$

The first one is a consequence of the inequality $(\varrho_j \pm \varrho_k)^2 \geq 0$, whence follows $\varrho_j^2 + \varrho_k^2 \geq 2|\varrho_j \varrho_k|$ and

$$\int_m \left(\varrho_j^2 + \varrho_k^2\right) \, \mathrm{d}m \geq 2 \int_m |\varrho_j \varrho_k| \, \mathrm{d}m \geq 2 \left| \int_m \varrho_j \varrho_k \, \mathrm{d}m \right| \,. \tag{3.30}$$

This is the inequality (3.28). The inequality (3.29) is derived from (3.22):

$$\begin{aligned} J_{ii}J_{jj} - J_{ij}^2 &= \left(\sum_{k=1}^{3} J_k a_{ki}^2\right)\left(\sum_{k=1}^{3} J_k a_{kj}^2\right) - \left(\sum_{k=1}^{3} J_k a_{ki} a_{kj}\right)^2 \\ &= J_1 J_2 (a_{1i}a_{2j} - a_{2i}a_{1j})^2 + J_2 J_3 (a_{2i}a_{3j} - a_{3i}a_{2j})^2 \\ &\quad + J_3 J_1 (a_{3i}a_{1j} - a_{1i}a_{3j})^2 \geq 0 \,. \end{aligned} \tag{3.31}$$

This inequality proves that in the second Eq. (3.25) each of the three terms on the left-hand side is individually nonnegative.

Problem 3.2. Under which conditions are the equality signs valid in (3.27), (3.28) and (3.29)?

Problem 3.3. Equation (3.21) for the transformation of moments and products of inertia is particularly simple if the two vector bases related by the matrix $\underline{A}^{12}$ have one base vector in common, say $\mathbf{e}_3^1 = \mathbf{e}_3^2$. Show that in this special case (3.21) can be interpreted geometrically by what is known as Mohr's circle.

Problem 3.4. In Fig. 3.3 a homogeneous solid tetrahedron of density ϱ with three mutually orthogonal edges of lengths ℓ, $\ell \tan\gamma$ and $\ell \tan\gamma$ is shown. Calculate, first, from triple integrals the moments and products of inertia in base $\underline{\mathbf{e}}$ with respect to the origin of this base. Then, determine the location of the body center of mass and calculate the central inertia matrix $\underline{J}^{\mathrm{C}}$ in the same base.

3.4 Angular Momentum Theorem

Newton's second axiom for translational motions finds its complement in the angular momentum theorem as the basic law governing rotational motions:

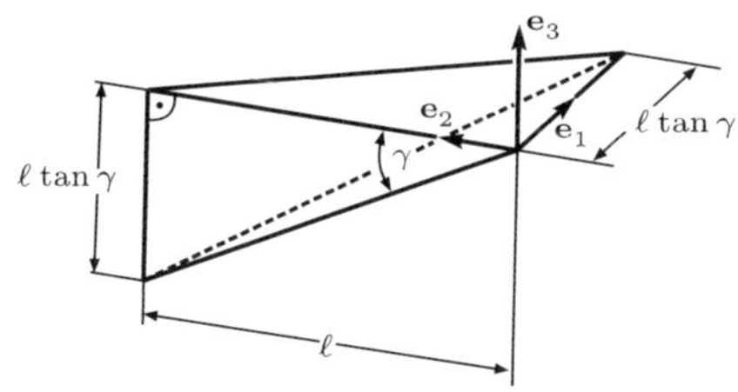

Fig. 3.3. Tetrahedron with three mutually orthogonal edges

Theorem 3.1. *For an arbitrary material system the absolute time derivative (i.e. the time derivative in an inertial reference base) of the absolute angular momentum with respect to a reference point 0 fixed in inertial space equals the resultant torque with respect to the same reference point:*

$$\dot{\mathbf{L}}^0 = \mathbf{M}^0 \,. \tag{3.32}$$

This law was first formulated as an axiom by Euler[3]. It cannot be derived from Newton's axioms.

The particular form of this law for a rigid body is obtained by substituting for $\mathbf{L}^0$ the expression given in (3.15). The time derivative in inertial space is obtained by applying (2.74) in combination with the product rule. The leading term of $\mathbf{L}^0$ has been shown to be $\mathbf{r}_\mathrm{A} \times \dot{\mathbf{r}}_\mathrm{C} m$ (see (3.13)). This explains the first two terms below.

$$\begin{aligned}\dot{\mathbf{L}}^0 = {} & m[\mathbf{r}_\mathrm{A} \times \ddot{\mathbf{r}}_\mathrm{C} + \dot{\mathbf{r}}_\mathrm{A} \times (\dot{\mathbf{r}}_\mathrm{A} + \boldsymbol{\omega} \times \boldsymbol{\varrho}_\mathrm{C}) + (\boldsymbol{\omega} \times \boldsymbol{\varrho}_\mathrm{C}) \times \dot{\mathbf{r}}_\mathrm{A} + \boldsymbol{\varrho}_\mathrm{C} \times \ddot{\mathbf{r}}_\mathrm{A}] \\ & + \mathsf{J}^\mathrm{A} \cdot \dot{\boldsymbol{\omega}} + \boldsymbol{\omega} \times \mathsf{J}^\mathrm{A} \cdot \boldsymbol{\omega} \,. \end{aligned} \tag{3.33}$$

Of the five terms in brackets the second is zero and the next two cancel each other. Consider next the resultant torque $\mathbf{M}^0$ on the body with respect to the point 0. If $\mathbf{F}$ is the resultant external force on the body and $\mathbf{M}^\mathrm{A}$ the resultant external torque with respect to A then $\mathbf{M}^0 = \mathbf{M}^\mathrm{A} + \mathbf{r}_\mathrm{A} \times \mathbf{F}$ or in view of Newton's law

$$m\ddot{\mathbf{r}}_\mathrm{C} = \mathbf{F} \tag{3.34}$$

$\mathbf{M}^0 = \mathbf{M}^\mathrm{A} + \mathbf{r}_\mathrm{A} \times \ddot{\mathbf{r}}_\mathrm{C} m$. When this together with (3.33) is substituted into (3.32) the angular momentum theorem for a rigid body is obtained in the final form

$$m\boldsymbol{\varrho}_\mathrm{C} \times \ddot{\mathbf{r}}_\mathrm{A} + \mathsf{J}^\mathrm{A} \cdot \dot{\boldsymbol{\omega}} + \boldsymbol{\omega} \times \mathsf{J}^\mathrm{A} \cdot \boldsymbol{\omega} = \mathbf{M}^\mathrm{A} \,. \tag{3.35}$$

The steps leading to this equation remain valid if the reference point A in Fig. 3.1 is not fixed on the body but moves relative to it. The only change that has to be made is to interpret $\ddot{\mathbf{r}}_\mathrm{A}$ as absolute acceleration not of A but of the body-fixed point which momentarily coincides with A. The angular momentum theorem takes its simplest form

$$\mathsf{J}^\mathrm{A} \cdot \dot{\boldsymbol{\omega}} + \boldsymbol{\omega} \times \mathsf{J}^\mathrm{A} \cdot \boldsymbol{\omega} = \mathbf{M}^\mathrm{A} \tag{3.36}$$

[3] For the history of the law see Truesdell [81].

if as reference point A either the body center of mass is chosen ($\boldsymbol{\varrho}_{\rm C} = \mathbf{0}$) or a point (if it exists) for which $\ddot{\mathbf{r}}_{\rm A} = \mathbf{0}$ or a point (if it exists) for which $\boldsymbol{\varrho}_{\rm C}$ and $\ddot{\mathbf{r}}_{\rm A}$ are permanently parallel to one another. The case $\ddot{\mathbf{r}}_{\rm A} = \mathbf{0}$ applies when a body-fixed point A is also fixed in inertial space. The case $\boldsymbol{\varrho}_{\rm C}$ permanently parallel $\ddot{\mathbf{r}}_{\rm A}$ is very special. When the body is a homogeneous rolling circular cylinder then the point of contact is a point A with this property.

From now on the superscript A will be omitted. The coordinate formulation of (3.36) in a body-fixed base is then

$$\underline{J}\,\underline{\dot{\omega}} + \underline{\tilde{\omega}}\,\underline{J}\,\underline{\omega} = \underline{M}\ . \tag{3.37}$$

Using, in particular, principal axes of inertia as directions for the base vectors this matrix equation is equivalent to

$$\left.\begin{aligned} J_1\dot{\omega}_1 - (J_2 - J_3)\omega_2\omega_3 &= M_1\ ,\\ J_2\dot{\omega}_2 - (J_3 - J_1)\omega_3\omega_1 &= M_2\ ,\\ J_3\dot{\omega}_3 - (J_1 - J_2)\omega_1\omega_2 &= M_3\ . \end{aligned}\right\} \tag{3.38}$$

These are Euler's equations of motion for a single rigid body. They can be integrated in closed form in a few special cases only. Mathematical problems arise for two reasons. One is the nonlinearity of the left-hand side of the equations. The other is the generally complicated form of the right-hand side expressions. Three types of problems can be distinguished. In the first and simplest case the torque coordinates M_1, M_2 and M_3 are known functions of ω_1, ω_2, ω_3 and t (and possibly of $\dot{\omega}_1$, $\dot{\omega}_2$, $\dot{\omega}_3$). Physically, this means that the source of the torque $\mathbf{M}$ is rotating together with the body. A typical example is the torque caused by the reaction of a rocket engine which is mounted on a missile and which moves relative to the missile according to some prescribed function of time. In such cases the rigid body is said to be self-excited. All problems not being of this type can be divided into two classes. The first class comprises problems in which M_1, M_2 and M_3 depend not only on ω_1, ω_2, ω_3 and t but also on generalized coordinates which describe the angular orientation of the body in some external reference base. To give an example gravity acting on a body which is suspended as a pendulum causes a torque whose coordinates are functions of the direction of the vertical in the principal axes system. The dependence of M_1, M_2 and M_3 on such generalized coordinates causes a mathematical coupling between Euler's equations and kinematic differential equations describing the angular orientation of the body. These are (2.104) or (2.109) or (2.112) or (2.119) depending on the choice of generalized coordinates. Still more complicated are problems in which M_1, M_2 and M_3 depend also on the location and velocity of the body center of mass. This dependency provides a coupling with Newton's law (3.34). Examples for this most general case are motions of airplanes and ships.

Problem 3.5. In Fig. 3.4 an inhomogeneous circular cylinder of radius R, mass m and moment of inertia $J^{\rm C}$ about an axis through the center of mass C is shown. The center of mass is located at the radius b. The cylinder is rolling without slipping on

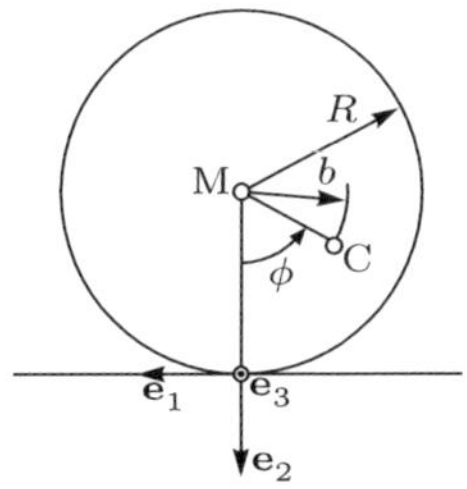

Fig. 3.4. Inhomogeneous rolling cylinder with center of mass C

a horizontal plane. Formulate the equation of motion for the angular coordinate ϕ by using as reference point A in (3.35) (i) the center of mass C, (ii) the geometric center M and (iii) the point of contact with the plane.

3.5 Principle of Virtual Power

The principal of virtual power, also referred to as Jourdain's principle, has the general form[4]

$$\int_m \delta\dot{\mathbf{r}} \cdot (\ddot{\mathbf{r}}\,\mathrm{d}m - \mathrm{d}\mathbf{F}) = 0\,. \tag{3.39}$$

It is valid for any material system (axiom). The integral is taken over the total system mass. The vector $\mathbf{r}$ is the radius vector of the mass particle $\mathrm{d}m$ measured from a point fixed in inertial space. The second derivative $\ddot{\mathbf{r}}$ is the absolute acceleration of the mass particle, and $\delta\dot{\mathbf{r}}$ is the variation of its velocity $\dot{\mathbf{r}}$. This variation is understood to be an arbitrary, infinitesimally small increment of $\dot{\mathbf{r}}$ which is imposed upon the system at a fixed time $t = \text{const}$ and in a fixed position of the system ($\mathbf{r} = \text{const}$ for all mass particles) and which is, furthermore, compatible with all velocity constraints of the system. The quantity $\mathrm{d}\mathbf{F}$ represents the total force acting on the mass element $\mathrm{d}m$. It can be the weight $\mathbf{g}\,\mathrm{d}m$ or a finite external force or an internal force such as the force of a spring. Constraint forces acting in frictionless contacts of rigid bodies have zero virtual power.

In what follows the principle of virtual power is formulated for a single rigid body. Let A be an arbitrary body-fixed point. Using again the vectors shown in Fig. 3.1 one has $\mathbf{r} = \mathbf{r}_{\mathrm{A}} + \boldsymbol{\varrho}$ and $\delta\dot{\mathbf{r}} = \delta\dot{\mathbf{r}}_{\mathrm{A}} + \delta\boldsymbol{\omega} \times \boldsymbol{\varrho}$. With these expressions the principle is written in the form

$$\int_m (\delta\dot{\mathbf{r}}_{\mathrm{A}} + \delta\boldsymbol{\omega} \times \boldsymbol{\varrho}) \cdot [(\ddot{\mathbf{r}}_{\mathrm{A}} + \ddot{\boldsymbol{\varrho}})\,\mathrm{d}m - \mathrm{d}\mathbf{F}] = 0\,. \tag{3.40}$$

The symbols for dot and cross multiplication can be interchanged. The integral $\int_m \mathrm{d}\mathbf{F}$ represents the resultant external force $\mathbf{F}$ on the body. Integrating

[4] For relationships between the principles of d'Alembert (virtual work), Jourdain (virtual power) and Gauss (least action) see Pars [56].

term by term one gets

$$\delta\dot{\mathbf{r}}_{\rm A}\cdot[(\ddot{\mathbf{r}}_{\rm A}+\ddot{\boldsymbol{\varrho}}_{\rm C})m-\mathbf{F}]+\delta\boldsymbol{\omega}\cdot\left[m\boldsymbol{\varrho}_{\rm C}\times\ddot{\mathbf{r}}_{\rm A}+\int_m\boldsymbol{\varrho}\times\ddot{\boldsymbol{\varrho}}\,{\rm d}m-\int_m\boldsymbol{\varrho}\times{\rm d}\mathbf{F}\right]=0\,. \tag{3.41}$$

The sum $\ddot{\mathbf{r}}_{\rm A}+\ddot{\boldsymbol{\varrho}}_{\rm C}$ represents the absolute acceleration $\ddot{\mathbf{r}}_{\rm C}$ of the body center of mass. The last integral is the resultant external torque $\mathbf{M}^{\rm A}$ with respect to A. The first integral is the absolute time derivative of $\int_m\boldsymbol{\varrho}\times\dot{\boldsymbol{\varrho}}\,{\rm d}m=\int_m\boldsymbol{\varrho}\times(\boldsymbol{\omega}\times\boldsymbol{\varrho})\,{\rm d}m$ and, hence, according to (3.14), of $\mathsf{J}^{\rm A}\cdot\boldsymbol{\omega}$. With this expression the principle of virtual power for a single rigid body gets the final form

$$\delta\dot{\mathbf{r}}_{\rm A}\cdot(m\ddot{\mathbf{r}}_{\rm C}-\mathbf{F})+\delta\boldsymbol{\omega}\cdot(m\boldsymbol{\varrho}_{\rm C}\times\ddot{\mathbf{r}}_{\rm A}+\mathsf{J}^{\rm A}\cdot\dot{\boldsymbol{\omega}}+\boldsymbol{\omega}\times\mathsf{J}^{\rm A}\cdot\boldsymbol{\omega}-\mathbf{M}^{\rm A})=0\,. \tag{3.42}$$

If the body is unconstrained then $\delta\dot{\mathbf{r}}_{\rm A}$ and $\delta\boldsymbol{\omega}$ are independent variations. The principle then yields the two equations

$$m\ddot{\mathbf{r}}_{\rm C}=\mathbf{F}\,,\qquad m\boldsymbol{\varrho}_{\rm C}\times\ddot{\mathbf{r}}_{\rm A}+\mathsf{J}^{\rm A}\cdot\dot{\boldsymbol{\omega}}+\boldsymbol{\omega}\times\mathsf{J}^{\rm A}\cdot\boldsymbol{\omega}=\mathbf{M}^{\rm A}\,. \tag{3.43}$$

These are Newton's law (3.34) and Euler's law of angular momentum (3.35), respectively. Thus, the principle of virtual power contains these two basic laws as special cases. Much more important are applications in cases when the variations $\delta\dot{\mathbf{r}}_{\rm A}$ and $\delta\boldsymbol{\omega}$ are not independent. As an example see Problem 3.6. The full potential of the principle will be seen in Sect. 5.4 when it is used for the formulation of equations of motion of complicated multibody systems. There, the special form will be used, when as point A the center of mass C is chosen. This is the form

$$\delta\dot{\mathbf{r}}_{\rm C}\cdot(m\ddot{\mathbf{r}}_{\rm C}-\mathbf{F})+\delta\boldsymbol{\omega}\cdot(\mathsf{J}^{\rm C}\cdot\dot{\boldsymbol{\omega}}+\boldsymbol{\omega}\times\mathsf{J}^{\rm C}\cdot\boldsymbol{\omega}-\mathbf{M}^{\rm C})=0\,. \tag{3.44}$$

Problem 3.6. Two straight lines forming an angle α are frictionless guides for two pegs $\rm P_1$ and $\rm P_2$ which are fixed on a rigid body of mass m and central moment of inertia $J^{\rm C}$ (Fig. 3.5). The body center of mass C is constrained to move in the plane of the guides. The body is subject to external forces $\mathbf{F}_1$ and $\mathbf{F}_2$ which are applied to the pegs in the direction of the guides. Derive from (3.42) an equation of motion for the variable ϕ and with parameters α, β, $a=|\mathbf{a}|$, $\varrho_{\rm C}=|\boldsymbol{\varrho}_{\rm C}|$, m and $J^{\rm C}$. Let $\rm P_1$ be the reference point A for $\mathsf{J}^{\rm A}$ and $\mathbf{M}^{\rm A}$. Consider the special case $\alpha=\pi/2$, $\beta=0$, $\varrho_{\rm C}=a/2$.

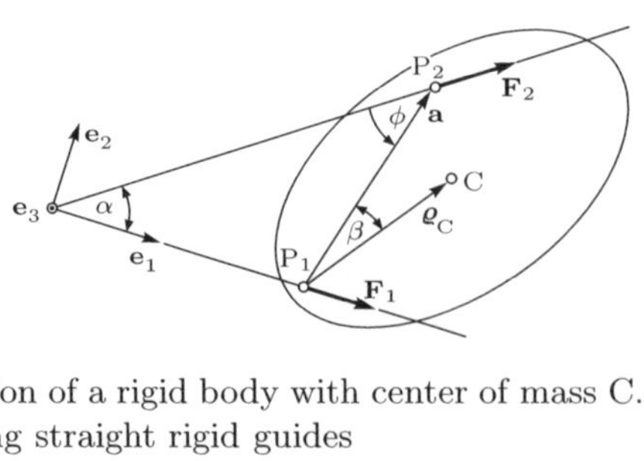

Fig. 3.5. Plane motion of a rigid body with center of mass C. The body-fixed pegs $\rm P_1$ and $\rm P_2$ move along straight rigid guides

4

Classical Problems of Rigid Body Mechanics

In this chapter some of the rare rigid body problems are considered in which equations of motion can be integrated in closed form. With the exception of the gyrostat in Sects. 4.6 and 4.7 these problems are treated also in other books on rigid body mechanics.

4.1 Unsymmetric Torque-Free Rigid Body

In the absence of external torques the equations of motion in the forms (3.36) and (3.38) read

$$\mathsf{J} \cdot \dot{\boldsymbol{\omega}} + \boldsymbol{\omega} \times \mathsf{J} \cdot \boldsymbol{\omega} = \mathbf{0} \tag{4.1}$$

and

$$\left.\begin{aligned} J_1\dot{\omega}_1 - (J_2 - J_3)\omega_2\omega_3 &= 0\,, \\ J_2\dot{\omega}_2 - (J_3 - J_1)\omega_3\omega_1 &= 0\,, \\ J_3\dot{\omega}_3 - (J_1 - J_2)\omega_1\omega_2 &= 0\,, \end{aligned}\right\} \tag{4.2}$$

respectively. Reference point for the moments of inertia is the body center of mass. These equations apply, for instance, to celestial bodies which are isolated from any external torque. They describe also the motion of a body which is supported without friction at its center of mass (provided torques caused by air resistance can be neglected). Such a support is approximately realized by a system of gimbals of the kind shown in Fig. 2.2. Strictly speaking, this system consists of three kinematically coupled bodies. The influence of the gimbals can, however, often be neglected (in Sect. 4.5 this influence will be the subject of investigation). Equation (4.1) has two algebraic first integrals which are obtained through scalar multiplication by $\boldsymbol{\omega}$ and by $\mathsf{J} \cdot \boldsymbol{\omega}$, respectively:

$$\left.\begin{aligned} \boldsymbol{\omega} \cdot (\mathsf{J} \cdot \dot{\boldsymbol{\omega}} + \boldsymbol{\omega} \times \mathsf{J} \cdot \boldsymbol{\omega}) &= \boldsymbol{\omega} \cdot \mathsf{J} \cdot \dot{\boldsymbol{\omega}} = \frac{1}{2}\frac{\mathrm{d}}{\mathrm{d}t}(\boldsymbol{\omega} \cdot \mathsf{J} \cdot \boldsymbol{\omega}) = 0\,, \\ \mathsf{J} \cdot \boldsymbol{\omega} \cdot (\mathsf{J} \cdot \dot{\boldsymbol{\omega}} + \boldsymbol{\omega} \times \mathsf{J} \cdot \boldsymbol{\omega}) &= \mathsf{J} \cdot \boldsymbol{\omega} \cdot \mathsf{J} \cdot \dot{\boldsymbol{\omega}} = \frac{1}{2}\frac{\mathrm{d}}{\mathrm{d}t}(\mathsf{J} \cdot \boldsymbol{\omega})^2 = 0\,. \end{aligned}\right\} \tag{4.3}$$

From these equations it follows that

$$\left.\begin{aligned}\boldsymbol{\omega}\cdot\mathsf{J}\cdot\boldsymbol{\omega} &= 2T = \text{const}\,,\\ (\mathsf{J}\cdot\boldsymbol{\omega})^2 &= L^2 = \text{const}\end{aligned}\right\} \tag{4.4}$$

or in terms of coordinates in the principal axes frame of reference

$$\sum_{i=1}^{3} J_i\omega_i^2 = 2T\,, \tag{4.5}$$

$$\sum_{i=1}^{3} J_i^2\omega_i^2 = L^2 = 2DT\,. \tag{4.6}$$

The quantities T and L represent the kinetic energy of rotation and the magnitude of the absolute angular momentum, respectively. In (4.6) a parameter D with the physical dimension of moment of inertia has been introduced. The use of $2T$ and $2DT$ instead of $2T$ and L^2 as parameters simplifies subsequent formulations. Only the general case of three different principal moments of inertia will be considered. Without loss of generality it is assumed that

$$J_3 < J_2 < J_1\,. \tag{4.7}$$

The body-fixed base vector in the principal axis associated with J_i is called $\mathbf{e}_i$ $(i = 1, 2, 3)$. Equations (4.5) and (4.6) define two ellipsoids which are fixed on the body and which are both geometric locus of the angular velocity vector $\boldsymbol{\omega}$. The vector is, therefore, confined to the line of intersection of the two ellipsoids. These lines are called polhodes.

4.1.1 Polhodes. Permanent Rotations

An investigation of geometric properties of the polhodes contributes to an understanding of the dynamic behavior of the torque-free rigid body. It is useful to think of the energy ellipsoid as given and to imagine that the angular momentum ellipsoid is "blown up" by increasing the parameter D so that on the invariable energy ellipsoid the family of all physically realizable polhodes is generated. This family corresponds to a certain interval of D values for which the angular momentum ellipsoid lies neither entirely inside nor entirely outside the energy ellipsoid. The minimum and maximum D values are found by multiplying (4.5) with J_1 and with J_3, respectively, and by subtracting both equations separately from (4.6). In the resulting equations

$$J_2(J_1 - J_2)\omega_2^2 + J_3(J_1 - J_3)\omega_3^2 = 2T(J_1 - D)\,, \tag{4.8}$$

$$J_1(J_1 - J_3)\omega_1^2 + J_2(J_2 - J_3)\omega_2^2 = 2T(D - J_3) \tag{4.9}$$

the left-hand side expressions are nonnegative so that the inequalities

$$J_3 \le D \le J_1 \tag{4.10}$$

must be satisfied. Thus, J_3 and J_1 are the extreme values of D for which the equations of motion have real solutions.

Of particular interest are degenerate polhodes which consist of singular points. In such points the ellipsoids have a common tangential plane. Each singular point marks a solution $\boldsymbol{\omega} \equiv \boldsymbol{\omega}^* =$ const of the equations of motion. This particular state of motion is referred to as permanent rotation. From (4.1) it is seen that a solution $\boldsymbol{\omega} \equiv \boldsymbol{\omega}^* =$ const is possible only if either $\boldsymbol{\omega}^*$ equals zero (in this trivial case the body is not rotating; both ellipsoids degenerate into a single point) or $\boldsymbol{\omega}^*$ and the angular momentum $\mathbf{L}^* = \mathsf{J} \cdot \boldsymbol{\omega}^*$ are collinear. In matrix form this latter condition yields the coordinate equation $\underline{J}\,\underline{\omega}^* = \lambda \underline{\omega}^*$ with an unknown factor λ, i.e. the eigenvalue problem

$$(\underline{J} - \lambda \underline{I})\underline{\omega}^* = \underline{0} \,. \tag{4.11}$$

This equation is identical in form with (3.23) which led to principal moments and principal axes of inertia. From this identity it follows that the eigenvectors of (4.11), i.e. the axes of permanent rotations, are identical with the principal axes of inertia. It is now possible to specify the particular values $D = D^*$ which cause the angular momentum ellipsoid to touch the energy ellipsoid in singular points. In a state of permanent rotation with an angular velocity of magnitude ω^* about the principal axis $\mathbf{e}_i$ $(i = 1, 2, 3)$ the integrals of motion are $2T = J_i \omega^{*2}$ and $2D^*T = J_i^2 \omega^{*2}$, whence follows

$$D^* = J_i \,. \tag{4.12}$$

Consider, again, the entire family of polhodes. A clear picture is obtained if the polhodes are seen in projections along principal axes. The projection along $\mathbf{e}_i$ $(i = 1, 2, 3)$ requires the elimination of the coordinate ω_i from (4.5) and (4.6). For the projections along $\mathbf{e}_1$ and $\mathbf{e}_3$ this has been done already. The resulting equations are (4.8) and (4.9). In a similar manner the projection along $\mathbf{e}_2$ yields

$$J_1(J_1 - J_2)\omega_1^2 - J_3(J_2 - J_3)\omega_3^2 = 2T(D - J_2) \,. \tag{4.13}$$

Because of the inequalities $J_3 < J_2 < J_1$ and $J_3 \le D \le J_1$ (4.8) and (4.9) represent families of ellipses whereas (4.13) represents a family of hyperbolas. The asymptotes of the hyperbolas correspond to $D = J_2$. In Figs. 4.1a to c all three projections of polhodes for one and the same set of D values are illustrated. Solid lines represent the contour ellipses of the energy ellipsoid. Of the polhode ellipses and hyperbolas only those parts are relevant which lie inside these contour ellipses. All three projections together produce an image of the three-dimensional pattern of polhodes. A perspective view is shown in Fig. 4.1d.

The results just obtained can be summarized as follows. Each of the parameter values $D = D^* = J_i$ $(i = 1, 2, 3)$ specifies the axis of permanent rotation which coincides with the principal axis $\mathbf{e}_i$. The value $D = D^* = J_2$

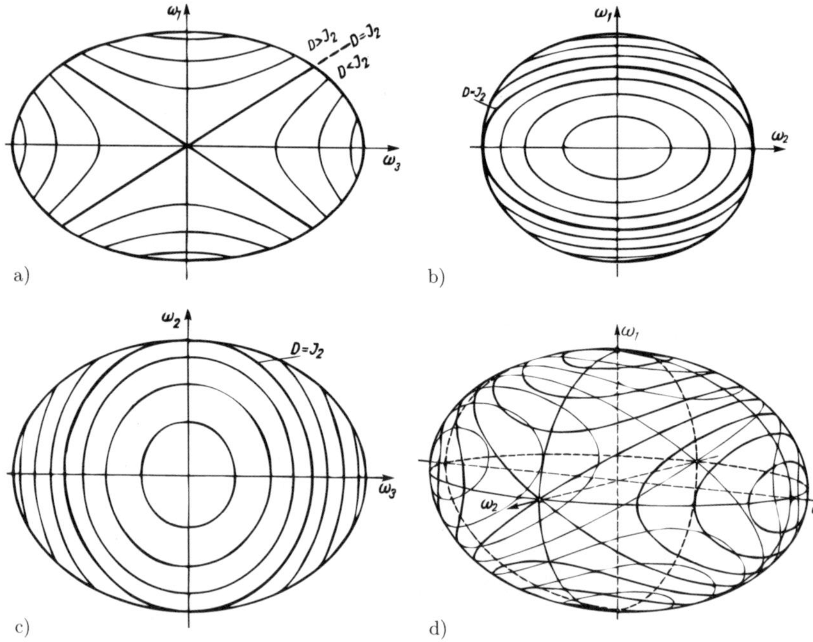

Fig. 4.1. Polhodes on the energy ellipsoid of an unsymmetric rigid body seen in projections along principal axes **(a, b, c)** and in perspective **(d)**. The figures are based on the parameters $J_1 = 7$, $J_2 = 5$, $J_3 = 3$, $D = 3$, 3.3, 3.9, 4.5, 5, 5.5, 6.1, 6.7 and 7 (the same physical unit for all quantities)

specifies in addition, two particular polhodes which intersect one another in the points of permanent rotation about the axis $\mathbf{e}_2$ and which separate all other polhodes into four families. Two families corresponding to $D < J_2$ envelop the axis $\mathbf{e}_3$ and the others corresponding to $D > J_2$ envelop the axis $\mathbf{e}_1$. The separating polhodes are called separatrices.

4.1.2 Poinsot's Geometric Interpretation of the Motion

So far the integrals of motion have been used for characterizing the geometric locus of the angular velocity vector on the body. The integrals can also be used for an interpretation of the motion of the body relative to inertial space. This interpretation is due to Poinsot. The energy equation states that the scalar product of angular velocity $\boldsymbol{\omega}$ and angular momentum $\mathbf{L} = \mathsf{J} \cdot \boldsymbol{\omega}$ is constant. Since the magnitude and the direction of $\mathbf{L}$ are both constant it follows that the projection of $\boldsymbol{\omega}$ onto this invariable direction of $\mathbf{L}$ is constant. Figure 4.2a illustrates this situation. The vector $\boldsymbol{\omega}$ is confined to an invariable plane perpendicular to $\mathbf{L}$. From this it follows that any (finite or infinitesimally small) increment $\Delta\boldsymbol{\omega}$ between two arbitrary moments of time is perpendicular to $\mathbf{L}$. Hence, $\mathbf{L} \cdot \Delta\boldsymbol{\omega} = 0$. This equation defines the invariable plane. The

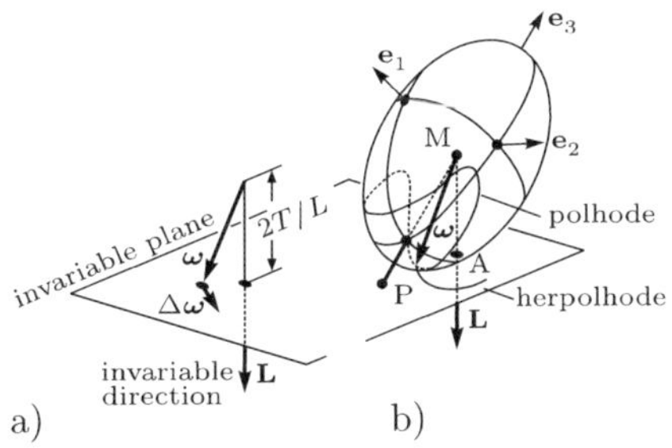

Fig. 4.2. Poinsot's interpretation of the motion of a torque-free rigid body

vector $\boldsymbol{\omega}$ is also located on the energy ellipsoid defined by (4.6). The total differential of this equation has the form $\sum_{i=1}^{3} J_i\omega_i\,\mathrm{d}\omega_i = \mathbf{L}\cdot\mathrm{d}\boldsymbol{\omega} = 0$. It defines the tangential plane of the energy ellipsoid at the point with coordinates $\omega_1, \omega_2, \omega_3$. Comparison with the equation for the invariable plane reveals that both planes are parallel. Moreover, since both are geometric locus of $\boldsymbol{\omega}$ the planes coincide with one another. This situation is illustrated in Fig. 4.2b. The point of contact between the ellipsoid and the invariable plane is located on the instantaneous axis of rotation. Poinsot's interpretation of the motion can be summarized as follows. *The body is moving as if its energy ellipsoid were rolling without slipping on the invariable plane the geometric center M of the ellipsoid being fixed in inertial space a distance* $\overline{AM} = 2T/L$ *above this plane.* During this rolling motion the contact point traces in the invariable plane another curve called herpolhode. Properties of herpolhodes are discussed by Grammel [22] and Magnus [51].

4.1.3 Solution of Euler's Equations of Motion

Euler [15] gave the following closed-form solutions of the dynamic equations of motion[1]. First, (4.9) and (4.8) are solved for ω_1 and ω_3, respectively, as functions of ω_2:

$$\omega_1^2 = \frac{J_2(J_2 - J_3)}{J_1(J_1 - J_3)}\left(a^2 - \omega_2^2\right), \qquad \omega_3^2 = \frac{J_2(J_1 - J_2)}{J_3(J_1 - J_3)}\left(b^2 - \omega_2^2\right). \tag{4.14}$$

In these expressions a^2 and b^2 are the non-negative constants

$$a^2 = \frac{2T(D - J_3)}{J_2(J_2 - J_3)}, \qquad b^2 = \frac{2T(J_1 - D)}{J_2(J_1 - J_2)} \tag{4.15}$$

which satisfy the relationship

$$a^2 - b^2 = \frac{2T(J_1 - J_3)(D - J_2)}{J_2(J_1 - J_2)(J_2 - J_3)}. \tag{4.16}$$

[1] In the same volume (pp. 315–317) Euler angles are introduced (1760).

Substitution of (4.14) into the second of Eqs. (4.2) and separation of the variables yield

$$\int \frac{\mathrm{d}\omega_2}{\sqrt{(a^2-\omega_2^2)(b^2-\omega_2^2)}} = s_2(t-t_0)\sqrt{\frac{(J_1-J_2)(J_2-J_3)}{J_1J_3}} \tag{4.17}$$

where s_2 is an abbreviation for the resultant sign of the two square roots. This sign will be determined later. The integral on the left-hand side is an elliptic integral of the first kind. In reducing it to a Legendre normal form three cases have to be distinguished:

a) $a^2 < b^2$ or $D < J_2$
b) $a^2 = b^2$ or $D = J_2$
c) $a^2 > b^2$ or $D > J_2$.

Cases a) and c) correspond to polhodes which envelop the axes $\mathbf{e}_3$ and $\mathbf{e}_1$, respectively, and case b) corresponds to the separatrices (see Fig. 4.1). Consider, first, case a). Equation (4.17) is rewritten in the form

$$\int \frac{\mathrm{d}\omega_2/a}{\sqrt{(1-\omega_2^2/a^2)(1-\omega_2^2/b^2)}} = s_2 b(t-t_0)\sqrt{\frac{(J_1-J_2)(J_2-J_3)}{J_1J_3}} \tag{4.18}$$

or

$$\int \frac{\mathrm{d}x}{\sqrt{(1-x^2)(1-k^2x^2)}} = s_2\tau \tag{4.19}$$

with the abbreviations

$$x = \frac{\omega_2}{a}\,, \qquad k = \frac{a}{b}\,, \qquad \tau = (t-t_0)\sqrt{\frac{2T(J_1-D)(J_2-J_3)}{J_1J_2J_3}}\,. \tag{4.20}$$

The solution has the form $x = s_2 \operatorname{sn}\tau$ or

$$\omega_2 = s_2\sqrt{\frac{2T(D-J_3)}{J_2(J_2-J_3)}}\operatorname{sn}\tau \tag{4.21}$$

(for elliptic integrals and Jacobian elliptic functions see Tölke [80]). When this is substituted into (4.14) and use is made of the addition theorems $\operatorname{sn}^2\tau + \operatorname{cn}^2\tau = 1$ and $\operatorname{dn}^2\tau + k^2\operatorname{sn}^2\tau = 1$ solutions for ω_1 and ω_3 are obtained in the forms

$$\omega_1 = s_1\sqrt{\frac{2T(D-J_3)}{J_1(J_1-J_3)}}\operatorname{cn}\tau\,, \qquad \omega_3 = s_3\sqrt{\frac{2T(J_1-D)}{J_3(J_1-J_3)}}\operatorname{dn}\tau\,. \tag{4.22}$$

The quantities s_1 and s_3 are as yet undetermined signs of the respective square roots. The missing relationship between s_1, s_2 and s_3 is found when (4.21) and (4.22) are substituted back into the second of Eqs. (4.2). Taking into account the relationship $\mathrm{d}\operatorname{sn}\tau/\mathrm{d}\tau = \operatorname{cn}\tau\operatorname{dn}\tau$ one gets $s_2 = -s_1s_3$ or

$$s_1s_2s_3 = -1\,. \tag{4.23}$$

Altogether four combinations of signs satisfying this relationship are possible. This result is in accordance with the fact that for each value of the parameter D two separate polhodes exist and that on each of them ω_2 passes through zero in two different points (see Fig. 4.1d).

Case c) for $D > J_2$ is treated in a similar manner. The results are

$$\left.\begin{aligned} \omega_1 &= s_1\sqrt{\frac{2T(D-J_3)}{J_1(J_1-J_3)}}\,\operatorname{dn}\tau\,, \qquad \omega_2 = s_2\sqrt{\frac{2T(J_1-D)}{J_2(J_1-J_2)}}\,\operatorname{sn}\tau\,, \\ \omega_3 &= s_3\sqrt{\frac{2T(J_1-D)}{J_3(J_1-J_3)}}\,\operatorname{cn}\tau \end{aligned}\right\} \tag{4.24}$$

with the modulus $k = b/a$ and the argument

$$\tau = (t-t_0)\sqrt{\frac{2T(J_1-J_2)(D-J_3)}{J_1J_2J_3}}\,. \tag{4.25}$$

The signs s_1, s_2 and s_3 satisfy (4.23), again.

It is unnecessary to integrate case b) since the solutions for the cases a) and c) converge both in the limit $D \to J_2$ toward the same result

$$\left.\begin{aligned} \omega_1 &= s_1\sqrt{\frac{2T(J_2-J_3)}{J_1(J_1-J_3)}}\,\frac{1}{\cosh\tau}\,, \qquad \omega_2 = s_2\sqrt{\frac{2T}{J_2}}\,\tanh\tau\,, \\ \omega_3 &= s_3\sqrt{\frac{2T(J_1-J_2)}{J_3(J_1-J_3)}}\,\frac{1}{\cosh\tau} \end{aligned}\right\} \tag{4.26}$$

which, therefore, represents the solution for case b). These formulas show that the motion of $\boldsymbol{\omega}$ along a separatrix is aperiodic. For $\tau \to \infty$ ω_1 and ω_3 tend toward zero and ω_2 approaches asymptotically the value $s_2\sqrt{2T/J_2}$. This represents a permanent rotation about the axis $\mathbf{e}_2$.

Problem 4.1. Determine from the second Eq. (4.2) the sense of direction in which $\boldsymbol{\omega}$ traces the polhodes in Fig. 4.1d.

4.1.4 Solution of the Kinematic Differential Equations

The last part of the problem is to show how the body is moving in inertial space. For this purpose Euler angles are used as generalized coordinates. According to Fig. 2.1 the intermediate angle θ is measured between two axes one of which is fixed in the reference base $\underline{\mathbf{e}}^1$ (here inertial space) and the other is fixed on the body. It is convenient to choose as axis fixed in inertial space the direction of the angular momentum $\mathbf{L}$ since this is the only significant axis. On the body the axis $\mathbf{e}_3$ is an appropriate choice for polhodes which envelop this axis, i.e. for $D < J_2$. In the case $D > J_2$ the polhodes envelop the axis $\mathbf{e}_1$. Therefore, this axis will be chosen. Consider first the case $D < J_2$ which is illustrated in Fig. 4.2. In this figure θ is the angle PMA and

ψ is measured in the invariable plane between the straight line $\overline{PA}$ and some inertially fixed reference line through A. It is unnecessary to solve kinematic differential equations of the form (2.109). The problem can be simplified substantially if use is made of the fact that the angular momentum $\mathbf{L}$ has constant magnitude and constant direction in inertial space. In the principal axes frame $\mathbf{L}$ has coordinates $L_i = J_i\omega_i$ $(i = 1, 2, 3)$. These coordinates are expressed as functions of Euler angles and of time derivatives of Euler angles. Using the notation of Fig. 2.1 $\mathbf{L}$ has the direction of $\mathbf{e}_3^1$. Its coordinates in the principal axes frame are, therefore, found in the third column of the direction cosine matrix $\underline{A}^{21}$ in (2.5). The results are the equations

$$J_1\omega_1 = L\sin\theta\sin\phi\ , \qquad J_2\omega_2 = L\sin\theta\cos\phi\ , \qquad J_3\omega_3 = L\cos\theta\ . \quad (4.27)$$

They yield without integration

$$\cos\theta = \frac{J_3}{L}\,\omega_3\ , \qquad \tan\phi = \frac{J_1\omega_1}{J_2\omega_2}\ . \quad (4.28)$$

Only the angle ψ is not found directly. Its time derivative is according to the third Eq. (2.108)

$$\dot{\psi} = \frac{\omega_3 - \dot{\phi}}{\cos\theta} = \frac{L}{J_3}\left(1 - \frac{\dot{\phi}}{\omega_3}\right)\ . \quad (4.29)$$

The expressions for $\cos\theta$ and $\tan\phi$ become with (4.21), (4.22) and (4.23) and with $L^2 = 2DT$

$$\cos\theta = s_3\sqrt{\frac{J_3(J_1 - D)}{D(J_1 - J_3)}}\,\mathrm{dn}\,\tau\ , \qquad \tan\phi = -s_3\sqrt{\frac{J_1(J_2 - J_3)}{J_2(J_1 - J_3)}}\,\frac{\mathrm{cn}\,\tau}{\mathrm{sn}\,\tau}\ . \quad (4.30)$$

From this follows by differentiation

$$\dot{\phi} = s_3\cos^2\phi\sqrt{\frac{J_1(J_2 - J_3)}{J_2(J_1 - J_3)}}\,\frac{\mathrm{dn}\,\tau}{\mathrm{sn}^2\,\tau}\,\frac{\mathrm{d}\tau}{\mathrm{d}t}\ . \quad (4.31)$$

Expressing $\cos^2\phi$ through $\tan^2\phi$ and $\mathrm{dn}\,\tau$ through ω_3 this is given the form

$$\dot{\phi} = \omega_3\left(\frac{J_2}{J_2 - J_3}\,\mathrm{sn}^2\,\tau + \frac{J_1}{J_1 - J_3}\,\mathrm{cn}^2\,\tau\right)^{-1}\ . \quad (4.32)$$

With this expression (4.29) takes the form

$$\dot{\psi} = \frac{L}{J_3}\,\frac{[J_3/(J_2 - J_3)]\,\mathrm{sn}^2\,\tau + [J_3/(J_1 - J_3)]\,\mathrm{cn}^2\,\tau}{[J_2/(J_2 - J_3)]\,\mathrm{sn}^2\,\tau + [J_1/(J_1 - J_3)]\,\mathrm{cn}^2\,\tau}\ . \quad (4.33)$$

The results show that θ, $\dot{\phi}$ and $\dot{\psi}$ are periodic functions of time. The sign of $\dot{\phi}$ equals that of ω_3, and $\dot{\psi}$ is always positive (clockwise rotation about $\mathbf{L}$). If the function $\psi(t)$ is desired (4.33) is rewritten in the form

$$\psi - \psi_0 = a\tau + b\int\frac{\mathrm{d}\tau}{\mathrm{sn}^2\,\tau + c^2} \quad (4.34)$$

with new constants a, b and c. The integral has the normal form of an elliptic integral of the third kind (see Tölke [80]).

The solution for $\theta(t)$ can be used to answer the question whether the orientation of the axis $\mathbf{e}_3$ in inertial space during a permanent rotation about this axis is stable or not. For reasons of symmetry of the polhodes on the energy ellipsoid only the case $s_3 = +1$ need be considered. The elliptic function $\mathrm{dn}\,\tau$ has the lower bound $\sqrt{1-k^2}$ where k is the modulus given by (4.20). This yields for θ the inequality

$$\cos^2\theta \geq (1-k^2)\,\frac{J_3(J_1-D)}{D(J_1-J_3)} = \frac{J_3(J_2-D)}{D(J_2-J_3)}\;. \tag{4.35}$$

For motions close to a permanent rotation about the axis $\mathbf{e}_3$ D is slightly larger than J_3. Setting $D = J_3 + \delta$ with $\delta \ll J_3$ one obtains the inequality

$$\sin^2\theta \leq \frac{J_2}{J_3(J_2-J_3)}\,\delta\;. \tag{4.36}$$

This indicates that by choosing appropriate initial conditions $\theta(t)$ can be kept smaller than any given arbitrarily small angle. Thus, the orientation in inertial space of the axis of permanent rotation is stable.

Arguments similar to those just used lead to solutions also in the case $D > J_2$. Now, θ is defined to be the angle between $\mathbf{L}$ and $\mathbf{e}_1$. More precisely, imagine that the body-fixed axes $\mathbf{e}_1$, $\mathbf{e}_2$, $\mathbf{e}_3$ are given the new names $\mathbf{e}_3'$, $\mathbf{e}_1'$, $\mathbf{e}_2'$, respectively, and that the Euler angles ψ, θ, ϕ specify the orientation of this base in the inertial reference base. This explains the following equations which replace (4.27).

$$J_1\omega_1 = L\cos\theta\;,\qquad J_2\omega_2 = L\sin\theta\sin\phi\;,\qquad J_3\omega_3 = L\sin\theta\cos\phi\;. \tag{4.37}$$

From these equations the following results are derived:

$$\left.\begin{aligned}
&\cos\theta = s_1\sqrt{\frac{J_1(D-J_3)}{D(J_1-J_3)}}\,\mathrm{dn}\,\tau\;,\quad \tan\phi = -s_1\sqrt{\frac{J_2(J_1-J_3)}{J_3(J_1-J_2)}}\,\frac{\mathrm{sn}\,\tau}{\mathrm{cn}\,\tau}\;,\\
&\dot\phi = -\omega_1\left(\frac{J_2}{J_1-J_2}\,\mathrm{sn}^2\tau + \frac{J_3}{J_1-J_3}\,\mathrm{cn}^2\tau\right)^{-1}\;,\\
&\dot\psi = \frac{\omega_1-\dot\phi}{\cos\theta} = \frac{L}{J_1}\,\frac{[J_1/(J_1-J_2)]\,\mathrm{sn}^2\tau + [J_1/(J_1-J_3)]\,\mathrm{cn}^2\tau}{[J_2/(J_1-J_2)]\,\mathrm{sn}^2\tau + [J_3/(J_1-J_3)]\,\mathrm{cn}^2\tau}\;.
\end{aligned}\right\} \tag{4.38}$$

The only difference compared with the case $D < J_2$ is that now $\dot\phi$ and ω_1 have opposite signs. In every other respect the results are qualitatively the same. In particular, it is found that the orientation of the axis of permanent rotation $\mathbf{e}_1$ in inertial space is stable.

Finally, the stability behavior of permanent rotations about the axis $\mathbf{e}_2$ is investigated. For this purpose θ is defined to be the angle between $\mathbf{L}$ and $\mathbf{e}_2$. This time, the body-fixed axes $\mathbf{e}_1$, $\mathbf{e}_2$, $\mathbf{e}_3$ are given the names $\mathbf{e}_2'$, $\mathbf{e}_3'$, $\mathbf{e}_1'$, respectively, and ψ, θ, ϕ specify the orientation of this base in the inertial reference base. The Eqs. (4.37) are replaced by $J_2\omega_2 = L\cos\theta$, $J_3\omega_3 = L\sin\theta\sin\phi$, $J_1\omega_1 = L\sin\theta\cos\phi$. The first equation in combination with (4.21) and (4.24) yields

$$\cos\theta = \begin{cases} s_2\sqrt{\dfrac{J_2(D-J_3)}{D(J_2-J_3)}}\,\operatorname{sn}\tau & (D < J_2) \\ s_2\sqrt{\dfrac{J_2(J_1-D)}{D(J_1-J_2)}}\,\operatorname{sn}\tau & (D > J_2) \end{cases} \tag{4.39}$$

For $D \to J_2$ both square roots tend toward unity while sn τ changes periodically between $+1$ and -1. This means periodic changes of θ between $\approx +90°$ and $\approx -90°$ indicating instability of the permanent rotation about the axis $\mathbf{e}_2$.

Next, the case $D = J_2$ is considered. Both expressions yield $\cos\theta = s_2\tanh\tau$ whence follows $\dot\theta = -\dot\tau/\cosh\tau$. Moreover, with (4.26) $\tan\phi = J_3\omega_3/(J_1\omega_1) = \text{const}$ and also $\dot\psi = (\omega_2 - \dot\phi)/\cos\theta = \sqrt{2T/J_2} = \text{const}$. Imagine a sphere of Radius R fixed in inertial space with its center at the body center of mass and on this sphere the path generated by the point of intersection with the axis $\mathbf{e}_2$. The coordinates ψ and $\lambda = \pi/2 - \theta$ of this point are interpreted as geographic longitude and geographic latitude, respectively ($\mathbf{L}$ playing the role of the polar axis). The path is traced with velocity coordinates $\dot\lambda = -R\dot\theta = R\dot\tau/\cosh\tau$ due north and $R\dot\psi\sin\theta = R\dot\psi\sqrt{1-\tanh^2\tau} = R\dot\psi/\cosh\tau$ due east. The ratio of the two coordinates is constant, whence follows that the path is a curve of constant heading (a loxodrome) spiraling toward the north pole.

4.2 Symmetric Torque-Free Rigid Body

The solution developed in the previous section becomes particularly simple if the body under consideration has two equal principal moments of inertia as is often the case in engineering applications. It is left to the reader to adapt the general solution to this special case. Here, it is preferred to start again from the equations of motion and to develop from them the special solution directly. It is assumed that the principal axis $\mathbf{e}_3$ is the symmetry axis of the body so that $J_1 = J_2 \neq J_3$. The principal moment of inertia J_3 can be either smaller or larger than J_1 (the trivial case $J_1 = J_2 = J_3$ will not be considered). With these assumptions Euler's equations reduce to

$$\left.\begin{aligned} J_1\dot\omega_1 - (J_1 - J_3)\omega_2\omega_3 &= 0\,, \\ J_1\dot\omega_2 - (J_3 - J_1)\omega_3\omega_1 &= 0\,, \\ J_3\dot\omega_3 &= 0\,. \end{aligned}\right\} \tag{4.40}$$

They yield at once

$$\omega_3 \equiv \omega_{30} = \text{const} \tag{4.41}$$

and by substituting this into the first two equations

$$\dot{\omega}_1 - \nu\omega_2 = 0\,, \qquad \dot{\omega}_2 + \nu\omega_1 = 0 \tag{4.42}$$

where ν is the constant

$$\nu = \omega_{30}\,\frac{J_1 - J_3}{J_1}\,. \tag{4.43}$$

The differential equations have the first integral $\omega_1^2 + \omega_2^2 = \Omega^2 = \text{const}$. The general solution for initial values ω_{10} and ω_{20} has the form

$$\left.\begin{aligned} \omega_1 &= \omega_{10}\cos\nu(t-t_0) + \omega_{20}\sin\nu(t-t_0) = \Omega\sin\nu(t-t_0')\,,\\ \omega_2 &= \omega_{20}\cos\nu(t-t_0) - \omega_{10}\sin\nu(t-t_0) = \Omega\cos\nu(t-t_0')\,. \end{aligned}\right\} \tag{4.44}$$

Next, the kinematic equations are considered. Again, Euler angles are used with θ being the angle between the symmetry axis and the inertially fixed angular momentum vector $\mathbf{L} = J_1\boldsymbol{\Omega} + J_3\boldsymbol{\omega}_{30}$. Equations (4.27) – (4.29) are valid, again, so that

$$\cos\theta = \frac{\omega_{30}J_3}{L}\,, \qquad \tan\phi = \frac{\omega_1}{\omega_2} = \tan\nu(t-t_0') \quad \rightarrow \quad \phi = \nu(t-t_0')\,, \tag{4.45}$$

$$\dot{\psi} = \frac{\omega_{30} - \dot{\phi}}{\cos\theta} = \frac{\omega_{30}J_3}{J_1\cos\theta} = \frac{L}{J_1}\,. \tag{4.46}$$

Thus, θ, $\dot{\phi}$ and $\dot{\psi}$ turn out to be constant. For a better understanding of these results Poinsot's interpretation of the motion is considered again. The energy ellipsoid in Fig. 4.2 is now an ellipsoid of revolution with the symmetry axis $\mathbf{e}_3$. The polhodes are, therefore, circles (and so are the herpolhodes). The axis $\mathbf{e}_3$ is moving around a circular cone whose axis is the angular momentum vector $\mathbf{L}$. The angular velocity of this motion around the cone, called nutation angular velocity, is $\dot{\psi}$. The angular velocity vector $\boldsymbol{\omega}$ of the body lies in the plane spanned by the symmetry axis and the angular momentum $\mathbf{L}$. Relative to the body the vector $\boldsymbol{\omega}$ is moving around a circular cone which is defined by the polhode. Relative to inertial space the vector $\boldsymbol{\omega}$ is also moving around a circular cone. The axis of this cone is the angular momentum $\mathbf{L}$. Following Sect. 2.2.3 (Fig. 2.8) the motion of the body can be visualized as rolling motion without slipping of the body-fixed cone on the inertially fixed cone. In Fig. 4.3a,b the cone swept out by $\mathbf{e}_3$ and the two cones swept out by $\boldsymbol{\omega}$ are illustrated. Only their projections onto the plane spanned by $\mathbf{e}_3$, $\boldsymbol{\omega}$ and $\mathbf{L}$ are shown. The ellipses represent contours of the energy ellipsoid. Two figures are necessary since rod-shaped bodies with $J_1 > J_3$ show another behavior than disc-shaped bodies with $J_1 < J_3$. For the former $\dot{\phi}$ is positive and for the latter negative (see (4.43)). Note that an observer of the motion can see the cone described by the symmetry axis $\mathbf{e}_3$ but that the other two cones are invisible. Magnus [50] describes an experimental trick which renders the

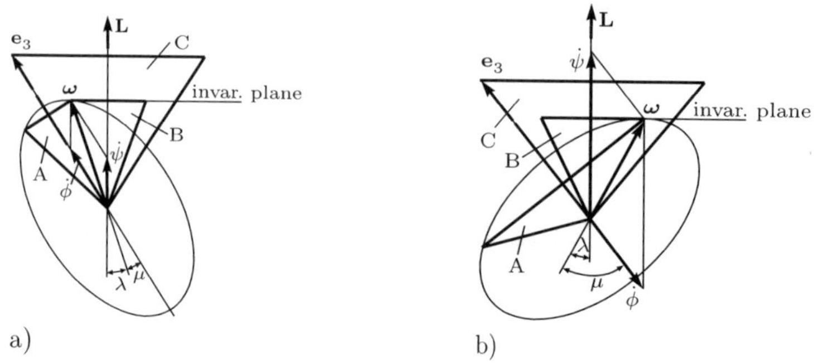

Fig. 4.3. The body-fixed $\boldsymbol{\omega}$-cone A is rolling on the space-fixed $\boldsymbol{\omega}$-cone B while the symmetry axis $\mathbf{e}_3$ generates the space-fixed cone C. The energy ellipsoid intersects cone A in a polhode and rolls on the invariable plane. Figure **(a)** belongs to a rod-shaped body and **(b)** to a disc-shaped body

motion of $\boldsymbol{\omega}$ around the body-fixed cone visible. It is left to the reader to verify that the angles λ and μ are related by the equation

$$\frac{\sin\lambda}{\sin\mu} = \frac{|J_1 - J_3|}{J_3}\cos\theta\ . \tag{4.47}$$

In engineering applications motions of symmetric bodies usually differ very little from permanent rotations about the symmetry axis. For such motions ω_{30} is practically identical with $\omega = |\boldsymbol{\omega}|$, and θ is very small. The nutation angular velocity is then approximately

$$\dot{\psi} \approx \frac{J_3}{J_1}\,\omega_{30} \qquad \text{for} \quad \theta \ll 1\ . \tag{4.48}$$

4.3 Self-Excited Symmetric Rigid Body

The subject of this section is a symmetric rigid body with principal moments of inertia J_1, $J_2 = J_1$ and $J_3 \neq J_1$ which is under the action of a torque whose coordinates in the principal axes frame are given functions of time. Euler's equations of motion have the form

$$\left.\begin{aligned} J_1\dot{\omega}_1 - (J_1 - J_3)\omega_2\omega_3 &= M_1(t)\ ,\\ J_1\dot{\omega}_2 - (J_3 - J_1)\omega_3\omega_1 &= M_2(t)\ ,\\ J_3\dot{\omega}_3 \qquad\qquad\qquad &= M_3(t)\ . \end{aligned}\right\} \tag{4.49}$$

We begin with the simple case in which $M_3(t)$ is identically zero. Then, $\omega_3 =$ const, and the first two equations reduce to

$$\dot{\omega}_1 - \nu\omega_2 = m_1(t)\ , \qquad \dot{\omega}_2 + \nu\omega_1 = m_2(t) \tag{4.50}$$

with the constant $\nu = \omega_3(J_1 - J_3)/J_1$ and with functions $m_1(t) = M_1(t)/J_1$ and $m_2(t) = M_2(t)/J_1$. By introducing the complex quantities $\omega^* = \omega_1 + \mathrm{i}\omega_2$

and $m^* = m_1 + \mathrm{i}m_2$ these equations are combined in the single complex equation

$$\dot{\omega}^* + \mathrm{i}\nu\omega^* = m^*(t)\ . \tag{4.51}$$

It has the general solution

$$\omega^*(t) = \mathrm{e}^{-\mathrm{i}\nu t}\left[\omega_0^* + \int_0^t m^*(\tau)\mathrm{e}^{\mathrm{i}\nu\tau}\,\mathrm{d}\tau\right]. \tag{4.52}$$

When this is split, again, into real and imaginary parts the solutions for $\omega_1(t)$ and $\omega_2(t)$ are obtained:

$$\left.\begin{aligned}\omega_1(t) &= \omega_{10}\cos\nu t + \omega_{20}\sin\nu t\\ &\quad + \int_0^t [m_1(\tau)\cos\nu(\tau - t) + m_2(\tau)\sin\nu(\tau - t)]\,\mathrm{d}\tau\ ,\\ \omega_2(t) &= \omega_{20}\cos\nu t - \omega_{10}\sin\nu t\\ &\quad + \int_0^t [m_1(\tau)\sin\nu(\tau - t) + m_2(\tau)\cos\nu(\tau - t)]\,\mathrm{d}\tau\ .\end{aligned}\right\} \tag{4.53}$$

Next, the general case with $M_3(t) \neq 0$ is investigated. From the third of Eqs. (4.49) the solution for $\omega_3(t)$ is obtained

$$\omega_3(t) = \omega_{30} + \frac{1}{J_3}\int_0^t M_3(\tau)\,\mathrm{d}\tau\ . \tag{4.54}$$

An auxiliary variable $\alpha(t)$ is now introduced by the equation

$$\alpha(t) = \int_0^t \omega_3(\tau)\,\mathrm{d}\tau\ . \tag{4.55}$$

This variable is a known function of time. For the inverse function $t(\alpha)$ a closed-form expression may not exist. However, it is available at least numerically. In the first two of Euler's equations $\dot{\omega}_1$ and $\dot{\omega}_2$ are expressed in the form

$$\dot{\omega}_i = \frac{\mathrm{d}\omega_i}{\mathrm{d}\alpha}\,\dot{\alpha} = \omega_i'\,\omega_3 \qquad (i = 1, 2) \tag{4.56}$$

where the prime denotes differentiation with respect to α. With these expressions the two equations of motion take the form

$$\omega_1' - \nu\omega_2 = m_1(\alpha)\ , \qquad \omega_2' + \nu\omega_1 = m_2(\alpha) \tag{4.57}$$

with the constant $\nu = (J_1 - J_3)/J_1$ and with functions

$$m_1(\alpha) = \frac{M_1(t(\alpha))}{J_1\omega_3(t(\alpha))}\ , \qquad m_2(\alpha) = \frac{M_2(t(\alpha))}{J_1\omega_3(t(\alpha))}\ . \tag{4.58}$$

These equations are identical with (4.50) except that α is the independent variable instead of t. The solution has, therefore, the form (4.53) provided t is replaced everywhere by $\alpha(t)$ and m_1 and m_2 are the functions of α defined above.

4.4 Symmetric Heavy Top

A rigid body is considered which is supported in inertial space at a single point which is not the body center of mass. The body is subject to gravity only. In the literature this system is known as heavy top. The general solution of its equations of motion is not known. It is known only for the special case in which the body is inertia-symmetric and in which, furthermore, the support point is located on the symmetry axis. In Fig. 4.4 such a symmetric heavy top is shown in a position in which the support point is at a lower level than the center of mass. This system the solutions of which were found by Lagrange is the subject of the following investigation.

The torque caused by gravity is a function of the orientation of the body in inertial space. This functional relationship has the consequence that Euler's equations (3.38) for ω_1, ω_2, ω_3 are coupled with kinematic differential equations which relate $\boldsymbol{\omega}$ to generalized coordinates. For this reason Euler's equations will not be used. Instead, it is preferred to establish second-order differential equations for appropriately chosen generalized coordinates and to solve these equations. On the body as well as in inertial space there exists one physically significant direction each, namely the symmetry axis on the body and the vertical line of action of gravity in inertial space. This suggests the use of Euler angles as generalized coordinates with θ being the angle between these two directions. The Euler angles ψ, θ and ϕ are defined as in Figs. 2.1 and 2.2. They relate a body-fixed base to a base fixed in inertial space. In Fig. 4.4 the body-fixed base is not shown. Shown are the base $\underline{\mathbf{e}}^1$ which is fixed in inertial space and the base $\underline{\mathbf{e}}^{2'}$ of Figs. 2.1 and 2.2. Its base vector $\mathbf{e}_3^{2'} = \mathbf{e}_3^2$ lies in the symmetry axis, and $\mathbf{e}_1^{2'}$ is always perpendicular to the vertical $\mathbf{e}_3^1$. The absolute angular velocity of this base differs from the absolute angular velocity $\boldsymbol{\omega}$ of the body by a component along the symmetry axis which is equal to $\dot{\phi}\mathbf{e}_3^2$. For reasons of symmetry the body moments of

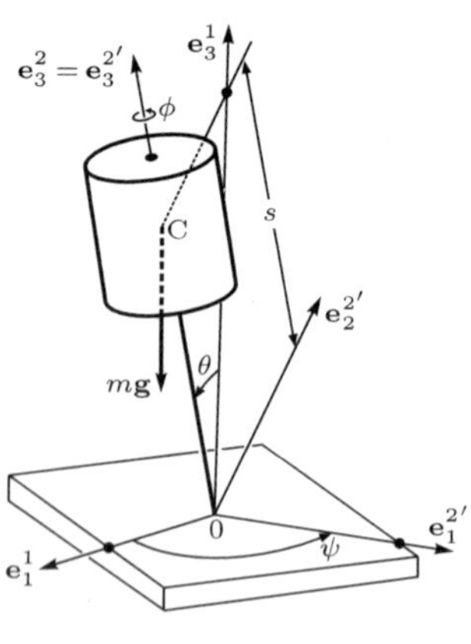

Fig. 4.4. Symmetric heavy top with coordinates ψ, θ, ϕ

inertia in base $\underline{\mathbf{e}}^{2'}$ are constant in spite of the motion of this base relative to the body.

Equations of motion are established from the angular momentum theorem in the general form (cf. (3.32)):

$$\dot{\mathbf{L}}^0 = \mathbf{M}^0 . \tag{4.59}$$

As reference point 0 in inertial space the support point of the body is chosen. Using the basic Eq. (2.74) the absolute time derivative $\dot{\mathbf{L}}^0$ is expressed in terms of the time derivative in base $\underline{\mathbf{e}}^{2'}$. Let $\boldsymbol{\Omega}$ be the angular velocity of this base. Then, the angular momentum theorem becomes

$$\frac{^{(2')}\mathrm{d}}{\mathrm{d}t}\mathbf{L}^0 + \boldsymbol{\Omega} \times \mathbf{L}^0 = \mathbf{M}^0 . \tag{4.60}$$

For the components of $\boldsymbol{\Omega}$, $\mathbf{L}^0$ and $\mathbf{M}^0$ in base $\underline{\mathbf{e}}^{2'}$ Fig. 4.4 yields the expressions

$$\boldsymbol{\Omega} = \dot{\theta}\mathbf{e}_1^{2'} + \dot{\psi}\sin\theta\mathbf{e}_2^{2'} + \dot{\psi}\cos\theta\mathbf{e}_3^{2'} , \tag{4.61}$$

$$\mathbf{L}^0 = J_1\dot{\theta}\mathbf{e}_1^{2'} + J_1\dot{\psi}\sin\theta\mathbf{e}_2^{2'} + J_3(\dot{\phi} + \dot{\psi}\cos\theta)\mathbf{e}_3^{2'} , \tag{4.62}$$

$$\mathbf{M}^0 = mgs\sin\theta\mathbf{e}_1^{2'} . \tag{4.63}$$

The quantity $(\dot{\phi}+\dot{\psi}\cos\theta)$ represents the angular velocity component ω_3 of the body along the symmetry axis. With these vector components (4.60) yields directly the desired scalar differential equations of motion:

$$J_1\ddot{\theta} + [J_3(\dot{\phi} + \dot{\psi}\cos\theta) - J_1\dot{\psi}\cos\theta]\dot{\psi}\sin\theta - mgs\sin\theta = 0 , \tag{4.64}$$

$$J_1\ddot{\psi}\sin\theta + 2J_1\dot{\psi}\dot{\theta}\cos\theta - J_3\dot{\theta}(\dot{\phi} + \dot{\psi}\cos\theta) = 0 , \tag{4.65}$$

$$J_3\frac{\mathrm{d}}{\mathrm{d}t}(\dot{\phi} + \dot{\psi}\cos\theta) = 0 . \tag{4.66}$$

The equations have three first integrals. The first one is

$$\dot{\phi} + \dot{\psi}\cos\theta \equiv \omega_3 = \text{const} . \tag{4.67}$$

This integral of motion can be concluded more directly from the third Euler equation which in the case $J_1 = J_2$ and $M_3 = 0$ reduces to $\dot{\omega}_3 = 0$ (cf. (4.40)). With this integral of motion (4.64) and (4.65) become

$$J_1\ddot{\theta} + \left(J_3\omega_3 - J_1\dot{\psi}\cos\theta\right)\dot{\psi}\sin\theta - mgs\sin\theta = 0 , \tag{4.68}$$

$$J_1\ddot{\psi}\sin\theta + 2J_1\dot{\psi}\dot{\theta}\cos\theta - J_3\omega_3\dot{\theta} = 0 . \tag{4.69}$$

These equations furnish two more first integrals. When the second equation is multiplied by $\sin\theta$ it can be written in the form

$$\frac{\mathrm{d}}{\mathrm{d}t}\left(J_1\dot{\psi}\sin^2\theta + J_3\omega_3\cos\theta\right) = 0 . \tag{4.70}$$

If, on the other hand, (4.68) is multiplied by $\dot{\theta}$ and (4.69) by $\dot{\psi}\sin\theta$ and both expressions are summed it is found that

$$\frac{\mathrm{d}}{\mathrm{d}t}\left[J_1\left(\dot{\psi}^2\sin^2\theta+\dot{\theta}^2\right)+2mgs\cos\theta\right]=0\,. \qquad (4.71)$$

The two integrals of motion revealed by these equations can be found directly without equations of motion. Since gravity does not produce a torque about the vertical $\mathbf{e}_3^1$ the angular momentum component in this direction has a constant magnitude L. Equation (4.62) and the geometric relationships shown in Fig. 4.4 yield for L the expression

$$J_1\dot{\psi}\sin^2\theta+J_3\omega_3\cos\theta=L=\text{ const}\,. \qquad (4.72)$$

This equation is equivalent to (4.70). The system is conservative so that its total energy E is constant. This yields

$$J_1\left(\omega_1^2+\omega_2^2\right)+J_3\omega_3^2+2mgs\cos\theta=2E \qquad (4.73)$$

or with $\omega_1=\Omega_1$ and $\omega_2=\Omega_2$ from (4.61)

$$J_1\left(\dot{\psi}^2\sin^2\theta+\dot{\theta}^2\right)+2mgs\cos\theta=2E-J_3\omega_3^2=\text{ const}\,. \qquad (4.74)$$

This is equivalent to (4.71).

Before the general solution to the problem is developed two special types of motion are considered which can be produced by a proper choice of initial conditions. One is the common plane pendulum motion with $\omega_3\equiv 0$ in which θ is the only time dependent variable. In this case the equations of motion reduce to $J_1\ddot{\theta}-mgs\sin\theta=0$, and of the three integrals of motion only the energy equation $J_1\dot{\theta}^2+2mgs\cos\theta=2E$ is not trivial. These two equations represent, indeed, the differential equation and the energy integral, respectively, of a plane physical pendulum (note that, normally, $\varphi=\pi-\theta$ is used as variable). The second special type of motion is characterized by a time independent angle $\theta\equiv\theta_0$. With this condition (4.69) yields $\dot{\psi}=$ const and, furthermore, (4.67) leads to $\dot{\phi}=$ const. This geometrically simple motion is referred to as regular precession. The symmetry axis of the body is moving with a constant precession angular velocity $\dot{\psi}$ around a circular cone the axis of which is the vertical $\mathbf{e}_3^1$. Equation (4.68) with $\ddot{\theta}=0$ represents a quadratic equation for $\dot{\psi}$ which has the solutions

$$\dot{\psi}_{1,2}=\begin{cases}\dfrac{J_3\omega_3}{2J_1\cos\theta_0}\left(1\pm\sqrt{1-\dfrac{4J_1mgs\cos\theta_0}{J_3^2\omega_3^2}}\right) & (\cos\theta_0\neq 0)\,,\\[2ex] \dfrac{mgs}{J_3\omega_3} & (\cos\theta_0=0)\,.\end{cases} \qquad (4.75)$$

At this point it should be noted that all results obtained thus far are valid also for the special case $s=0$, i.e. for a symmetric body which is supported at its center of mass and which is, therefore, torque-free. From Sect. 4.2 it is known that under these conditions only two types of motion can occur, namely a permanent rotation about the symmetry axis and nutations with a nutation angular velocity $\dot{\psi}$ given by (4.46). Equation (4.75) yields for $s = 0$ the two solutions

$$\dot{\psi}_1 = \frac{J_3\omega_3}{J_1 \cos\theta_0}\,, \qquad \dot{\psi}_2 = 0\,. \tag{4.76}$$

These are, indeed, the nutation angular velocity and a result for $\dot{\psi}_2$ which in the case $\omega_3 \neq 0$ can be interpreted as a permanent rotation about the symmetry axis.

Next, the general case of (4.75) with $s \neq 0$ is considered. If the body is hanging ($\cos\theta_0 < 0$) both roots $\dot{\psi}_1$ and $\dot{\psi}_2$ are real for any value of θ_0. In positions with $\cos\theta_0 > 0$ for which Fig. 4.4 shows an example regular precessions are possible only if the body angular velocity ω_3 is large enough so as to render the expression under the square root positive. In the upright position $\theta_0 = 0$ this is the condition $\omega_3^2 \geq 4J_1 mgs/J_3^2$.

In Fig. 4.5 the relationship between $\dot{\psi}_1$, $\dot{\psi}_2$ and ω_3 is schematically illustrated for various values of the parameter θ_0. For rapidly spinning bodies the roots tend toward

$$\lim_{\omega_3\to\infty} \dot{\psi}_1 = \frac{J_3\omega_3}{J_1\cos\theta_0}\,, \qquad \lim_{\omega_3\to\infty} \dot{\psi}_2 = \frac{mgs}{J_3\omega_3}\,. \tag{4.77}$$

One of the asymptotic solutions is proportional to ω_3 and represents a fast regular precession while the other is proportional to $1/\omega_3$ and represents a slow regular precession. The fast regular precession angular velocity is identical with the nutation angular velocity of a torque-free symmetric rigid body and also with $\dot{\psi}_1$ for $s = 0$ while the slow one is identical with the exact solution for $\dot{\psi}$ in the case $\cos\theta_0 = 0$.

We now turn to the general solution of the equations of motion. It is deduced from the integrals of motion. Equation (4.72) yields

$$\dot{\psi} = \frac{L - J_3\omega_3\cos\theta}{J_1\sin^2\theta}\,. \tag{4.78}$$

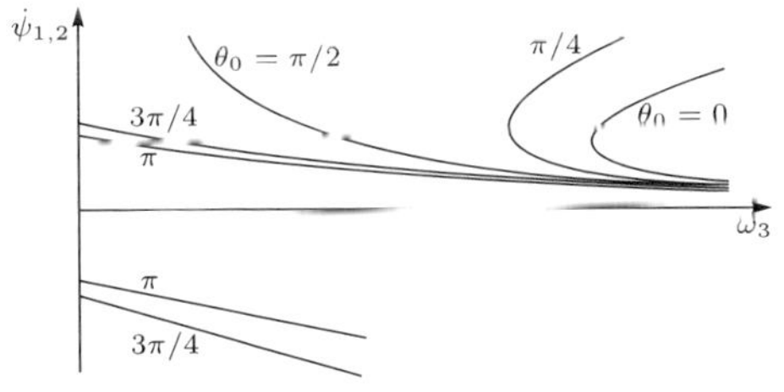

Fig. 4.5. Angular velocities $\dot{\psi}_1$ and $\dot{\psi}_2$ of regular precessions as functions of ω_3 and θ_0

Substitution into (4.74) results in the differential equation for θ

$$J_1\dot{\theta}^2 = 2E - J_3\omega_3^2 - 2mgs\cos\theta - \frac{(L - J_3\omega_3\cos\theta)^2}{J_1\sin^2\theta} \,. \tag{4.79}$$

As soon as its solution $\theta(t)$ is known $\psi(t)$ and $\phi(t)$ are found by simple integration from (4.78) and (4.67). With the new variable

$$u = \cos\theta \,, \qquad \dot{u} = -\dot{\theta}\sin\theta \tag{4.80}$$

(4.79) takes after simple manipulations the form

$$\dot{u}^2 = \frac{\left(2E - J_3\omega_3^2 - 2mgsu\right)\left(1 - u^2\right)}{J_1} - \frac{(L - J_3\omega_3 u)^2}{J_1^2} \,. \tag{4.81}$$

The expression on the right-hand side is a cubic polynomial in u. About the location of its roots the following statements can be made. For $u = +1$ and also for $u = -1$ the polynomial has negative values. In the limit $u \to +\infty$ it tends toward plus infinity. It has, therefore, at least one real root $u_3 > 1$. Because of (4.80) only the interval $|u| \le 1$ is of interest. Since for real solutions $\dot{u}^2(u)$ must be nonnegative somewhere in this interval the polynomial must have either two real roots or one real double root in the interval. For parameter combinations corresponding to real solutions the diagram of the function $\dot{u}^2(u)$ has, therefore, a form of the kind shown in Fig. 4.6. The roots u_1 and u_2 have either equal or opposite signs. That both cases are physically possible is demonstrated by the two special types of motion studied earlier. For pendulum motions the amplitude of θ can be chosen such that the sign of $u(t) = \cos\theta(t)$ is either always negative or alternating. For a regular precession u is constant.

It is assumed that the roots u_1, u_2 and u_3 are ordered as shown in Fig. 4.6 ($u_1 \le u_2 < u_3$). In terms of these roots (4.81) reads

$$\dot{u}^2 = \frac{2mgs}{J_1}(u - u_1)(u - u_2)(u - u_3) \,. \tag{4.82}$$

When u is replaced by the new variable v defined by

$$u = u_1 + (u_2 - u_1)v^2 \tag{4.83}$$

this equation becomes after simple manipulations

$$\dot{v}^2 = \frac{mgs}{2J_1}(u_3 - u_1)(1 - v^2)(1 - k^2v^2) \tag{4.84}$$

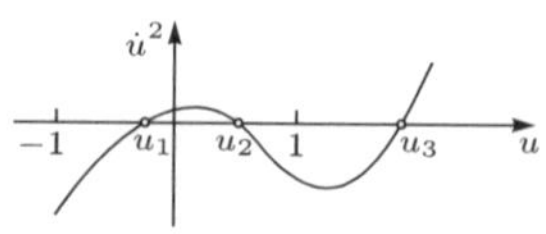

Fig. 4.6. Graph of the function $\dot{u}^2(u)$

with

$$0 \le k^2 = \frac{u_2 - u_1}{u_3 - u_1} \le 1 \, . \tag{4.85}$$

Separation of the variables leads to an elliptic integral of the first kind with modulus k

$$\int_{v_0}^{v} \frac{\mathrm{d}\bar{v}}{\sqrt{(1-\bar{v}^2)(1-k^2\bar{v}^2)}} = (t - t_0)\sqrt{\frac{(u_3 - u_1)mgs}{2J_1}} = \tau \, . \tag{4.86}$$

This equation has the solution $v = \operatorname{sn}\tau$. The solution for $\theta(t)$ is, therefore, in view of (4.83) and (4.80),

$$\cos\theta = \cos\theta_1 + (\cos\theta_2 - \cos\theta_1)\operatorname{sn}^2\tau \, . \tag{4.87}$$

The constants θ_1 and θ_2 are determined through $\cos\theta_1 = u_1$ and $\cos\theta_2 = u_2$, respectively. They represent the minimum and maximum values of $\theta(t)$. From (4.78) and (4.67) also $\dot{\psi}$ and $\dot{\phi}$ are obtained as functions of t:

$$\dot{\psi} = \frac{L - J_3\omega_3\cos\theta}{J_1(1-\cos^2\theta)} \, , \qquad \dot{\phi} = \omega_3 - \dot{\psi}\cos\theta \, . \tag{4.88}$$

The quantities θ, $\dot{\psi}$ and $\dot{\phi}$ are, thus, shown to be elliptic functions of time. The period of these functions is half the period of $\operatorname{sn}\tau$, i.e. $2K(k)$ in the variable τ and

$$t_{\mathrm{p}} = K\sqrt{\frac{8J_1}{mgs(u_3 - u_1)}} \tag{4.89}$$

in the variable t, K being the complete elliptic integral

$$K(k) = \int_0^1 \frac{\mathrm{d}v}{\sqrt{(1-v^2)(1-k^2v^2)}} \, . \tag{4.90}$$

The superposition of periodic changes in $\theta(t)$ onto a precession about the vertical with a (periodically changing) angular velocity $\dot{\psi}(t)$ is best visualized as follows. Imagine a sphere with the center at the support point of the body. The intersection point of the symmetry axis of the body with this sphere generates paths which render the periodic changes of $\theta(t)$ and of $\dot{\psi}(t)$ clearly

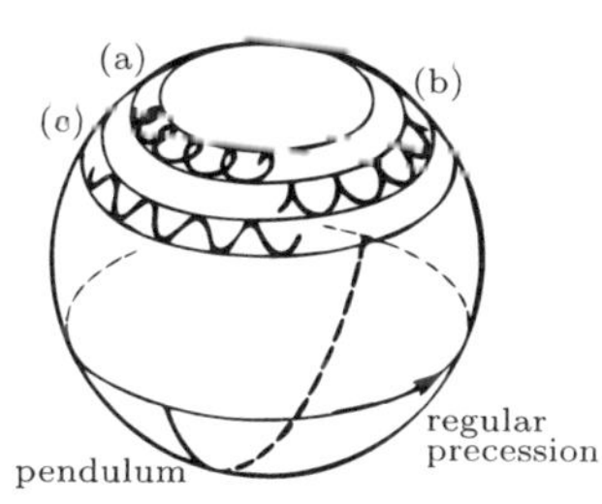

Fig. 4.7. Paths generated by the symmetry axis on a sphere surrounding the support point of the heavy top

visible. In Fig. 4.7 characteristic features of all physically realizable types of paths are illustrated schematically. The two lower paths belong to the special types of motion described as pendulum motion and regular precession. The upper three paths represent general cases of motion in which $\dot{\psi}(t)$ is changing either between negative and positive values (path a) or between zero and a positive maximum (b) or between two positive extreme values (c). The periodic nodding motion of the top with $\theta(t)$ which is superimposed on the precessional motion is called nutation. For a more detailed discussion of the solutions the reader is referred to Arnold/Maunder [3] and Magnus [51].

Many gyroscopic instruments represent, in principle, a symmetric heavy top (see Magnus [51]). Such instruments are operated under special conditions. For one thing, they are in rapid rotation. By this is meant that the part $J_3\omega_3^2/2$ of the kinetic energy is very large compared with the potential energy mgs. Second, such instruments are set into motion in such a way that initially the precession angular velocity $\dot{\psi}$ is very small compared with ω_3. Usually, θ and ψ are held fixed until the body has reached its full spin. Only then the constraints on θ and ψ are lifted keeping the initial values of $\dot{\theta}$ and $\dot{\psi}$ as small as possible. Under these conditions motions are observed which can hardly be distinguished from regular precessions. In reality, these motions are governed by (4.87) and (4.88). The nutation amplitude $(\theta_2 - \theta_1)/2$ is extremely small, however, and $\theta(t)$ is oscillating rapidly. The precession angular velocity $\dot{\psi}(t)$ is very small, and it appears to be constant although it is undergoing rapid oscillations. Such motions are called pseudo-regular precessions. Their characteristic properties can be developed from the general solution by means of approximation formulas. For this purpose it is assumed that the top is started with the initial conditions $\theta(0) = \theta_1$, $\dot{\theta}(0) = 0$, $\dot{\psi}(0) = \dot{\psi}_1$ and ω_3 (constant throughout the motion). The angular momentum integral and the energy integral are (see (4.72) and (4.74))

$$\left.\begin{aligned} L &= J_1\dot{\psi}_1\left(1-u_1^2\right) + J_3\omega_3 u_1 \;, \\ 2E &= J_1\dot{\psi}_1^2\left(1-u_1^2\right) + 2mgsu_1 + J_3\omega_3^2 \;. \end{aligned}\right\} \tag{4.91}$$

These expressions have to be substituted into the cubic in (4.81). Because of the initial condition $\dot{\theta}(0) = 0$ the angle θ_1 is one of the two extreme values of $\theta(t)$. This means that $u_1 = \cos\theta_1$ is a root of the cubic. Division by $(u - u_1)$ leads after some algebraic manipulations to

$$\dot{u}^2 = \frac{2mgs}{J_1}\left(u-u_1\right)\left\{u^2 - 2au - 1 + 2a\left[u_1 + b\left(1-u_1^2\right)\left(2 - bu_1 - b\right)\right]\right\} \tag{4.92}$$

where a and b are the dimensionless quantities

$$a = \frac{J_3^2\omega_3^2}{4mgsJ_1}\;, \qquad b = \frac{J_1\dot{\psi}_1}{J_3\omega_3}\;. \tag{4.93}$$

The quadratic function of u in curled brackets has the roots

$$u_{2,3} = a \mp \sqrt{a^2 + 1 - 2a\left[u_1 + b\left(1 - u_1^2\right)\left(2 - bu_1 - b\right)\right]} \,. \tag{4.94}$$

Under the assumed conditions the quantity a is much larger than one, and the absolute value of b is much smaller than one. This allows the approximations

$$u_{2,3} \approx a \mp \sqrt{a^2 + 1 - 2a\left[u_1 + 2b\left(1 - u_1^2\right)\right]} \,. \tag{4.95}$$

A Taylor expansion up to second-order terms yields

$$u_2 \approx u_1 - \left(1 - u_1^2\right)\left(\frac{1}{2a} - 2b\right) , \qquad u_3 \approx 2a \,. \tag{4.96}$$

With this result for u_2 and with the Taylor formula $\cos\theta_2 \approx \cos\theta_1 - (\theta_2 - \theta_1)\sin\theta_1$ the nutation amplitude $(\theta_2 - \theta_1)/2$ becomes approximately

$$\theta_2 - \theta_1 \approx \left(\frac{1}{2a} - 2b\right)\sin\theta_1 = \frac{2J_1}{J_3\omega_3}\left(\frac{mgs}{J_3\omega_3} - \dot{\psi}_1\right)\sin\theta_1 \,. \tag{4.97}$$

This is, indeed, a very small quantity. It becomes zero if the initial value $\dot{\psi}_1$ equals the angular velocity of a slow regular precession (see (4.77)). The modulus k of the elliptic functions determined by (4.85) is very small compared with unity as can be seen from (4.96) and (4.93). The complete elliptic integral $K(k)$ is, therefore, approximately $\pi/2$. This yields for the period length of the functions $\theta(t)$ and $\dot{\psi}(t)$ the approximation

$$t_{\mathrm{p}} \approx \frac{\pi}{2}\sqrt{\frac{8J_1}{mgsu_3}} = \frac{2\pi J_1}{J_3\omega_3} \,. \tag{4.98}$$

The corresponding circular frequency $2\pi/t_{\mathrm{p}} \approx \omega_3 J_3/J_1$ is very large. It equals the nutation angular velocity of a torque-free symmetric body in the case of small nutation amplitudes (see (4.48)). Finally, an approximation formula for $\dot{\psi}$ is developed from (4.88) and (4.91):

$$\dot{\psi} \approx \frac{J_1\dot{\psi}_1(1 - u_1^2) + J_3\omega_3(u_1 - u)}{J_1(1 - u_1^2)} = \dot{\psi}_1 + \frac{J_3\omega_3}{J_1(1 - u_1^2)}(u_1 - u) \,. \tag{4.99}$$

For the difference $u_1 - u = \cos\theta_1 - \cos\theta$ (4.87) yields (with the approximation $\operatorname{sn}\tau \approx \sin\tau$ valid for $k \approx 0$) $u_1 - u \approx (u_1 - u_2)\sin^2\tau$. For $(u_1 - u_2)$ (4.96) is used. This leads to

$$u_1 - u \approx \left(1 - u_1^2\right)\left(\frac{1}{2a} - 2b\right)\sin^2\tau \tag{4.100}$$

and, furthermore, to

$$\dot{\psi} \approx \dot{\psi}_1 + 2\left(\frac{mgs}{J_3\omega_3} - \dot{\psi}_1\right)\sin^2\tau = \frac{mgs}{J_3\omega_3} - \left(\frac{mgs}{J_3\omega_3} - \dot{\psi}_1\right)\cos 2\tau \,. \tag{4.101}$$

This result shows that $\dot{\psi}$ is oscillating about the mean value $mgs/(J_3\omega_3)$ which is the angular velocity of a slow regular precession. The amplitude of the oscillation is zero if the initial value $\dot{\psi}_1$ equals this mean value. The approximation formulas just established confirm the initial statement that motions of a rapidly rotating symmetric heavy top differ only slightly from regular precessions. The motion can be interpreted as superposition of a fast nutation with very small amplitude $(\theta_2 - \theta_1)/2$ onto a slow regular precession.

4.5 Symmetric Heavy Body in a Cardan Suspension

Figure 4.8 depicts a symmetric rigid rotor in a two-gimbal suspension. All three rotation axes intersect at the point 0. This point is the center of mass of the rotor as well as of the inner gimbal. In a position when the three rotation axes are mutually perpendicular these axes represent principal axes of inertia of the rotor and of the inner gimbal. The axis of the outer gimbal is mounted in a vertical position. To the symmetry axis of the rotor, at a distance s from 0, a point mass m is attached. Such a system has many features in common with a symmetric heavy top (see Fig. 4.4). The only differences are the presence of gimbals with inertia properties and the constraint torque normal to the outer gimbal axis which is transmitted to the system by the bearings on this axis.

In Fig. 4.8 the inertial base $\underline{\mathbf{e}}^1$, the base $\underline{\mathbf{e}}^{2'}$ fixed on the inner gimbal and the Euler angles ψ, θ and ϕ are identical with those of Fig. 2.2 and also with those of Fig. 4.4. As reference point for angular momenta and torques the point 0 is chosen. If there is no friction in the bearings which is supposed to be the case then the resultant torque on the rotor has no component along the rotor axis. The coordinate ω_3 in this direction of the absolute rotor angular velocity is, therefore, constant. This constitutes a first integral of motion. It

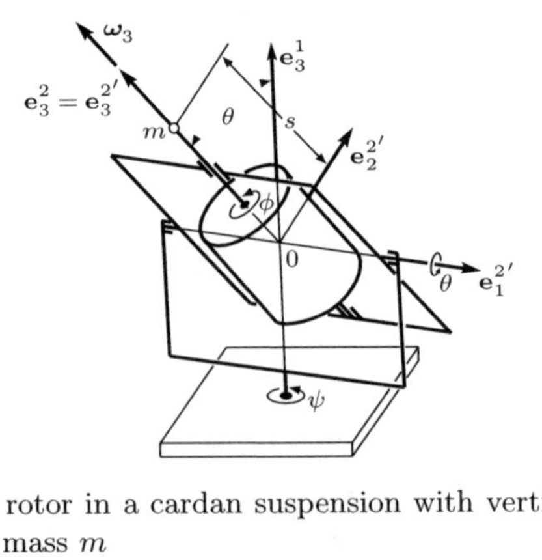

Fig. 4.8. Symmetric rotor in a cardan suspension with vertical outer gimbal axis and with offset point mass m

is known already from the heavy top (see (4.67)):

$$\dot{\psi}\cos\theta + \dot{\phi} \equiv \omega_3 = \text{const} . \tag{4.102}$$

The resultant torque on the entire system, composed of the constraint torque in the outer gimbal axis and of the torque caused by the weight of the point mass has no vertical component. Hence, the coordinate L in this direction of the absolute angular momentum of the entire system is a constant. Note that this first integral exists only if the axis of the outer gimbal is mounted vertically. A third first integral expresses the fact that the total energy E of the system is constant. These three first integrals are the same on which the analysis of the symmetric heavy top was based. In the present case the two gimbals contribute to the angular momentum coordinate L and to the energy E. In order to formulate L and E the absolute angular velocities of the three bodies are expressed. Figure 4.8 yields

$$\left.\begin{aligned} &\text{outer gimbal:} && \boldsymbol{\omega}^{(1)} = \dot{\psi}\mathbf{e}_3^1 , \\ &\text{inner gimbal:} && \boldsymbol{\omega}^{(2)} = \dot{\theta}\mathbf{e}_1^{2'} + \dot{\psi}\sin\theta\mathbf{e}_2^{2'} + \dot{\psi}\cos\theta\mathbf{e}_3^{2'} , \\ &\text{rotor:} && \boldsymbol{\omega}^{(r)} = \dot{\theta}\mathbf{e}_1^{2'} + \dot{\psi}\sin\theta\mathbf{e}_2^{2'} + \omega_3\mathbf{e}_3^{2'} . \end{aligned}\right\} \tag{4.103}$$

Let $J_3^{(1)}$ be the moment of inertia of the outer gimbal about its vertical axis. In base $\underline{\mathbf{e}}^{2'}$ the inner gimbal has principal moments of inertia $J_1^{(2)}, J_2^{(2)}, J_3^{(2)}$ and rotor and point mass together have principal moments of inertia J_1, $J_2 = J_1$, J_3. With this notation, the outer gimbal has the angular momentum $\mathbf{L}^{(1)} = J_3^{(1)}\dot{\psi}\mathbf{e}_3^1$. Inner gimbal, rotor and point mass together have in base $\underline{\mathbf{e}}^{2'}$ the angular momentum components

$$\begin{aligned} \mathbf{L}^{(2)} + \mathbf{L}^{(r)} = {} & \left(J_1^{(2)} + J_1\right)\dot{\theta}\mathbf{e}_1^{2'} + \left(J_2^{(2)} + J_1\right)\dot{\psi}\sin\theta\mathbf{e}_2^{2'} \\ & + \left(J_3^{(2)}\dot{\psi}\cos\theta + J_3\omega_3\right)\mathbf{e}_3^{2'} . \end{aligned} \tag{4.104}$$

The constant vertical coordinate L of the total angular momentum is (see Fig. 4.8)

$$\begin{aligned} L &= J_3^{(1)}\dot{\psi} + \sin\theta\left(J_2^{(2)} + J_1\right)\dot{\psi}\sin\theta + \cos\theta\left(J_3^{(2)}\dot{\psi}\cos\theta + J_3\omega_3\right) \\ &= \dot{\psi}\left[J_1 + J_2^{(2)} + J_3^{(1)} - \left(J_1 + J_2^{(2)} - J_3^{(2)}\right)\cos^2\theta\right] \\ &\quad + J_3\omega_3\cos\theta . \end{aligned} \tag{4.105}$$

The total energy of the system is calculated from the angular velocities and from the position of the mass m:

$$\begin{aligned} 2E &= J_3^{(1)}\dot{\psi}^2 + \left(J_1^{(2)} + J_1\right)\dot{\theta}^2 + \left(J_2^{(2)} + J_1\right)\dot{\psi}^2\sin^2\theta + J_3^{(2)}\dot{\psi}^2\cos^2\theta \\ &\quad + J_3\omega_3^2 + 2mgs\cos\theta \\ &= \dot{\psi}^2\left[J_1 + J_2^{(2)} + J_3^{(1)} - \left(J_1 + J_2^{(2)} - J_3^{(2)}\right)\cos^2\theta\right] \\ &\quad + \left(J_1 + J_1^{(2)}\right)\dot{\theta}^2 + J_3\omega_3^2 + 2mgs\cos\theta . \end{aligned} \tag{4.106}$$

Equations (4.105) and (4.106) correspond to (4.72) and (4.74), respectively, for the symmetric heavy top. They are identical with these equations if all gimbal moments of inertia are equal to zero. The general solution to the problem is achieved by the same approach that was used for the symmetric heavy top. Equation (4.105) is solved for $\dot{\psi}$

$$\dot{\psi} = \frac{L - J_3\omega_3\cos\theta}{J_1 + J_2^{(2)} + J_3^{(1)} - \left(J_1 + J_2^{(2)} - J_3^{(2)}\right)\cos^2\theta} . \tag{4.107}$$

Substitution into (4.106) yields for θ the differential equation

$$\begin{aligned}\left(J_1 + J_1^{(2)}\right)\dot{\theta}^2 &= 2E - J_3\omega_3^2 - 2mgs\cos\theta \\ &\quad - \frac{\left(L - J_3\omega_3\cos\theta\right)^2}{J_1 + J_2^{(2)} + J_3^{(1)} - \left(J_1 + J_2^{(2)} - J_3^{(2)}\right)\cos^2\theta} .\end{aligned} \tag{4.108}$$

This corresponds to (4.79). As before, θ is substituted by the new variable $u = \cos\theta$, $\dot{u} = -\dot{\theta}\sin\theta$. This results in the differential equation for u

$$\begin{aligned}\dot{u}^2 &= \frac{\left(2E - J_3\omega_3^2 - 2mgsu\right)\left(1 - u^2\right)}{J_1 + J_1^{(2)}} \\ &\quad - \frac{(L - J_3\omega_3 u)^2(1 - u^2)}{\left(J_1 + J_1^{(2)}\right)\left[J_1 + J_2^{(2)} + J_3^{(1)} - \left(J_1 + J_2^{(2)} - J_3^{(2)}\right)u^2\right]}\end{aligned} \tag{4.109}$$

If all gimbal moments of inertia are zero this is identical with (4.81) for the symmetric heavy top. The gimbal inertia has the effect that the right-hand side expression is a fractional rational function of u instead of a cubic polynomial. The solution has, therefore, a completely different character. It cannot be expressed in terms of known special functions. The equation is not even soluble in the special case when no point mass and, thus, no torque caused by gravity is present. For more details about the dynamic effects of suspension systems the reader is referred to Magnus [51], Arnold/Maunder [3] and Saidov [70].

4.6 Gyrostat. General Considerations

A gyrostat is a mechanical system which is composed of more than one body and, yet, has the rigid body property that its moments and products of inertia are time independent constants. In its simplest form a gyrostat consists of two bodies as shown in Fig. 4.9. A symmetric rigid rotor is supported in rigid bearings on another rigid body called the carrier.

We shall first establish equations of motion for this particular system. Later it will be simple to formulate equations for gyrostats which consist of more than two bodies. About the central principal moments of inertia of the

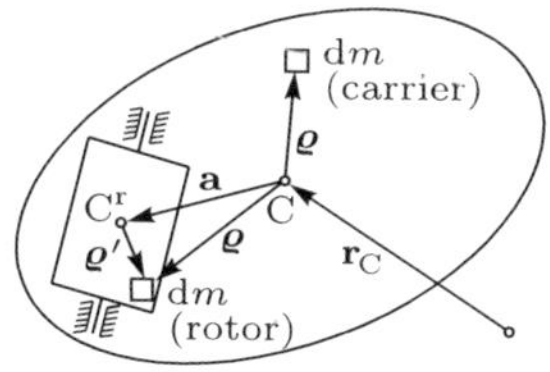

Fig. 4.9. Radius vectors of mass particles on the carrier and the rotor of a gyrostat

composite system and about the orientation of the rotor axis relative to the principal axes of inertia no special assumptions are made. Let $\boldsymbol{\omega}$ denote the absolute angular velocity of the carrier and $\boldsymbol{\omega}_{\mathrm{rel}}$ the angular velocity of the rotor relative to the carrier. In Fig. 4.9 the points C and $\mathrm{C^r}$ designate the composite system center of mass and the rotor center of mass, respectively. The point 0 is fixed in inertial space. The figure also shows representative mass particles of the carrier and the rotor. Both have position vectors $\mathbf{r}_{\mathrm{C}}+\boldsymbol{\varrho}$. Their absolute velocities are different, however, namely $\mathbf{v}_{\mathrm{C}}+\boldsymbol{\omega}\times\boldsymbol{\varrho}$ (carrier) and $\mathbf{v}_{\mathrm{C}}+\boldsymbol{\omega}\times\boldsymbol{\varrho}+\boldsymbol{\omega}_{\mathrm{rel}}\times\boldsymbol{\varrho}'$ (rotor). With these vectors the absolute angular momentum of the total system with respect to the point 0 is the sum of two integrals, one over the mass m^{c} of the carrier and the other over the mass m^{r} of the rotor:

$$\begin{aligned}\mathbf{L}^0 &= \int_{m^{\mathrm{c}}} (\mathbf{r}_{\mathrm{C}}+\boldsymbol{\varrho})\times(\mathbf{v}_{\mathrm{C}}+\boldsymbol{\omega}\times\boldsymbol{\varrho})\,\mathrm{d}m \\ &\quad + \int_{m^{\mathrm{r}}} (\mathbf{r}_{\mathrm{C}}+\boldsymbol{\varrho})\times[\mathbf{v}_{\mathrm{C}}+\boldsymbol{\omega}\times\boldsymbol{\varrho}+\boldsymbol{\omega}_{\mathrm{rel}}\times\boldsymbol{\varrho}']\,\mathrm{d}m\,. \qquad (4.110)\end{aligned}$$

The parts common to both integrals are combined under one integral over the total system mass $m=m^{\mathrm{c}}+m^{\mathrm{r}}$. In the remaining integral the vector $\boldsymbol{\varrho}$ is expressed as $\mathbf{a}+\boldsymbol{\varrho}'$. This results in the expression

$$\begin{aligned}\mathbf{L}^0 &= \int_{m} (\mathbf{r}_{\mathrm{C}}+\boldsymbol{\varrho})\times(\mathbf{v}_{\mathrm{C}}+\boldsymbol{\omega}\times\boldsymbol{\varrho})\,\mathrm{d}m \\ &\quad + \int_{m^{\mathrm{r}}} (\mathbf{r}_{\mathrm{C}}+\mathbf{a}+\boldsymbol{\varrho}')\times(\boldsymbol{\omega}_{\mathrm{rel}}\times\boldsymbol{\varrho}')\,\mathrm{d}m \qquad (4.111)\end{aligned}$$

From (3.15) it is known that the first integral is $\mathbf{r}_{\mathrm{C}}\times\mathbf{v}_{\mathrm{C}}m+\mathsf{J}\cdot\boldsymbol{\omega}$ with J being the central inertia tensor of the entire system composed of carrier and rotor. The contribution of the constant vector $\mathbf{r}_{\mathrm{C}}+\mathbf{a}$ to the second integral is zero because of $\int_{m_{\mathrm{r}}}\boldsymbol{\varrho}'\,\mathrm{d}m=\mathbf{0}$. Thus, the integral is $\mathsf{J}^{\mathrm{r}}\cdot\boldsymbol{\omega}_{\mathrm{rel}}$ where J^{r} is the rotor inertia tensor with respect to $\mathrm{C^r}$. The inertia tensors J and J^{r} have both constant coordinates in a vector base fixed on the carrier. The vector $\mathsf{J}^{\mathrm{r}}\cdot\boldsymbol{\omega}_{\mathrm{rel}}$ is the rotor angular momentum relative to the carrier. It is abbreviated $\mathbf{h}$. Since the rotor axis is a principal axis of inertia of the rotor the vector $\mathbf{h}$ has the direction of this axis. Thus, $\mathbf{L}^0$ has the final form

$$\mathbf{L}^0=\mathbf{r}_{\mathrm{C}}\times\mathbf{v}_{\mathrm{C}}m+\mathsf{J}\cdot\boldsymbol{\omega}+\mathbf{h}\,. \qquad (4.112)$$

With this expression the angular momentum theorem (3.32) yields the equation of motion

$$\mathsf{J} \cdot \dot{\boldsymbol{\omega}} + \overset{\circ}{\mathbf{h}} + \boldsymbol{\omega} \times (\mathsf{J} \cdot \boldsymbol{\omega} + \mathbf{h}) = \mathbf{M} \tag{4.113}$$

which represents a generalization of Euler's equation (3.36) for the rigid body. The symbol $\overset{\circ}{\mathbf{h}}$ serves as an abbreviation for the time derivative of $\mathbf{h}$ in a base fixed on the carrier. Also $\overset{\circ}{\mathbf{h}}$ has the direction of the rotor axis. The vector $\mathbf{M}$ is the resultant external torque with respect to C.

A case of considerable importance in engineering is the special case where the angular velocity of the rotor relative to the carrier is a prescribed function of time. Then, $\mathbf{h}(t)$ and $\overset{\circ}{\mathbf{h}}(t)$ are vectors along the rotor axis with magnitudes which are known as functions of time. The rotor has no degree of freedom of its own, and (4.113) together with kinematic differential equations for the carrier fully describes the system. The term $-\overset{\circ}{\mathbf{h}}(t)$ can be treated as if it were an external torque:

$$\mathsf{J} \cdot \dot{\boldsymbol{\omega}} + \boldsymbol{\omega} \times (\mathsf{J} \cdot \boldsymbol{\omega} + \mathbf{h}(t)) = \mathbf{M} - \overset{\circ}{\mathbf{h}}(t)\ . \tag{4.114}$$

In the simplest case of this kind the relative angular velocity of the rotor is constant. The equation then reads

$$\mathsf{J} \cdot \dot{\boldsymbol{\omega}} + \boldsymbol{\omega} \times (\mathsf{J} \cdot \boldsymbol{\omega} + \mathbf{h}) = \mathbf{M}\ . \tag{4.115}$$

In another case of importance in engineering the component along the rotor axis of the torque which is exerted on the rotor by the carrier is a prescribed function of time. In this case the rotor has one degree of freedom of motion of its own so that one additional scalar equation of motion is needed. This equation is established as follows. The angular momentum theorem for the rotor alone reads

$$\frac{\mathrm{d}}{\mathrm{d}t} \left[\mathsf{J}^{\mathrm{r}} \cdot (\boldsymbol{\omega} + \boldsymbol{\omega}_{\mathrm{rel}})\right] = \mathbf{M}^{\mathrm{r}} \tag{4.116}$$

with $\mathbf{M}^{\mathrm{r}}$ being the resultant torque on the rotor with respect to the center of mass C^{r}. Carrying out the differentiation in a carrier-fixed base we get

$$\mathsf{J}^{\mathrm{r}} \cdot \dot{\boldsymbol{\omega}} + \overset{\circ}{\mathbf{h}} + \boldsymbol{\omega} \times (\mathsf{J}^{\mathrm{r}} \cdot \boldsymbol{\omega} + \mathbf{h}) = \mathbf{M}^{\mathrm{r}}\ . \tag{4.117}$$

It is reasonable to assume that the external torque $\mathbf{M}$ which acts on the gyrostat as a whole does not contribute to the component of $\mathbf{M}^{\mathrm{r}}$ along the rotor axis. This component is then caused by interaction from the carrier alone and by assumption it is a given function of time. Let it be called $M^{\mathrm{r}}(t)$. If $\mathbf{u}$ is a unit vector along the rotor axis the desired scalar equation of motion is obtained by scalar multiplication of (4.117) by $\mathbf{u}$:

$$\mathbf{u} \cdot \left(\mathsf{J}^{\mathrm{r}} \cdot \dot{\boldsymbol{\omega}} + \overset{\circ}{\mathbf{h}}\right) = \mathbf{u} \cdot \mathbf{M}^{\mathrm{r}} = M^{\mathrm{r}}(t)\ . \tag{4.118}$$

Note that the third term on the left in (4.117) does not give a contribution since $\mathbf{u}$ and $\mathbf{h}$ are collinear and since for a symmetric body the vectors $\mathbf{u}$, $\boldsymbol{\omega}$

and $\mathsf{J}^{\mathrm{r}}\cdot\boldsymbol{\omega}$ are coplanar. The equation can be integrated once. For this purpose it is given the form

$$\frac{\mathrm{d}}{\mathrm{d}t}\left[\mathbf{u}\cdot(\mathsf{J}^{\mathrm{r}}\cdot\boldsymbol{\omega}+\mathbf{h})\right]=M^{\mathrm{r}}(t) \tag{4.119}$$

which is identical since the product $(\boldsymbol{\omega}\times\mathbf{u})\cdot(\mathsf{J}^{\mathrm{r}}\cdot\boldsymbol{\omega}+\mathbf{h})$ is zero. Integration then yields

$$\mathbf{u}\cdot(\mathsf{J}^{\mathrm{r}}\cdot\boldsymbol{\omega}+\mathbf{h})=\int M^{\mathrm{r}}(t)\,\mathrm{d}t=L^{\mathrm{r}}(t) \tag{4.120}$$

where the known function $L^{\mathrm{r}}(t)$ represents the coordinate along the rotor axis of the absolute angular momentum. In order to describe the motion completely it is necessary to supplement (4.113) and (4.120) by kinematic differential equations which relate $\boldsymbol{\omega}$ to generalized coordinates for the angular orientation of the carrier.

Having described the simplest possible type of gyrostat it is now a straightforward procedure to formulate equations of motion for gyrostats in which more than one rotor is mounted on the carrier. Let there be $m+n$ rotors altogether and let them be identified by an index i which is attached to all rotor quantities. It is assumed that for the rotors labeled $i=1,\ldots,m$ the axial torque component $M_i^{\mathrm{r}}(t)$ is a given function of time whereas for the remaining rotors $i=m+1,\ldots,m+n$ the coordinates of $\mathbf{h}_i(t)$ in a carrier-fixed base are given functions of time. The equations of motion consist of a single vector equation of the kind of (4.113) and of a set of m scalar equations of the form (4.120), one for each of the rotors $i=1,\ldots,m$:

$$\mathsf{J}\cdot\dot{\boldsymbol{\omega}}+\sum_{i=1}^{m}\overset{\circ}{\mathbf{h}}_i+\boldsymbol{\omega}\times\left[\mathsf{J}\cdot\boldsymbol{\omega}+\sum_{i=1}^{m}\mathbf{h}_i+\sum_{i=m+1}^{m+n}\mathbf{h}_i(t)\right]=\mathbf{M}-\sum_{i=m+1}^{m+n}\overset{\circ}{\mathbf{h}}_i(t)\ , \tag{4.121}$$

$$\mathbf{u}_i\cdot(\mathsf{J}_i^{\mathrm{r}}\cdot\boldsymbol{\omega}+\mathbf{h}_i)=\int M_i^{\mathrm{r}}(t)\,\mathrm{d}t=L_i^{\mathrm{r}}(t)\qquad(i=1,\ldots,m)\ . \tag{4.122}$$

The scalar equations are equivalent to

$$\mathbf{u}_i\cdot\left(\mathsf{J}_i^{\mathrm{r}}\cdot\dot{\boldsymbol{\omega}}+\overset{\circ}{\mathbf{h}}_i\right)=M_i^{\mathrm{r}}(t)\qquad(i=1,\ldots,m)\ . \tag{4.123}$$

This set of equations can be simplified substantially. To begin with, (4.122) and (4.123) are reformulated. Since $\mathbf{u}_i$ is an eigenvector of $\mathsf{J}_i^{\mathrm{r}}$ the product $\mathbf{u}_i\cdot\mathsf{J}_i^{\mathrm{r}}\cdot\boldsymbol{\omega}$ can be rewritten in the form $J_i^{\mathrm{r}}\mathbf{u}_i\cdot\boldsymbol{\omega}$ where J_i^{r} is the principal moment of inertia of the ith rotor about the rotor axis. Furthermore, $\mathbf{u}_i\cdot\overset{\circ}{\mathbf{h}}_i=|\overset{\circ}{\mathbf{h}}_i|$. Multiplication of (4.123) by $\mathbf{u}_i$ yields, therefore, the vector equation

$$J_i^{\mathrm{r}}\mathbf{u}_i\mathbf{u}_i\cdot\dot{\boldsymbol{\omega}}+\overset{\circ}{\mathbf{h}}_i=\mathbf{u}_iM_i^{\mathrm{r}}(t)\qquad(i=1,\ldots,m) \tag{4.124}$$

in which $J_i^{\mathrm{r}}\mathbf{u}_i\mathbf{u}_i$ is an inertia tensor. In a similar manner multiplication of (4.122) by $\boldsymbol{\omega}\times\mathbf{u}_i$ produces the vector equation

$$\boldsymbol{\omega}\times(J_i^{\mathrm{r}}\mathbf{u}_i\mathbf{u}_i)\cdot\boldsymbol{\omega}+\boldsymbol{\omega}\times\mathbf{h}_i=\boldsymbol{\omega}\times\mathbf{u}_iL_i^{\mathrm{r}}(t)\qquad(i=1,\ldots,m)\ . \tag{4.125}$$

Equations (4.124) and (4.125) are summed over $i = 1, \ldots, m$, and both sums are subtracted from (4.121):

$$\begin{aligned}&\left(\mathsf{J} - \sum_{i=1}^{m} J_i^{\mathrm{r}} \mathbf{u}_i \mathbf{u}_i\right) \cdot \dot{\boldsymbol{\omega}} \\ &+ \boldsymbol{\omega} \times \left[\left(\mathsf{J} - \sum_{i=1}^{m} J_i^{\mathrm{r}} \mathbf{u}_i \mathbf{u}_i\right) \cdot \boldsymbol{\omega} + \sum_{i=1}^{m} \mathbf{u}_i L_i^{\mathrm{r}}(t) + \sum_{i=m+1}^{m+n} \mathbf{h}_i(t)\right] \\ &= M - \sum_{i=1}^{m} \mathbf{u}_i M_i^{\mathrm{r}}(t) - \sum_{i=m+1}^{m+n} \overset{\circ}{\mathbf{h}}_i(t) \,. \end{aligned} \tag{4.126}$$

For abbreviation the new quantities are introduced:

$$\mathsf{J}^* = \mathsf{J} - \sum_{i=1}^{m} J_i^{\mathrm{r}} \mathbf{u}_i \mathbf{u}_i \,, \qquad \mathbf{h}^*(t) = \sum_{i=1}^{m} \mathbf{u}_i L_i^{\mathrm{r}}(t) + \sum_{i=m+1}^{m+n} \mathbf{h}_i(t) \,. \tag{4.127}$$

Because of the identity $M_i^{\mathrm{r}}(t) = \mathrm{d}L_i^{\mathrm{r}}(t)/\,\mathrm{d}t$ one has

$$\overset{\circ}{\mathbf{h}}{}^*(t) = \sum_{i=1}^{m} \mathbf{u}_i M_i^{\mathrm{r}}(t) + \sum_{i=m+1}^{m+n} \overset{\circ}{\mathbf{h}}_i(t) \,. \tag{4.128}$$

In a carrier-fixed base the tensor J^* has constant coordinates, and $\mathbf{h}^*(t)$ as well as $\overset{\circ}{\mathbf{h}}{}^*(t)$ have coordinates which are known functions of time. In terms of these quantities (4.126) becomes

$$\mathsf{J}^* \cdot \dot{\boldsymbol{\omega}} + \boldsymbol{\omega} \times [\mathsf{J}^* \cdot \boldsymbol{\omega} + \mathbf{h}^*(t)] = \mathbf{M} - \overset{\circ}{\mathbf{h}}{}^*(t) \,. \tag{4.129}$$

This equation, together with kinematic differential equations for the carrier, fully describes the motion of the carrier. The equation has the same form as (4.114) for a gyrostat with a single rotor and with given functions $\mathbf{h}(t)$ and $\overset{\circ}{\mathbf{h}}(t)$. Note, however, that in contrast to $\mathbf{h}(t)$ in (4.114) the vector $\mathbf{h}^*(t)$ in (4.129) does not have, in general, a fixed direction in the carrier. The two equations are, therefore, not equivalent. They are equivalent only in special cases such as the following:

(i) For arbitrary m and n all rotor axes are aligned parallel.
(ii) $m = 1$ and $n = 0$. Thus, a gyrostat with a single rotor with given torque $M^{\mathrm{r}}(t)$ and a gyrostat with a single rotor with given relative angular momentum $\mathbf{h}(t)$ are equivalent.
(iii) $m = 0$, n arbitrary and $|\mathbf{h}_i(t)| = \lambda_i h(t)$ for $i = 1, \ldots, n$ with constants $\lambda_1, \ldots, \lambda_n$ and with an arbitrary function $h(t)$. These conditions are fulfilled if all rotors are connected by gear wheels.
(iv) m and n arbitrary, $M_i^{\mathrm{r}}(t) \equiv 0$ for $i = 1, \ldots, m$ and $|\mathbf{h}_i(t)| = \mathrm{const}$ for $i = m+1, \ldots, m+n$.

Carrier bodies with built-in rotors are not the only multibody systems with time independent moments of inertia. This property is conserved if the carrier, in addition to rotors, contains cavities which are completely filled with homogeneous fluids. Various technical instruments and vehicles with rotating engines, with fuel tanks and with hydraulic systems represent gyrostats of this nature. Their dynamic behavior has been studied by Moiseev/Rumiancev [52].

Problem 4.2. In the absence of external torques (4.129) represents the equation of a self-excited rigid body with the driving torque $-\boldsymbol{\omega} \times \mathbf{h}^*(t) - \overset{\circ}{\mathbf{h}}{}^*(t)$. Integrate the equations of motion in closed form in the following case. A gyrostat with one rotor has principal moments of inertia $J_1 = J_2 \neq J_3$. The axis of the rotor whose moment of inertia about the symmetry axis is J^{r} is mounted parallel to the principal axis about which J_3 is measured. The carrier transmits a torque $M^{\mathrm{r}}(t)$ to the rotor about the rotor axis. Solve the equation also if the rotor torque has the form $M^{\mathrm{r}}(t) - ah$ ($a > 0$, const) where $M^{\mathrm{r}}(t)$ is, again, a given function of time and the term $-ah$ represents a viscous damping torque.

4.7 Torque-Free Gyrostat

In this section (4.129) is investigated in the special case when the external torque $\mathbf{M}$ on the gyrostat is identically zero and when, furthermore, the relative angular momentum $\mathbf{h}^*$ has constant coordinates in a carrier-fixed base. Omitting the asterisk we have the equation

$$\mathsf{J} \cdot \dot{\boldsymbol{\omega}} + \boldsymbol{\omega} \times (\mathsf{J} \cdot \boldsymbol{\omega} + \mathbf{h}) = \mathbf{0}\,. \tag{4.130}$$

It governs torque-free gyrostats whose rotors are operated under the condition $M_i^{\mathrm{r}}(t) \equiv 0$ for $i = 1, \ldots, m$ (arbitrary) and $|\mathbf{h}_i(t)| = \text{const}$ for $i = m + 1, \ldots, m + n$ (n arbitrary). For the sake of simplicity the equation will always be interpreted as that of a gyrostat with a single rotor of constant relative angular momentum $\mathbf{h}$.

Equation (4.130) possesses two algebraic integrals which are found by multiplication with $\boldsymbol{\omega}$ and with $\mathsf{J} \cdot \boldsymbol{\omega} + \mathbf{h}$, respectively:

$$\boldsymbol{\omega} \cdot \mathsf{J} \cdot \boldsymbol{\omega} = 2T = \text{const}\,, \qquad (\mathsf{J} \cdot \boldsymbol{\omega} + \mathbf{h})^2 = L^2 = \text{const}\,. \tag{4.131}$$

They are an energy integral and an angular momentum integral. Note that T is not the total kinetic energy but only that portion of it which remains when the carrier is rotating with its actual angular velocity $\boldsymbol{\omega}$ while the rotor is "frozen" in the carrier. The quantity L is the magnitude of the absolute angular momentum of the composite system. Resolved in the principal axes frame of the composite system (4.130) and (4.131) read

$$\left.\begin{aligned} J_1\dot{\omega}_1 - (J_2 - J_3)\omega_2\omega_3 + \omega_2 h_3 - \omega_3 h_2 = 0\,,\\ J_2\dot{\omega}_2 - (J_3 - J_1)\omega_3\omega_1 + \omega_3 h_1 - \omega_1 h_3 = 0\,,\\ J_3\dot{\omega}_3 - (J_1 - J_2)\omega_1\omega_2 + \omega_1 h_2 - \omega_2 h_1 = 0\,, \end{aligned}\right\} \tag{4.132}$$

$$\sum_{i=1}^{3} J_i \omega_i^2 = 2T \;, \tag{4.133}$$

$$\sum_{i=1}^{3} (J_i \omega_i + h_i)^2 = L^2 = 2DT \;. \tag{4.134}$$

The parameter D with the physical dimension of moment of inertia is introduced for convenience and also in order to show the similarity with (4.6) for the torque-free rigid body.

The following investigation will be concerned with the most general case in which all three principal moments of inertia are different from one another and, furthermore, all three angular momentum coordinates h_1, h_2, h_3 are different from zero[2]. Without loss of generality it is assumed that $J_3 < J_2 < J_1$. The same assumption was made in Sect. 4.1 on the dynamics of the torque-free rigid body.

4.7.1 Polhodes. Permanent Rotations

Equations (4.133) and (4.134) are analogous to the integrals of motion for a torque-free rigid body (see (4.5) and (4.6)) in that they define two ellipsoids which are fixed on the carrier and which, both, represent a geometric locus for the vector $\boldsymbol{\omega}$. The angular velocity is, therefore, confined to the lines of intersection of the ellipsoids. As for the rigid body, these lines are called polhodes. Their investigation is considerably more complicated than in the case of the rigid body since the centers of the ellipsoids are not coincident. This has the consequence, for instance, that projections of the polhodes along principal axes are not ellipses or hyperbolas but fourth-order curves. As in Sect. 4.1.1 it is imagined that the energy ellipsoid is given and that the angular momentum ellipsoid is "blown up" by increasing D. In this process the family of all physically realizable polhodes is produced on the energy ellipsoid. Of particular interest are, again, those degenerate polhodes which have singular points. These points mark permanent rotations with constant angular velocities $\boldsymbol{\omega} \equiv \boldsymbol{\omega}^* = \text{const}$. Each singular point has its particular value D^* of D which determines the corresponding angular momentum ellipsoid. The location of the singular points on the energy ellipsoid as well as relationships between $\boldsymbol{\omega}^*$, D^* and the system parameters J_i, h_i $(i = 1, 2, 3)$ and $2T$ are found from the differential equations and from the integrals of motion. According to (4.130) solutions $\boldsymbol{\omega} \equiv \boldsymbol{\omega}^* = \text{const}$ require that either $\boldsymbol{\omega}^* = \mathbf{0}$ or $\mathsf{J} \cdot \boldsymbol{\omega}^* + \mathbf{h}^* = \mathbf{0}$ or $\mathsf{J} \cdot \boldsymbol{\omega}^* + \mathbf{h}^* = \lambda^* \boldsymbol{\omega}^*$ with an as yet unknown scalar λ^*. The first two conditions represent trivial cases. In the first case, the energy ellipsoid degenerates into a single point. Only the rotor is moving, not the carrier. In the second case, the angular momentum ellipsoid degenerates

[2] For special cases not treated here and also for additional details about the general case see Wittenburg [96].

into a single point. The third condition implies that the ellipsoids are not degenerate and that they have a common tangential plane in the singular point marked by $\boldsymbol{\omega}^*$. This condition does not yield an eigenvalue problem as in the corresponding case for the rigid body (see (4.11)). Let u_1, u_2 and u_3 be the coordinates of the unit vector $\mathbf{u}$ along the rotor axis so that $h_i = hu_i$ $(i = 1, 2, 3)$. The third condition then yields

$$\omega_i^* = \frac{hu_i}{\lambda^* - J_i} \qquad (i = 1, 2, 3) . \tag{4.135}$$

The unknown λ^* is determined from the equation

$$\sum_{i=1}^{3} \frac{u_i^2 J_i}{(\lambda^* - J_i)^2} = \frac{2T}{h^2} = \frac{1}{J_0} \tag{4.136}$$

which is found by substituting ω_i^* into (4.133). The newly introduced constant parameter J_0 has the physical dimension of moment of inertia. Equation (4.136) represents a sixth-order equation for λ^*. Every real solution is associated with one permanent rotation. The value D^* of D which defines the corresponding angular momentum integral follows from (4.134):

$$D^* = J_0 {\lambda^*}^2 \sum_{i=1}^{3} \frac{u_i^2}{(\lambda^* - J_i)^2} . \tag{4.137}$$

In order to be able to make statements about the number of axes of permanent rotation the function

$$F(\lambda) = \sum_{i=1}^{3} \frac{u_i^2 J_i}{(\lambda - J_i)^2} \tag{4.138}$$

is considered (see Fig. 4.10). It has second-order poles at $\lambda = J_i$ $(i = 1, 2, 3)$. Furthermore, it is positive everywhere and it has exactly one minimum in each of the intervals $J_3 < \lambda < J_2$ and $J_2 < \lambda < J_1$ since $\mathrm{d}^2 F(\lambda)/\mathrm{d}\lambda^2 > 0$. The location of these minima is found from the condition $\mathrm{d}F(\lambda)/\mathrm{d}\lambda = 0$, i.e. from

$$\sum_{i=1}^{3} \frac{u_i^2 J_i}{(\lambda - J_i)^3} = 0 . \tag{4.139}$$

This is a sixth-order equation for λ which has exactly two real solutions. From these properties of the function $F(\lambda)$ the following conclusions can be drawn.

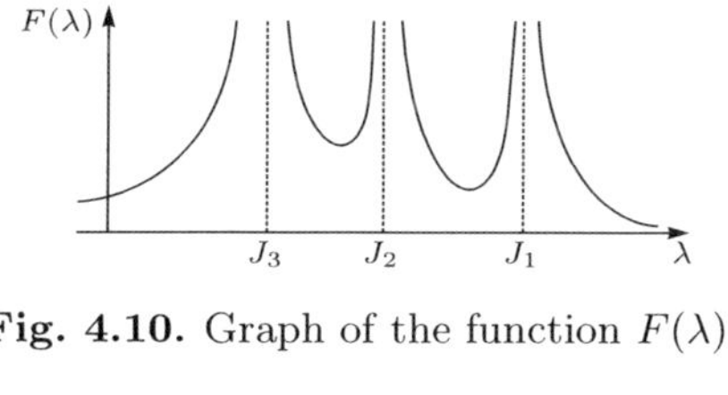

Fig. 4.10. Graph of the function $F(\lambda)$

Equation (4.136) has either six or four or two real solutions λ^* depending on the choice of system parameters. Double roots which are then also roots of (4.139) may occur. Every real solution is associated with a permanent angular velocity with coordinates calculated from (4.135). None of these coordinates can be zero so that axes of permanent rotation are neither parallel to principal axes of inertia nor parallel to planes spanned by two principal axes of inertia. Any other axis can – for given moments of inertia J_1, J_2 and J_3 – be made an axis of permanent rotation by a proper choice of u_1, u_2 and u_3. For a given set of parameters J_i, u_i $(i = 1, 2, 3)$ and J_0 no two angular velocity vectors $\boldsymbol{\omega}^*$ have opposite directions. It is interesting to look at two degenerate cases. For $h = 0$ the gyrostat is a rigid body. It should then have six axes of permanent rotation which coincide with the principal axes of inertia (three if permanent rotations of opposite directions are not counted separately). In the limit $h \to 0$, i.e. for $J_0 \to 0$ the roots of (4.136) tend toward double roots $\lambda^* = J_i$ $(i = 1, 2, 3)$. For each root two out of three coordinates of $\boldsymbol{\omega}^*$ are zero according to (4.135). This means that all six permanent angular velocity vectors have, indeed, directions of principal axes of inertia. In the second degenerate case h is infinitely large while $2T$ remains finite. This is the case of a gyrostat with a slowly rotating carrier and with an infinitely rapidly spinning rotor. The rotor then dominates the dynamic behavior of the gyrostat. It should, therefore, be expected that only two axes of permanent rotation exist and that these axes coincide with the symmetry axis of the rotor. This is, indeed, the case. In the limit $J_0 \to \infty$ (4.136) has only two real solutions, namely $\lambda^* \to \pm\infty$. For the corresponding permanent angular velocity coordinates the relationship $\omega_1^* : \omega_2^* : \omega_3^* = u_1 : u_2 : u_3$ is found from (4.135) which means that $\boldsymbol{\omega}^*$ is parallel to the rotor axis.

4.7.2 Solution of the Dynamic Equations of Motion

In this subsection closed-form solutions will be obtained for the differential equations (4.132) by a method which was suggested by Wangerin [88] and further developed by Wittenburg [96]. The approach leads to real functions $\omega_1(t)$, $\omega_2(t)$, $\omega_3(t)$. Another method developed by Volterra [84] (see also Wittenburg [96]) results in complex functions of time.

We start out from the integrals of motion. Multiply (4.133) by an as yet undetermined scalar λ with the dimension of moment of inertia and subtract (4.134). The result is written in the form

$$\sum_{i=1}^{3} J_i(\lambda - J_i)\left(\omega_i - \frac{hu_i}{\lambda - J_i}\right)^2 = 2T[f(\lambda) - D] \tag{4.140}$$

with

$$f(\lambda) = \lambda \left[1 + J_0 \sum_{i=1}^{3} \frac{u_i^2}{\lambda - J_i} \right] . \tag{4.141}$$

The function $f(\lambda)$ has first-order poles at $\lambda = J_i$ $(i = 1, 2, 3)$. From this it follows that in each of the intervals $J_3 < \lambda < J_2$ and $J_2 < \lambda < J_1$ all values from $-\infty$ to $+\infty$ are assumed at least once. For λ a value λ_0 is chosen which satisfies the conditions

$$f(\lambda_0) = D \qquad \text{and} \qquad J_3 < \lambda_0 < J_2 \, . \tag{4.142}$$

If several values exist which satisfy these conditions then one of them is chosen arbitrarily. The equation $f(\lambda_0) = D$ represents a fourth-order equation. When new variables w_i defined as

$$w_i = \omega_i - \frac{h u_i}{\lambda_0 - J_i} \qquad (i = 1, 2, 3) \tag{4.143}$$

are introduced (4.140) becomes

$$\left(\frac{w_1}{k_1 w_3} \right)^2 + \left(\frac{w_2}{k_2 w_3} \right)^2 = 1 \tag{4.144}$$

with real constants

$$k_1 = \sqrt{\frac{J_3(\lambda_0 - J_3)}{J_1(J_1 - \lambda_0)}} \, , \qquad k_2 = \sqrt{\frac{J_3(\lambda_0 - J_3)}{J_2(J_2 - \lambda_0)}} \, . \tag{4.145}$$

This equation defines a double cone of elliptic cross section whose axis is the w_3-axis (see Fig. 4.11). This cone, too, is a geometric locus of the polhodes, since its equation is a linear combination of the equations which define the ellipsoids.

Equation (4.144) for the cone is replaced by the parameter equations with parameter ϕ

$$w_1 = k_1 w_3 \sin \phi \, , \qquad w_2 = k_2 w_3 \cos \phi \, . \tag{4.146}$$

Through these equations in combination with (4.143) also ω_1 and ω_2 are functions of w_3 and ϕ. Substituting the expressions for ω_1 and ω_2 into the energy equation (4.133) one finds for w_3 the quadratic equation

$$a_1(\phi) w_3^2 - 2 a_2(\phi) w_3 + a_3 = 0 \tag{4.147}$$

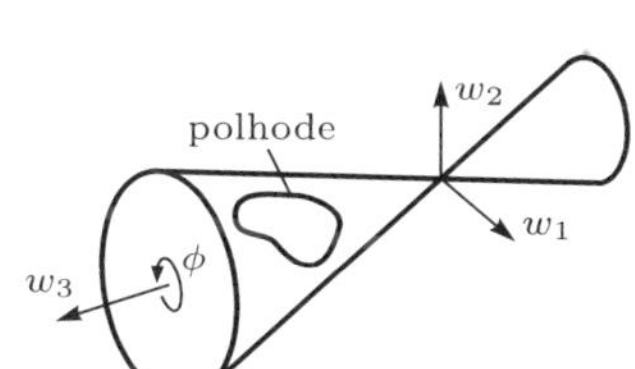

Fig. 4.11. The double cone of (4.144) with a polhode

with the coefficients

$$\left.\begin{aligned} a_1(\phi) &= J_1 k_1^2 \sin^2\phi + J_2 k_2^2 \cos^2\phi + J_3 > 0\,, \\ a_2(\phi) &= \frac{h_1 J_1 k_1}{J_1-\lambda_0}\sin\phi + \frac{h_2 J_2 k_2}{J_2-\lambda_0}\cos\phi + \frac{h_3 J_3}{J_3-\lambda_0}\,, \\ a_3 &= 2TJ_0\left[\sum_{i=1}^{3}\frac{u_i^2 J_i}{(\lambda_0-J_i)^2} - \frac{1}{J_0}\right]. \end{aligned}\right\} \tag{4.148}$$

The solutions for w_3 are

$$w_3(\phi) = \frac{a_2(\phi) \pm \sqrt{a_2^2(\phi) - a_1(\phi)a_3}}{a_1(\phi)}\,. \tag{4.149}$$

This equation in combination with (4.146) and (4.143) establishes all three angular velocity coordinates as functions of the single variable ϕ. For completing the solution it remains to be shown how ϕ is related to time. This is achieved by combining the two integrals of motion with one of the three differential equations of motion (4.132). First, the time derivative of the energy equation (4.133) is formulated:

$$J_1\omega_1\dot\omega_1 + J_2\omega_2\dot\omega_2 + J_3\omega_3\dot\omega_3 = 0\,. \tag{4.150}$$

In this equation $\dot\omega_1$ and $\dot\omega_2$ are replaced by the following expressions which result from (4.143) and (4.146):

$$\left.\begin{aligned} \dot\omega_1 = \dot w_1 &= k_1(\ \ w_3\dot\phi\cos\phi + \dot w_3\sin\phi) = k_1\left(\frac{w_2\dot\phi}{k_2} + \dot\omega_3\sin\phi\right), \\ \dot\omega_2 = \dot w_2 &= k_2(-w_3\dot\phi\sin\phi + \dot w_3\cos\phi) = k_2\left(\frac{-w_1\dot\phi}{k_1} + \dot\omega_3\cos\phi\right). \end{aligned}\right\} \tag{4.151}$$

This yields the equation

$$\begin{aligned} &\left(\frac{k_1}{k_2}\,J_1\omega_1 w_2 - \frac{k_2}{k_1}\,J_2\omega_2 w_1\right)\dot\phi \\ &+(J_1k_1\omega_1\sin\phi + J_2k_2\omega_2\cos\phi + J_3\omega_3)\dot\omega_3 = 0\,. \end{aligned} \tag{4.152}$$

The expression in front of $\dot\phi$ is rewritten by substituting for w_1 and w_2 (4.143) and for k_1 and k_2 (4.145). Following this, the third differential equation (4.132) is brought into play. The resulting expression reads:

$$\begin{aligned} &\frac{k_1}{k_2}\,J_1\omega_1 w_2 - \frac{k_2}{k_1}\,J_2\omega_2 w_1 \\ &= [-(J_1-J_2)\omega_1\omega_2 + \omega_1 h_2 - \omega_2 h_1]\sqrt{\frac{J_1J_2}{(J_1-\lambda_0)(J_2-\lambda_0)}} \\ &= -J_3\dot\omega_3\sqrt{\frac{J_1J_2}{(J_1-\lambda_0)(J_2-\lambda_0)}}\,. \end{aligned} \tag{4.153}$$

Next, the expression in front of $\dot{\omega}_3$ in (4.152) is rewritten. First, ω_1, ω_2 and ω_3 are expressed in terms of w_3 and ϕ with the help of (4.143) and (4.146). Following this, (4.148) and (4.149) are applied. The resulting expression reads:

$$\begin{aligned} &J_1 k_1 \omega_1 \sin\phi + J_2 k_2 \omega_2 \cos\phi + J_3 \omega_3 \\ &= w_3 \left(J_1 k_1^2 \sin^2\phi + J_2 k_2^2 \cos^2\phi + J_3 \right) \\ &\quad + \frac{h_1 J_1 k_1}{\lambda_0 - J_1} \sin\phi + \frac{h_2 J_2 k_2}{\lambda_0 - J_2} \cos\phi + \frac{h_3 J_3}{\lambda_0 - J_3} \\ &= w_3 a_1(\phi) - a_2(\phi) = \pm\sqrt{a_2^2(\phi) - a_1(\phi) a_3} \,. \end{aligned} \tag{4.154}$$

Substituting this together with (4.153) into (4.152) one gets, after separation of the variables, for ϕ the equation

$$\pm(t - t_0)\frac{1}{J_3}\sqrt{\frac{(J_1 - \lambda_0)(J_2 - \lambda_0)}{J_1 J_2}} = \int_{\phi_0}^{\phi} \frac{\mathrm{d}\bar{\phi}}{\sqrt{a_2^2(\bar{\phi}) - a_1(\bar{\phi}) a_3}} \,. \tag{4.155}$$

From (4.148) it is seen that the expression under the square root has the form

$$a_2^2 - a_1 a_3 = c_1 \sin^2\bar{\phi} + c_2 \cos^2\bar{\phi} + c_3 + c_4 \sin\bar{\phi} + c_5 \cos\bar{\phi} + c_6 \sin\bar{\phi}\cos\bar{\phi} \tag{4.156}$$

where $c_1, \ldots, c_6$ are abbreviations for constant coefficients. The integral is, therefore, an elliptic integral of the first kind. Its reduction to a normal form requires a sequence of substitutions of new variables in the course of which another fourth-order equation must be solved (the first one led to the root λ_0 of (4.142)). Additional difficulties arise from the necessity to distinguish between four different combinations of signs of certain constants in order to prevent the solution for ϕ from being a complex function of time. For details the reader is referred to Wittenburg [96]. At this place we content ourselves with the statement that the angular velocity coordinates ω_1, ω_2 and ω_3 are real elliptic functions of time.

A number of mathematical expressions in the analysis just presented are similar to other expressions which play a role in connection with permanent rotations. Compare, for example, (4.143) with (4.135) and the third Eq. (4.148) with (4.136). The similarity suggests an investigation of the question whether the root λ_0 of (4.142) can be identical with a root λ^* of (4.136) The function $f(\lambda)$ defined in (4.141) has the derivative with respect to λ

$$f'(\lambda) = \frac{\mathrm{d}f}{\mathrm{d}\lambda} = -J_0 \left[\sum_{i=1}^{3} \frac{u_i^2 J_i}{(\lambda - J_i)^2} - \frac{1}{J_0} \right] . \tag{4.157}$$

Comparison with (4.138) reveals the identity

$$f'(\lambda^*) = -J_0 \left[F(\lambda^*) - \frac{1}{J_0} \right] = 0 \,. \tag{4.158}$$

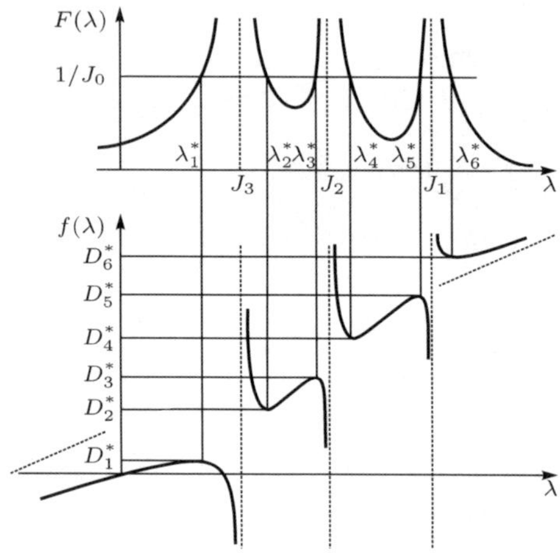

Fig. 4.12. Relationship between the functions $F(\lambda)$ and $f(\lambda)$

This relationship between $f(\lambda)$ and $F(\lambda)$ is illustrated in Fig. 4.12. The function $f(\lambda)$ is stationary for those values λ^* of λ which are roots of (4.136) and which determine the coordinates of permanent angular velocities. Furthermore, according to (4.142), $f(\lambda = \lambda^*)$ equals the parameter D^* which is associated with λ^* by (4.137). Figure 4.12 is based on a set of parameters for which six axes of permanent rotation exist. The roots λ^* are labeled in ascending order. Related quantities are identified by the respective indices. For parameter combinations for which only four axes of permanent rotation exist the function $f(\lambda)$ has stationary values in only one of the intervals $J_3 < \lambda < J_2$ and $J_2 < \lambda < J_1$. It has no stationary values in any of these intervals if only two axes of permanent rotation exist. In conclusion the following can be said. If a gyrostat is in a state of permanent rotation and if, in addition, the associated root λ^* of (4.136) lies in the interval $J_3 < \lambda^* < J_2$ then this λ^* is also a root of (4.142) and it can be used as λ_0 in all subsequent equations. Under no other condition is λ_0 a root of (4.142) and (4.136) simultaneously.

Using the identity $\lambda_0 = \lambda^*$ we can develop stability criteria for permanent rotations with $J_3 < \lambda^* < J_2$ (λ_2^* and λ_3^* in Fig. 4.12) on the basis of the exact solutions for the equations of motion. From (4.148) and (4.136) follows $a_3 = 0$ so that the solutions for $w_3(\phi)$ given by (4.149) are $w_3(\phi) \equiv 0$ and

$$w_3(\phi) = \frac{2a_2(\phi)}{a_1(\phi)} . \tag{4.159}$$

The first solution $w_3(\phi) \equiv 0$ yields, because of (4.146), $w_1 = w_2 = 0$ and, hence, with (4.143) $\omega_i = hu_i/(\lambda^* - J_i)$ $(i = 1, 2, 3)$. This represents the permanent rotation associated with λ^* (see (4.135)). From $w_1 = w_2 = w_3 = 0$ it follows that the apex of the cone of Fig. 4.11 lies on the energy ellipsoid. In addition to this singular point the cone and the energy ellipsoid intersect each

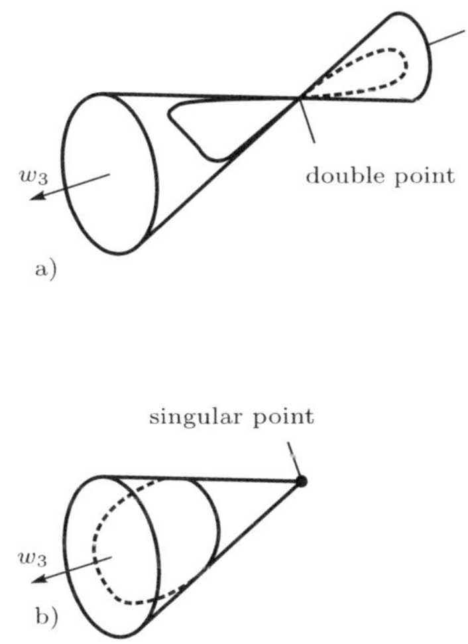

Fig. 4.13. Polhodes on the double cone associated with **(a)** unstable and **(b)** stable permanent rotations

other in a polhode which is described by (4.159). The form of this polhode decides whether the permanent rotation is stable or unstable. If the apex is a double point of the polhode (Fig. 4.13a) then this polhode represents a separatrix, and the permanent rotation is unstable. If, on the contrary, the polhode is a closed curve which is isolated from the singular point at the apex (Fig. 4.13b) then the permanent rotation is stable. The nature of the polhode is determined as follows. With $a_3 = 0$ (4.155) becomes

$$c(t - t_0) = \int_{\phi_0}^{\phi} \frac{\mathrm{d}\bar{\phi}}{s_1 \sin\bar{\phi} + s_2 \cos\bar{\phi} + s_3} \tag{4.160}$$

with

$$\left.\begin{aligned} c &= \sqrt{\frac{(J_1-\lambda^*)(J_2-\lambda^*)(\lambda^*-J_3)}{J_1J_2J_3}}\,, & s_1 &= \frac{h_1}{J_1-\lambda^*}\sqrt{\frac{J_1}{J_1-\lambda^*}}\,, \\ s_2 &= \frac{h_2}{J_2-\lambda^*}\sqrt{\frac{J_2}{J_2-\lambda^*}}\,, & s_3 &= \frac{h_3}{J_3-\lambda^*}\sqrt{\frac{J_3}{\lambda^*-J_3}}\,. \end{aligned}\right\} \tag{4.161}$$

The equation has the solution

$$\tan\frac{\phi}{2} = \begin{cases} \dfrac{s_1 - \sqrt{p}\tan\tau}{s_2 - s_3} & (p > 0)\,, \\[2ex] \dfrac{s_1 - \sqrt{-p}\tanh\tau}{s_2 - s_3} & (p < 0)\,, \\[2ex] \dfrac{s_1 + 2/\tau}{s_2 - s_3} & (p = 0) \end{cases} \tag{4.162}$$

where τ is a linear function of time and

$$p = s_3^2 - \left(s_1^2 + s_2^2\right) \ . \tag{4.163}$$

For $p > 0$ the solution is periodic. The polhode is then of the type shown in Fig. 4.13b, and the permanent rotation represented by the apex of the cone is stable. For $p < 0$ and also for $p = 0$ the solution is aperiodic. Using the expressions for $\tan \phi/2$ it is straightforward to show that in either case the function $a_2(\phi) = \text{const}\times(s_1 \sin\phi + s_2 \cos\phi + s_3)$ tends toward zero for $\tau \to \infty$. Because of (4.159) then also $w_3(\phi)$ tends toward zero. This means that the motion along the polhode approaches asymptotically the apex of the cone. The polhode is, therefore, of the type shown in Fig. 4.13a, and the permanent rotation is unstable. For the quantity p (4.163) and (4.161) yield

$$p = -\sum_{i=1}^{3} \frac{h_i^2 J_i}{(J_i - \lambda^*)^3} = -2h^2 \left.\frac{\mathrm{d}F}{\mathrm{d}\lambda}\right|_{\lambda=\lambda^*} \ . \tag{4.164}$$

Thus, the sign of the rate of change of the curve in the upper diagram of Fig. 4.12 determines the stability behavior. The permanent rotation belonging to the smaller root λ_2^* is stable and the one belonging to λ_3^* is unstable. Unstable is also the case of a double root $\lambda_2^* = \lambda_3^*$ in which p equals zero.

In an analogous way stability criteria can be developed which are associated with roots λ^* in the interval $J_2 < \lambda^* < J_1$. It has been shown that also these roots are roots of $f(\lambda) = D^*(\lambda^*)$. The details are left to the reader. It is, first, necessary, to develop new forms for (4.144)–(4.155) which are based on a root λ_0 of the equation $f(\lambda_0) = D$ in the interval $J_2 < \lambda_0 < J_1$. This can be done by a simple permutation of indices. Starting from the new equations it can be shown that permanent rotations for $J_2 < \lambda^* < J_1$ are stable if $F'(\lambda^*)$ is positive and unstable otherwise. Thus, the permanent rotation associated with λ_5^* in Fig. 4.12 is stable and the one associated with λ_4^* is unstable.

The stability of permanent rotations associated with the roots $\lambda_1^* < J_3$ and $\lambda_6^* > J_1$ cannot be investigated in a similar way. There is another and even simpler method, however. It is, first, noted that the values D^* belonging to these permanent rotations (D_1^* and D_6^* in Figs. 4.12a,b) are the smallest and the largest, respectively, of all values D^*. This can be proven as follows. For each value D^* belonging to one of the permanent rotations with $J_3 < \lambda^* < J_1$ a polhode is obtained which consists of a singular point and, in addition, of a closed curve (see Fig. 4.13b). As long as the angular momentum ellipsoid intersects the energy ellipsoid in a closed curve it is possible to deflate (or to inflate) the angular momentum ellipsoid by decreasing (increasing) the parameter D and still get an intersection curve. The intersection curve degenerates into a singular point when D reaches a certain minimum (maximum) value. These extreme values are the values D^* which belong to $\lambda^* < J_3$ and $\lambda^* > J_1$. For parameters D which are not in the interval $D_1^* \le D \le D_6^*$ no real polhodes exist. After these preparatory remarks the stability of the permanent rotations under consideration can be proven as follows. For $\lambda = \lambda^*$ (4.140) becomes in view of (4.135)

$$\sum_{i=1}^{3} J_i(\lambda^* - J_i)(\omega_i - \omega_i^*)^2 = 2T(D^* - D) . \qquad (4.165)$$

On the left-hand side $\Omega_i = \omega_i - \omega_i^*$ $(i = 1, 2, 3)$ represents the deviation of ω_i from the angular velocity of permanent rotation belonging to D^*. The function

$$V = \sum_{i=1}^{3} J_i(\lambda^* - J_i)\Omega_i^2 \qquad (4.166)$$

is negative definite for $\lambda_1^* < J_3$ and positive definite for $\lambda_6^* > J_1$. Furthermore, the total time derivative of V is zero since V is a linear combination of two integrals of motion. With these properties V is a Ljapunov function proving the stability of the two permanent rotations.

The stability analysis has provided us with information about the pattern of polhodes on the energy ellipsoid. In Fig. 4.14a the pattern is schematically illustrated. The separatrices belonging to the two unstable permanent rotations are the "figures eight" shown in solid lines. They divide the ellipsoid into five regions, namely the two "eyes" for each figure eight and the region between the two figures eight. The larger eye of one figure eight covers the entire reverse side and the outer part of the front side of the ellipsoid. The intersection points of the axes of permanent rotation carry the same indices as related quantities in Fig. 4.12. The polhodes shown as broken lines are associated with the permanent rotations number 2 and 5. The one which circles point 1 belongs to the solution in (4.162) for $p > 0$ while the solution for $p < 0$ belongs to the separatrix which passes through point 3. If the system parameters are chosen such that (4.136) has a double root $\lambda_2^* = \lambda_3^*$, then points 2 and 3 on the energy ellipsoid coincide, and also the polhode circling point 1 and the separatrix passing through point 3 coincide. The shape of such a particular polhode is drawn schematically in Fig. 4.14b. Such a polhode belongs to the solution in (4.162) for $p = 0$.

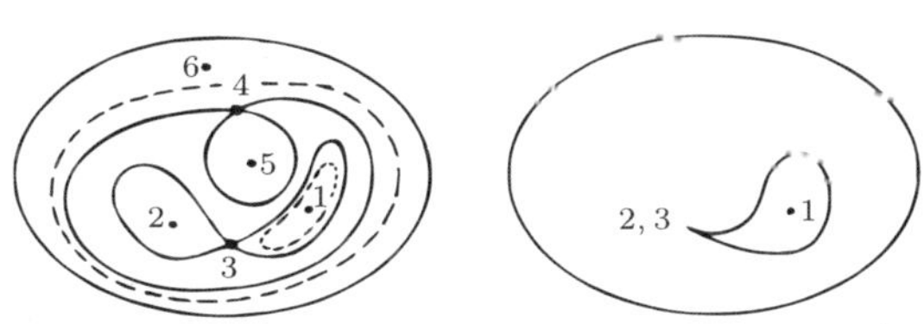

Fig. 4.14. The pattern of polhodes on the energy ellipsoid (schematically). **(a)** The general case of six different roots of (4.136). **(b)** A separatrix belonging to a real double root of (4.136)

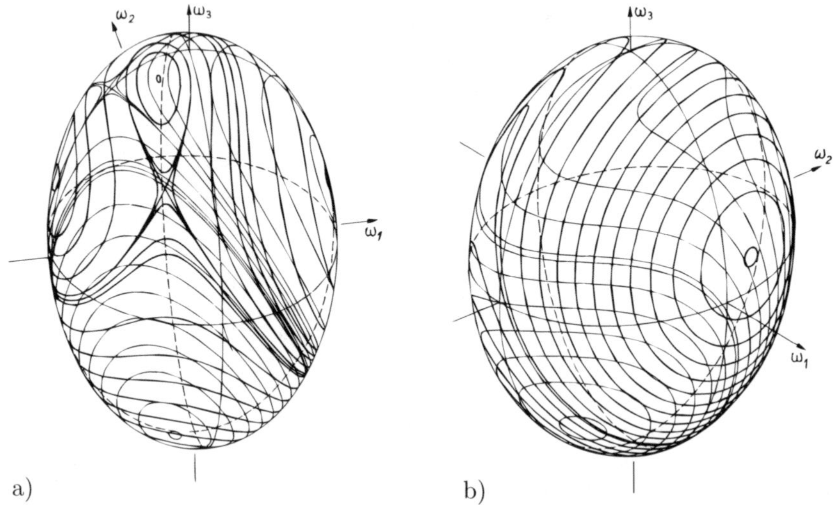

Fig. 4.15. Polhodes on the energy ellipsoid in the case of six **(a)** and of two axes of permanent rotation **(b)**. Both computer graphics are based on the parameters $J_1 = 7\,\mathrm{kg\,m^2}$, $J_2 = 5\,\mathrm{kg\,m^2}$, $J_3 = 3\,\mathrm{kg\,m^2}$, $J_0 = 0.48\,\mathrm{kg\,m^2}$ and $2T = 75\,\mathrm{kg\,m^2\,s^{-2}}$. Different are only the directions of $\mathbf{u}$. Its coordinates are 0.4, 0.1, $+\sqrt{0.83}$ in **(a)** and 0.6, 0.4, $+\sqrt{0.48}$ in **(b)**

In Fig. 4.15 parallel projections are shown of polhode families which were numerically calculated from the exact solutions of the equations of motion. The parameters for the two gyrostats in the figure differ only by the direction of the relative angular momentum $\mathbf{h}$ in the carrier. The moments of inertia J_1, J_2 and J_3, the magnitude of $\mathbf{h}$ and the energy constant $2T$ are the same in both cases. The gyrostat on the left has six axes and the other has two axes of permanent rotation.

5

General Multibody Systems

In the preceding chapter mechanical systems were investigated which consist either of a single rigid body or of several rigid bodies in some particularly simple geometric configuration. The important role they play in classical mechanics is due to the fact that their equations of motion can be integrated in closed form. This is not possible, in general, if a system is constructed of many rigid bodies in some arbitrary configuration. The engineer is confronted with an endless variety of such systems. To mention only a few examples one may think of linkages in machines, of steering mechanisms in cars, of railroad trains consisting of elastically connected cars, of a single railroad car with its undercarriage, of walking machines and manipulators etc. The assumption that the individual bodies of such systems are rigid is an idealization which may or may not be acceptable, depending largely on the kind of problem under investigation. Thus, in a crank-and-slider mechanism the seemingly rigid connecting rod has to be treated as elastic member when its forced bending vibrations are of concern. At the other extreme, the human body composed of obviously nonrigid members may well be treated as a system of interconnected rigid bodies when only its gross motion is of interest. In this chapter all bodies will be assumed rigid. This does not exclude springs and dampers interconnecting bodies. Those nonrigid elements must, however, be treated as massless.

5.1 Definition of Goals

The goal of this chapter is a minimal set of exact nonlinear differential equations of motion, of kinematic relationships, of energy expressions and other quantities required for investigations into the dynamics of multibody systems. Minimal means that the number n of variables and of equations equals the total degree of freedom of the system under consideration. This degree of freedom may be as small as one or large. Let $q_1, \ldots, q_n$ be a minimal set of suitably chosen variables. Then, on principle, it is possible to formulate equations of motion in the standard matrix form

$$\underline{A}\ddot{\underline{q}} = \underline{B} \tag{5.1}$$

where $\ddot{\underline{q}}$ is the column matrix of generalized accelerations $\ddot{q}_1, \ldots, \ddot{q}_n$. The nonlinearity of the problem has the effect that the matrix $\underline{A}$ depends on the variables $q_1, \ldots, q_n$ and that the column matrix $\underline{B}$ depends on $q_1, \ldots, q_n$ and on $\dot{q}_1, \ldots, \dot{q}_n$. The formalism to be developed should satisfy the following requirements. First, it should be applicable to such diverse mechanical systems as mentioned earlier. Second, it should give its user the freedom of choice of variables $q_1, \ldots, q_n$. In spite of this generality it should allow the generation of equations of motion, of kinematic relationships, of energy expressions etc. with only a minimum amount of labor. Classical methods do not satisfy these requirements. Take, for example, Lagrange's equations

$$\frac{\mathrm{d}}{\mathrm{d}t}\frac{\partial L}{\partial \dot{q}_k} - \frac{\partial L}{\partial q_k} = Q_k \qquad (k = 1, \ldots, n)\,. \tag{5.2}$$

The main disadvantage is that the Lagrangian L and its partial derivatives can only be formulated for a specific mechanical system and for a specific choice of generalized coordinates. Even then the amount of labor required for generating the matrices $\underline{A}$ and $\underline{B}$ is prohibitive. Another disadvantage is that large expressions resulting from the first term in (5.2) cancel identical expressions resulting from the second term. The larger a multibody system is the larger are these expressions. Which expressions cancel each other cannot be predicted so that double generation is unavoidable.

The new formalism to be developed is free of such disadvantages. First of all, it is applicable to arbitrary multibody systems and to arbitrary choices of generalized coordinates. This is achieved by the definition of a new parameter – a matrix with integer elements – which specifies the structure of a multibody system. The switch from, say, a four-body system forming a simple open chain to a five-body system having the structure of a star is effected by changing this matrix. The new formalism leads to simple expressions of such an explicit form that an automatic generation by a general-purpose computer program is possible. It is one of the goals of this chapter to enable the reader to write a program for dynamics simulations of a large class of engineering multibody systems.

This is not the only application of the formalism, however. The mathematical tools and notations used lead to expressions which can easily be interpreted in physical terms. As a consequence, purely analytical solutions can be found for some problems. This will be demonstrated in Sects. 5.7.3 and 5.7.4 and in Chap. 6.

5.2 Elements of Multibody Systems

A multibody system is composed of rigid bodies, of joints and of force elements. Another important element called the carrier body will be explained further below.

Joints and force elements have in common that they connect bodies and that they exert forces of equal magnitude and opposite direction on the two connected bodies. The difference between joints and force elements is the nature of these forces.

In a force element the force vector (direction and magnitude) is a *known* function of the positions and/or the velocities of the two bodies connected by the force element. The simplest force elements are springs and dampers. They are passive elements. Also active force elements (actuators) are admitted in which the force is determined by a *given* control law from observed position and velocity variables. The essential feature of force elements is that they do not create kinematical constraints.

In contrast, the force exerted on two bodies by a joint connecting these bodies is a pure kinematical constraint force. It is caused by frictionless rigid-body contacts. It cannot be expressed as function of position and velocity variables. Constraint forces do not enter the equations of motion because they have zero virtual work and zero virtual power. Note the following definition of joint. The joint connecting two bodies is the complete system of rigid-body contacts between these bodies. This definition has the consequence that two bodies cannot be connected by more than one joint. What this means is illustrated in Fig. 5.1. The two bodies are not connected by two spherical joints but by a single joint. This joint is a revolute joint with a single joint variable. Note also the following convention. A single joint cannot interconnect more than two bodies. What this means is illustrated in Figs. 5.2a,b. The three bodies in Fig. 5.2a are mounted on a single shaft. This shaft produces two revolute joints as is shown in Fig. 5.2b.

Depending on the nature of constraints the degree of freedom f of a joint is any number $1 \leq f \leq 5$. Figures 5.3a–e show as examples five joints which have, in this order, the degrees of freedom $f = 1, 2, 3, 4$ and 5. In Fig. 5.3c two plane surfaces, one on each body, are in contact. In Fig. 5.3d one of the

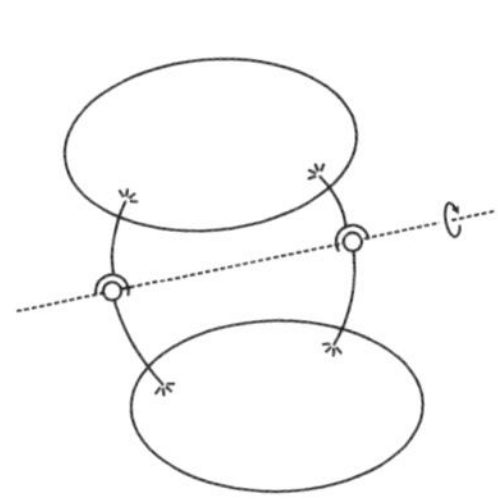

Fig. 5.1. Revolute joint connecting two bodies

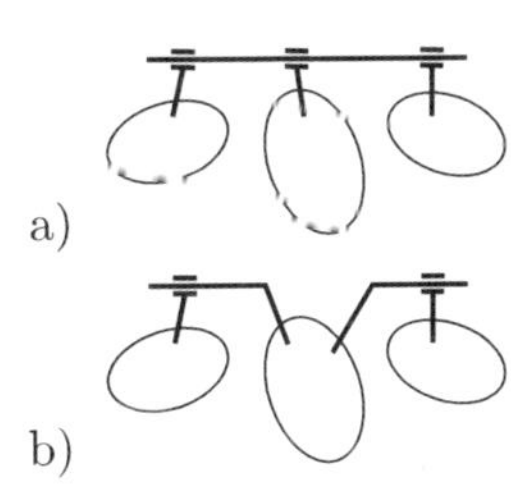

Fig. 5.2. The bodies in **(a)** are connected by two joints as is shown in **(b)**

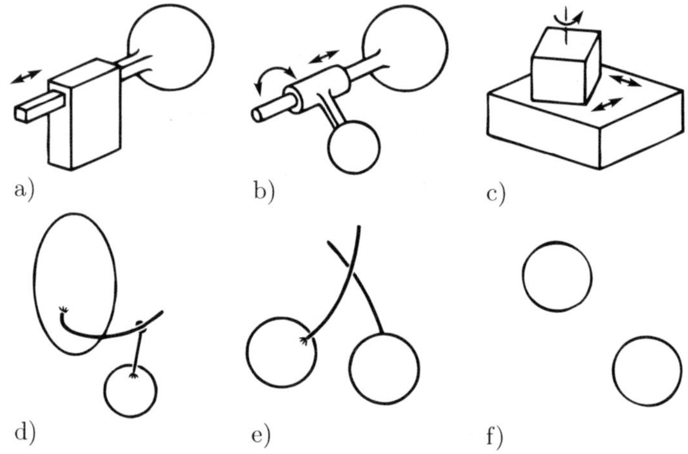

Fig. 5.3a–f. Joints with degrees of freedom 1, 2, 3, 4, 5 and 6

bodies is a pendulum whose suspension point is free to move along a guide which is fixed on the other body. In Fig. 5.3e each body has its own guide. The guides are constrained to touch each other but they are free to slip along each other.

In Fig. 5.3f two bodies without any material contact are shown. If it is decided to specify the position of one of the two bodies relative to the other by six variables then the two bodies are said to be connected by a *six-degree-of-freedom joint.* Such joints must be defined whenever without them the position of a single body or of a subsystem relative to the rest of the system would be unspecified.

The carrier body: Most multibody systems are connected by joints and/or by force elements to a frame which is fixed in inertial space. More general is the case when the system is connected to a moving carrier body the motion of which is *prescribed* as a function of time. Typical examples are the

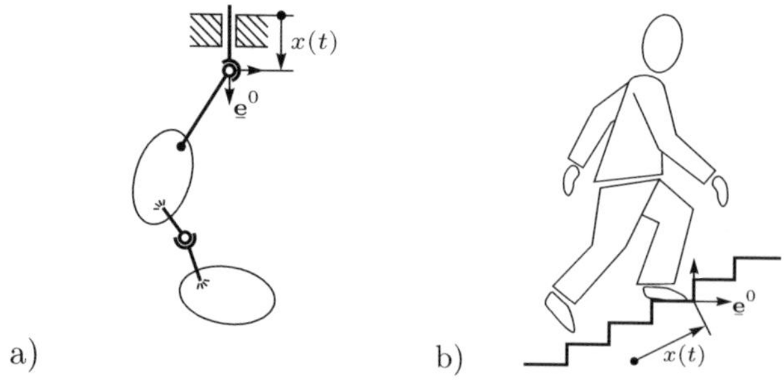

Fig. 5.4. Two systems with tree structure coupled to a carrier body, the motion of which is prescribed

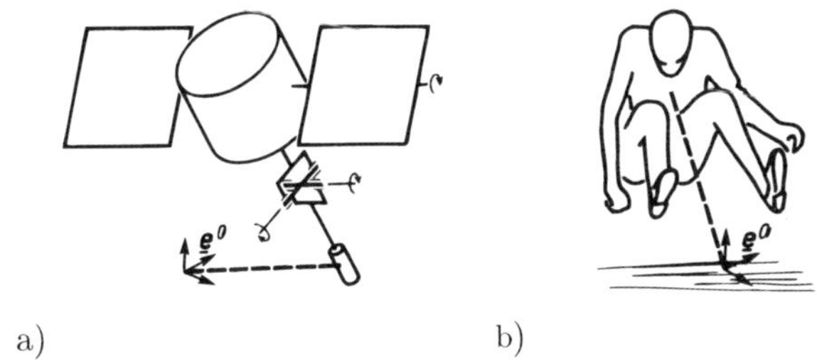

Fig. 5.5. Two systems with tree structure without kinematical constraints to inertial space

double-pendulum with a moving suspension point shown in Fig. 5.4a and the human figure on a moving escalator shown in Fig. 5.4b. It is obvious that the dimensions and inertia properties of the carrier body are irrelevant since its motion is prescribed. It is represented by a moving base $\underline{\mathbf{e}}^0$. The prescribed motion of $\underline{\mathbf{e}}^0$ as well as the properties of joints and of force elements between $\underline{\mathbf{e}}^0$ and the multibody system enter the equations of motion to be developed.

The multibody satellite in Fig. 5.5a and the human figure in Fig. 5.5b are examples of multibody systems without kinematical constraints to inertial space. For describing the position of such systems in inertial space it is necessary to define a six-degree-of-freedom joint connecting one arbitrarily chosen body of the system to a reference base $\underline{\mathbf{e}}^0$. In the two figures this joint is indicated by a broken line. Also in these cases the base $\underline{\mathbf{e}}^0$ is referred to as carrier body. When six-degree-of-freedom joints are taken into account then every multibody system is a *connected* system. By this is meant that between any two bodies of the system including body 0 there exists at least one path along a sequence of bodies and of joints such that no joint is passed more than once. A system is said to have *tree structure* if the path between *any two bodies* of the system is uniquely defined. The systems in Figs. 5.4a,b and 5.5a,b have tree structure. Tree-structured systems have the important property that the joint variables of all joints are kinematically unconstrained. This has the consequence that the total degree of freedom of the entire system equals the sum of the degrees of freedom of the individual joints.

In a system without tree structure the path between two bodies is not uniquely defined for all pairs of bodies. As an example, consider the system in Fig. 5.4b when both feet are in contact with the escalator. The legs, the trunk, the escalator and the connecting ankle, knee and hip joints form what is called a *closed kinematic chain*. The closure of this chain establishes constraints for the joint variables of the joints in the closed chain (and of these joints only). This has the consequence that the total degree of freedom of the entire system is smaller than the sum of the degrees of freedom of the individual joints. Constraint equations must be formulated for every closed kinematic chain individually. From these remarks it is seen that systems with tree structure are more easily analyzed than systems without tree structure.

Furthermore, it is seen that a system without tree structure can be analyzed by adding constraint equations to a system with tree structure. For this reason tree-structured systems are investigated first. They are the subject of Sect. 5.5. Section 5.6 is devoted to the formulation and to the incorporation of constraint equations for closed kinematic chains.

A final remark: Most multibody systems found in engineering have closed kinematic chains. However, systems with tree structure are not as exceptional as the Figs. 5.4a,b and 5.5a,b might suggest. In a road vehicle, for example, the engine is connected to the chassis not by joints but by bushings. A bushing is a force element with an internal force which is known as function of displacement and of velocity of the two connected bodies relative to one another. The same is true for the tires connecting the vehicle to the road. Neither bushings nor tires create closed kinematic chains.

5.3 Interconnection Structure of Multibody Systems

In this section mathematical tools are introduced for the description of interconnection structures of multibody systems (Wittenburg [98]). Because of their abstract nature these tools are applicable to interconnections by joints alone, to interconnections by force elements alone and also to interconnections by joints and by force elements in combination. In what follows joints and force elements in combination are considered. As illustrative example the system in Fig. 5.6 is used. The bodies are labeled $0, \ldots, n$ and the connections are labeled $1, \ldots, m$. In the example $n = 7$ and $m = 10$. Both labelings are arbitrary except that body 0 represents the carrier body. The connections labeled $1, \ldots, 7$ symbolize joints of unspecified nature whereas the connections 8, 9, 10 are drawn as force elements. For what follows this distinction is not important, however. The figure points to the fact that two bodies may be connected by more than a single force element.

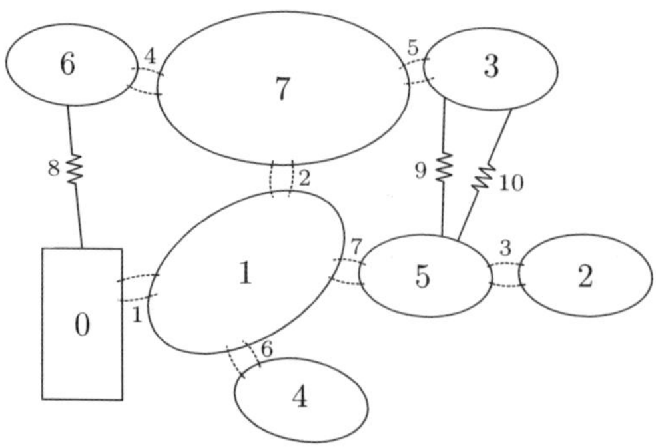

Fig. 5.6. Multibody system with joints and force elements. Carrier body 0

5.3.1 Directed System Graph. Associated Matrices

The basic idea is to display the interconnection structure of a multibody system by a graph. The graph consists of points called vertices and of lines connecting the vertices called arcs. The vertices $0, \ldots, n$ represent the bodies of the system, and the arcs $1, \ldots, m$ represent the connections. Since a graph displays neither locations nor physical properties of bodies or connections the vertices can be placed arbitrarily and the arcs can be drawn as straight or as curved lines. Figure 5.7a shows the graph for the system of Fig. 5.6. The graph is connected since the multibody system is connected by its joints (see the text following Figs. 5.5a,b).

To each arc of the graph an arbitrary sense of direction is assigned. It is indicated by the arrows in Fig. 5.7a. The resulting graph is called a directed graph. The sense of direction allows to distinguish the two vertices connected by an arc. This is necessary for two reasons. When formulating the kinematics of motion of two joint-connected bodies relative to one another it must be specified unambiguously which motion relative to which body is meant. Forces produced by a force element act with opposite signs on the two connected bodies. When formulating system dynamics it must be specified unambiguously on which body a force is acting with a positive sign and on which with a negative.

In the previous section it has been shown that the kinematics of multibody systems is simplest if the interconnection by joints is tree-structured. For this reason graphs with tree structure are given special attention. A graph is called tree-structured if between any two vertices there exists a unique minimal chain of arcs and vertices connecting the two vertices. This chain is called the path connecting the two vertices. In a tree-structured graph the identity $m = n$ holds. Proof: Starting with the single vertex 0 one must add one arc every time one vertex is added to the graph.

From a connected graph with $m > n$ arcs a graph with tree structure is produced by deleting $m - n$ suitably chosen arcs. In general, this can be done in more than one way. In Fig. 5.7a the arcs 8, 9 and 10 are deleted. These arcs are drawn with thin lines. The remaining arcs drawn with bold lines constitute what is called a *spanning tree* of the complete graph. In Fig. 5.7b this spanning tree is shown separately. The tree arcs are labeled $1, \ldots, n$ in an

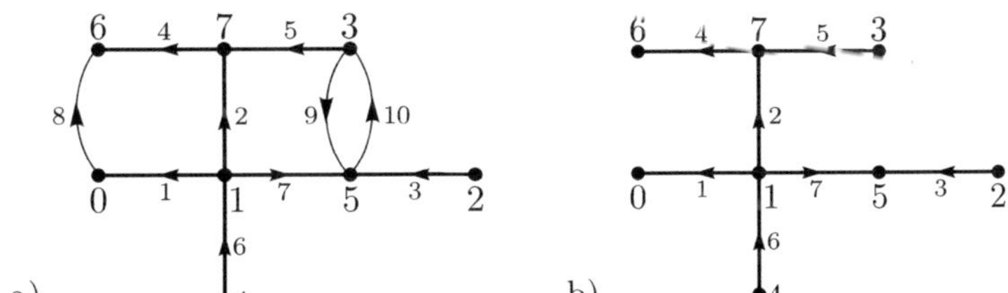

Fig. 5.7. Directed graph **(a)** and a spanning tree **(b)** for the system of Fig. 5.6

Table 5.1. Integer functions of the directed graph in Fig. 5.7a

a	1	2	3	4	5	6	7	8	9	10
$i^+(a)$	1	1	2	7	3	4	1	0	3	5
$i^-(a)$	0	7	5	6	7	1	5	6	5	3

arbitrary order (in the present case this desired labeling was arranged from the outset). The deleted arcs are called chords. They are labeled $n+1,\ldots,m$ in an arbitrary order. In order to simplify reading indices i and j refer to vertices, indices a and b to arcs (both tree arcs and chords) and the index c to chords alone.

In what follows the complete graph in Fig. 5.7a is considered, again. Each arc $a=1,\ldots,m$ is incident with two vertices. Let $i^+(a)$ and $i^-(a)$ be the labels of the two vertices at the starting point and at the terminating point of arc a, respectively. Thus, $i^+(a)$ and $i^-(a)$ are the names of two integer functions with integer arguments. Both functions can be read from the directed graph. For the directed graph of Fig. 5.7a the two functions are given in Table 5.1. Columns 1 to 7 are associated with the spanning tree. The directed graph is easily reconstructed from its functions $i^+(a)$ and $i^-(a)$ ($a=1,\ldots,m$). First, $n+1$ vertices labeled $0,\ldots,n$ are marked on a sheet of paper. Then, for every $a=1,\ldots,m$ an arc is drawn pointing from vertex $i^+(a)$ to vertex $i^-(a)$. The result of this procedure is the original directed graph.

In what follows matrices are defined for directed graphs. The first matrix called *incidence matrix* is defined for the complete graph. It has rows $0,\ldots,n$ and columns $1,\ldots,m$. The rows correspond to vertices and the columns to arcs. The matrix elements are denoted S_{ia} ($i=0,\ldots,n;\ a=1,\ldots,m$). They are defined as follows:

$$S_{ia}=\begin{cases}+1 & \text{(arc } a \text{ is incident with and pointing away from vertex } i)\\ -1 & \text{(arc } a \text{ is incident with and pointing toward vertex } i)\\ \;\;0 & \text{(arc } a \text{ is not incident with vertex } i)\end{cases}$$
$$(i=0,\ldots,n;\ a=1,\ldots,m)\,. \tag{5.3}$$

This can be expressed in the form

$$S_{ia}=\begin{cases}+1 & (i=i^+(a))\\ -1 & (i=i^-(a))\\ \;\;0 & (\text{else})\end{cases}\qquad (i=0,\ldots,n;\ a=1,\ldots,m)\,. \tag{5.4}$$

Still simpler is the formula employing the Kronecker delta

$$S_{ia}=\delta_{i,i^+(a)}-\delta_{i,i^-(a)}\qquad (i=0,\ldots,n;\ a=1,\ldots,m)\,. \tag{5.5}$$

The incidence matrix is partitioned into the row matrix $\underline{S}_0$ which corresponds to vertex 0 and the $(n\times m)$-matrix $\underline{S}$ composed of the elements S_{ia} ($i=1,\ldots,n;\ a=1,\ldots,m$). Both these matrices are further partitioned into submatrices $\underline{S}_{0\mathrm{t}}$ and $\underline{S}_{\mathrm{t}}$ associated with the spanning tree (columns $a = 1, \ldots, n$) and submatrices $\underline{S}_{0\mathrm{c}}$ and $\underline{S}_{\mathrm{c}}$ associated with chords (columns $a = n+1, \ldots, m$):

$$\begin{aligned} \underline{S}_0 &= [\ \underline{S}_{0\mathrm{t}} \quad \underline{S}_{\mathrm{c}}\ ]\,, \\ \underline{S} &= [\ \underline{S}_{\mathrm{t}} \quad \underline{S}_{\mathrm{c}}\ ]\,. \end{aligned} \tag{5.6}$$

Example: For the directed graph of Fig. 5.7a the matrices are

$$\underline{S}_0 = \left[\begin{array}{ccccccc|ccc} -1 & 0 & 0 & 0 & 0 & 0 & 0 & +1 & 0 & 0 \end{array}\right], \tag{5.7}$$

$$\underline{S} = \left[\begin{array}{ccccccc|ccc} +1 & +1 & 0 & 0 & 0 & -1 & +1 & 0 & 0 & 0 \\ 0 & 0 & -1 & 0 & 0 & 0 & 0 & 0 & 0 & 0 \\ 0 & 0 & 0 & 0 & +1 & 0 & 0 & 0 & +1 & -1 \\ 0 & 0 & 0 & 0 & 0 & +1 & 0 & 0 & 0 & 0 \\ 0 & 0 & +1 & 0 & 0 & 0 & -1 & 0 & -1 & +1 \\ 0 & 0 & 0 & -1 & 0 & 0 & 0 & -1 & 0 & 0 \\ 0 & -1 & 0 & +1 & -1 & 0 & 0 & 0 & 0 & 0 \end{array}\right]. \tag{5.8}$$

The submatrices to the left of the partitioning lines are associated with the spanning tree. From the definition (5.4) it follows that every column of $\underline{S}_0$ and $\underline{S}$ together contains exactly one element $+1$ and one element -1. Hence, the sum of rows $0, \ldots, n$ is a row of zeros. Expressed in matrix form this is the equation

$$\underline{S}_0 + \underline{1}^T \underline{S} = \underline{0} \tag{5.9}$$

with the row matrix $\underline{1}^T = [1\ 1\ \ldots\ 1]$.

In every row j ($j = 0, \ldots, n$) the number of nonzero elements equals the number of arcs which are incident with vertex j. If a row j of $\underline{S}$ has a single nonzero element S_{jb} then the vertex j is incident with arc b alone. This means that the vertex j is a terminal vertex of the graph. Example: The graph in Fig. 5.7a has the terminal vertices 2 and 4, and the spanning tree in Fig. 5.7b has the terminal vertices 2, 3, 4 and 6 (these are the rows of $\underline{S}_{\mathrm{t}}$ with a single nonzero element). The matrix $\underline{S}$ alone suffices to reconstruct the directed graph and, hence, the functions $i^+(a)$ and $i^-(a)$ $(a = 1, \ldots, m)$. The matrix $\underline{S}_0$ is not required, because a single nonzero element in a column b of $\underline{S}$ indicates that arc b is incident with vertex 0.

The second matrix called *path matrix* $\underline{T}$ is defined for tree-structured directed graphs only. Also this matrix has elements $+1$, -1 and zero. Like $\underline{S}_{\mathrm{t}}$ it is an $(n \times n)$-matrix. The elements are denoted T_{ai}. The letters a and i indicate that in this matrix rows correspond to arcs and columns to vertices The elements are defined as follows:

$$T_{ai} = \begin{cases} +1 & \text{(arc } a \text{ is on the path between vertices 0 and } i \\ & \ \text{and is directed toward vertex 0)} \\ -1 & \text{(arc } a \text{ is on the path between vertices 0 and } i \\ & \ \text{and is directed toward vertex } i) \\ 0 & \text{(arc } a \text{ is not on the path between vertices 0 and } i) \end{cases}$$

$$(i, a = 1, \ldots, n)\,. \tag{5.10}$$

There is no column corresponding to vertex 0. Example: The spanning tree in Fig. 5.7b has the path matrix

$$\underline{T} = \begin{bmatrix} +1 & +1 & +1 & +1 & +1 & +1 & +1 \\ 0 & 0 & -1 & 0 & 0 & -1 & -1 \\ 0 & +1 & 0 & 0 & 0 & 0 & 0 \\ 0 & 0 & 0 & 0 & 0 & -1 & 0 \\ 0 & 0 & +1 & 0 & 0 & 0 & 0 \\ 0 & 0 & 0 & +1 & 0 & 0 & 0 \\ 0 & -1 & 0 & 0 & -1 & 0 & 0 \end{bmatrix} . \tag{5.11}$$

From the definition (5.10) it follows that in every row of $\underline{T}$ all nonzero elements are identical. Every row has at least one nonzero element. If in row b the element T_{bj} is the only nonzero element then vertex j is a terminal vertex incident with arc b.

Since the spanning tree is uniquely determined by the matrix $\underline{S}_\mathrm{t}$ also the path matrix $\underline{T}$ is uniquely determined by $\underline{S}_\mathrm{t}$. Proposition 1: One matrix is the inverse of the other:

$$\underline{T} = \underline{S}_\mathrm{t}^{-1} . \tag{5.12}$$

Proposition 2: The matrices $\underline{T}$ and $\underline{S}_{0\mathrm{t}}$ are related through the equation

$$\underline{S}_{0\mathrm{t}}\underline{T} = -\underline{1}^T \tag{5.13}$$

(the matrix $\underline{1}^T = [1\ 1\ \ldots\ 1]$ is known from (5.9)).

For proving (5.12) it suffices to show that $\underline{T}\,\underline{S}_\mathrm{t}$ is the unit matrix. This product is an $(n \times n)$-matrix with elements $(\underline{T}\,\underline{S}_\mathrm{t})_{ab} = \sum_{i=1}^{n} T_{ai}S_{ib}$ $(a, b = 1, \ldots, n)$. According to (5.4) S_{ib} is equal to $+1$ for $i = i^+(b)$, equal to -1 for $i = i^-(b)$ and zero otherwise. Therefore, $(\underline{T}\,\underline{S}_\mathrm{t})_{ab} = T_{ai^+(b)} - T_{ai^-(b)}$. First, the case $b = a$ is considered. Arc a is either directed toward vertex 0 or away from vertex 0. If the former is true then $T_{a,i^+(a)} = 1$, $T_{a,i^-(a)} = 0$. If the latter is true then $T_{a,i^+(a)} = 0$, $T_{a,i^-(a)} = -1$. Hence, in either case $(\underline{T}\,\underline{S}_\mathrm{t})_{aa} = 1$. Next, the case $b \neq a$ is investigated. Consider the path between the vertices 0 and $i^+(b)$ and the path between the vertices 0 and $i^-(b)$. Arc a belongs either to both paths or to none of them. In either case, $T_{a,i^+(b)} = T_{a,i^-(b)}$ and, hence, $(\underline{T}\,\underline{S}_\mathrm{t})_{ab} = 0$. End of proof.

For proving (5.13) it must be shown that $\sum_{a=1}^{n} S_{0a}T_{ai} = -1$ for $i = 1, \ldots, n$. This is, indeed the case, since for every vertex i a single arc $b(i)$ satisfies the condition $S_{0b}T_{bi} \neq 0$ (arc b is on the path between vertices 0 and i and incident with vertex 0). Furthermore, $S_{0b}T_{bi} = -1$ independent of the sense of direction of this arc b. End of proof.

The existence of the path matrix $\underline{T}$ for tree-structured graphs and the two equations (5.12) and (5.13) are the mathematical reasons why multibody systems with tree structure are simpler than systems without tree structure.

In contrast to the matrix $\underline{S}_\mathrm{t}$ the matrix $\underline{T}$ is not easily determined directly from the two functions $i^+(a)$ and $i^-(a)$ of the spanning tree. It is equally difficult to reconstruct $i^+(a)$ and $i^-(a)$ from $\underline{T}$. An efficient method is described in Sect. 5.3.3 on regular labeling.

Problem 5.1. Give a direct proof for the statement $\underline{S}_{\rm t}\underline{T} = \underline{I}$ in (5.12).

Problem 5.2. Draw a tree-structured directed graph as follows. Arc a is directed from vertex a toward vertex 0 $(a = 1, \ldots, n)$. Determine the functions $i^+(a)$ and $i^-(a)$ and the matrices $\underline{S}_{0\rm t}$, $\underline{S}_{\rm t}$ and $\underline{T}$.

Problem 5.3. Delete in Fig. 5.7a arcs $1, \ldots, 7$ and give to arcs 8, 9, 10 the new labels 1, 2, 3, respectively. The *unconnected* directed graph thus defined has functions $i^+(a)$ and $i^-(a)$ $(a = 1, 2, 3)$ and an (8×3) incidence matrix with elements S_{ia} $(i = 0, \ldots, 7,\ a = 1, 2, 3)$ defined by (5.3). Determine this incidence matrix.

For connected directed graphs without tree structure two more matrices are defined. As illustrative example the graph in Fig. 5.7a is used, again. As before it is referred to as complete graph in contrast to its spanning tree shown in bold lines and separately in Fig. 5.7b.

Each arc of the spanning tree defines a *cutset* of the complete graph. It consists of the tree arc itself and of the minimal set of chords which must be cut in order to split the complete graph in two subgraphs. The cutset associated with arc a is also called cutset a. Example: The cutset 7 of the graph in Fig. 5.7a consists of the tree arc 7 and of the chords 9 and 10. A chord belonging to cutset a is said to be positively directed (in the cutset) if it points toward the same subgraph as arc a does. Otherwise it is negatively directed.

Each chord defines a *circuit* of the complete graph. It consists of the chord itself and of the minimal set of tree arcs creating a circuit. The circuit associated with chord c is also called circuit c. Example: The circuit 8 of the graph in Fig. 5.7a consists of the chord 8 and of the tree arcs 1, 2 and 4. A tree arc belonging to circuit c is said to be positively directed (in the circuit) if its sense of direction around the circuit is the same as that of chord c. Otherwise it is negatively directed.

After this introduction the $(n \times m)$ cutset matrix $\underline{P}$ and the $[(m-n) \times m]$ circuit matrix $\underline{U}$ are defined. Their elements are denoted P_{ab} and U_{ca}, respectively. Each row of $\underline{P}$ corresponds to a cutset and each row of $\underline{U}$ corresponds to a circuit. The columns of both matrices correspond to the arcs $a = 1, \ldots, m$ of the complete graph. The matrix elements are defined as follows:

$$P_{ab} = \begin{cases} +1 & \text{(arc } b \text{ belongs to cutset } a \text{ and is positively directed)} \\ -1 & \text{(arc } b \text{ belongs to cutset } a \text{ and is negatively directed)} \\ 0 & \text{(arc } b \text{ does not belong to cutset } a) \end{cases}$$
$$(a = 1, \ldots, n,\ b = 1, \ldots, m)\,, \tag{5.14}$$

$$U_{ca} = \begin{cases} +1 & \text{(arc } a \text{ belongs to circuit } c \text{ and is positively directed)} \\ -1 & \text{(arc } a \text{ belongs to circuit } c \text{ and is negatively directed)} \\ 0 & \text{(arc } a \text{ does not belong to circuit } c) \end{cases}$$
$$(c = n+1, \ldots, m;\ a = 1, \ldots, m)\,. \tag{5.15}$$

Like the matrices $\underline{S}_0$ and $\underline{S}$ also $\underline{P}$ and $\underline{U}$ are partitioned into submatrices $\underline{P}_t$, $\underline{P}_c$ and $\underline{U}_t$, $\underline{U}_c$ associated with the spanning tree and with chords, respectively. From the definitions it follows that $\underline{P}_t$ and $\underline{U}_c$ are both unit matrices (of different dimensions). Thus

$$\underline{P} = [\,\underline{I} \;\; \underline{P}_c]\,, \qquad \underline{U} = [\underline{U}_t \;\; \underline{I}\,]\,. \tag{5.16}$$

For the directed graph and its spanning tree shown in Figs. 5.7a and b the matrices are

$$\underline{P} = \left[\begin{array}{ccccccc|ccc} +1 & 0 & 0 & 0 & 0 & 0 & 0 & -1 & 0 & 0 \\ 0 & +1 & 0 & 0 & 0 & 0 & 0 & +1 & -1 & +1 \\ 0 & 0 & +1 & 0 & 0 & 0 & 0 & 0 & 0 & 0 \\ 0 & 0 & 0 & +1 & 0 & 0 & 0 & +1 & 0 & 0 \\ 0 & 0 & 0 & 0 & +1 & 0 & 0 & 0 & +1 & -1 \\ 0 & 0 & 0 & 0 & 0 & +1 & 0 & 0 & 0 & 0 \\ 0 & 0 & 0 & 0 & 0 & 0 & +1 & 0 & +1 & -1 \end{array}\right], \tag{5.17}$$

$$\underline{U} = \left[\begin{array}{ccccccc|ccc} +1 & -1 & 0 & -1 & 0 & 0 & 0 & +1 & 0 & 0 \\ 0 & +1 & 0 & 0 & -1 & 0 & -1 & 0 & +1 & 0 \\ 0 & -1 & 0 & 0 & +1 & 0 & +1 & 0 & 0 & +1 \end{array}\right]. \tag{5.18}$$

Between the matrices $\underline{P}$, $\underline{U}$, $\underline{S}$ and $\underline{T}$ there exist numerous relationships. First, the orthogonality relationship

$$\underline{S}\,\underline{U}^T = \underline{0}\,. \tag{5.19}$$

Proof: A single element of the product is

$$(\underline{S}\,\underline{U}^T)_{ic} = \sum_{a=1}^{m} S_{ia}U_{ca} = \sum_{a=1}^{n} S_{ia}U_{ca} + \sum_{a=n+1}^{m} S_{ia}\underbrace{U_{ca}}_{\delta_{ca}} = \sum_{a=1}^{n} S_{ia}U_{ca} + S_{ic} \tag{5.20}$$

($i = 1, \ldots, n;\ c = n+1, \ldots, m$). Two cases must be distinguished. Case 1: Vertex i is incident with chord c ($S_{ic} \neq 0$). Then, vertex i is incident with exactly one tree arc, and the sum over a is equal to $-S_{ic}$ independent of the senses of direction of chord c and of this single tree arc. Case 2: Vertex i is not incident with chord c ($S_{ic} = 0$). Then, vertex i is incident with two tree arcs, and the sum over a is equal to zero independent of whether these two tree arcs belong to circuit c or not. End of proof.

The next relationship is

$$\underline{T}\,\underline{S}_c = \underline{P}_c\,. \tag{5.21}$$

Proof: A single element of the product is

$$(\underline{T}\,\underline{S}_c)_{ac} = \sum_{i=1}^{n} T_{ai}S_{ic} = T_{a,i^+(c)} - T_{a,i^-(c)} \quad (a = 1, \ldots, n;\ c = n+1, \ldots, m)\,. \tag{5.22}$$

Two cases must be distinguished. Case 1: Chord c belongs to cutset a. Then, independent of the senses of direction of arc a and of chord c, $T_{a,i^+(c)} - T_{a,i^-(c)} = P_{ac} \neq 0$. Case 2: Chord c does not belong to cutset a. Then, $T_{a,i^+(c)} = T_{a,i^-(c)}$ and, hence, $T_{a,i^+(c)} - T_{a,i^-(c)} = 0 = P_{ac}$. End of proof.

Equations (5.12) and (5.21) together establish the relationship

$$\underline{T}\,\underline{S} = \underline{P}\,. \tag{5.23}$$

When this is postmultiplied by $\underline{U}^T$ one gets, due to (5.19), the equation

$$\underline{P}\,\underline{U}^T = \underline{0}\,. \tag{5.24}$$

This represents another orthogonality relationship. Using the partitioning of (5.16), the equation takes the form $\underline{U}_{\mathrm{t}}^T + \underline{P}_{\mathrm{c}} = \underline{0}$ or

$$\underline{U}_{\mathrm{t}}^T = -\underline{P}_{\mathrm{c}}\,. \tag{5.25}$$

The matrices $\underline{S}$, $\underline{P}$ and $\underline{U}$ were known to mathematicians for a long time (see Busacker/Saaty [12]). The path matrix $\underline{T}$ was first defined by Branin [8] for electrical networks and independently by Roberson/Wittenburg [66] for multibody systems.

5.3.2 Directed Graphs with Tree Structure

For graphs with tree structure a few more definitions are introduced. For every arc $a = 1, \ldots, n$ a number σ_a is defined:

$$\sigma_a = \begin{cases} +1 & \text{(arc } a \text{ is directed toward vertex 0)} \\ -1 & \text{(arc } a \text{ is directed away from vertex 0)} \end{cases} \qquad (a = 1, \ldots, n)\,. \tag{5.26}$$

Examples: The graph in Fig. 5.7b has $\sigma_3 = +1$ and $\sigma_4 = -1$.

For every pair of arcs a, b a set κ_{ab} of vertices is defined as follows. Cutting two arcs a and b $(a, b = 1, \ldots, n)$ results either in three subgraphs $(a \neq b)$ or in two subgraphs $(a = b)$. In the case $a \neq b$ κ_{ab} is the set of vertices of that subgraph which contains no vertex which is incident with arc b. In the case $a = b$ κ_{aa} is the set of vertices of that subgraph which does not contain vertex 0. Examples: In the graph of Fig. 5.7b κ_{25} is the set of vertices 0, 1, 2, 4, 5 and κ_{55} contains only vertex 3.

For arcs as well as for vertices weak ordering relationships are defined. For two arcs a and $b \neq a$ the ordering relationship arc $a <$ arc b means that arc a is on the path from vertex 0 to vertex $i_{i^+(b)}$ (and also on the path from vertex 0 to vertex $i_{i^-(b)}$). Note that two arcs a and b located on different branches of the tree as seen from vertex 0 satisfy neither the relationship arc $a <$ arc b nor the relationship arc $b <$ arc a.

Similarly, the relationship vertex i $<$ vertex j means that vertex i is on the path from vertex 0 to vertex j, but that it is not vertex j. In some places this is written in the short form $\mathrm{v}_i < \mathrm{v}_j$. Two vertices i and j located on different branches as seen from vertex 0 satisfy neither the relationship $\mathrm{v}_i < \mathrm{v}_j$ nor the relationship $\mathrm{v}_j < \mathrm{v}_i$.

Next, the inboard arc of a vertex and the inboard vertex of a vertex are defined. The inboard arc of a vertex $j \neq 0$ is the arc which is located on the path between the vertices 0 and j and which, furthermore, is incident with vertex j. The inboard vertex of a vertex $j \neq 0$ is the vertex which is connected with vertex j by the inboard arc of vertex j. Example: In the graph of Fig. 5.7b arc 7 and vertex 1 are the inboard arc and the inboard vertex, respectively, of vertex 5.

Problem 5.4. For the graph in Fig. 5.7b specify the sets κ_{52} and κ_{22}.

Problem 5.5. For the graph in Fig. 5.7b specify the sets of all vertices i which satisfy the following conditions (one at a time) for $k = 3$ and for $k = 5$
1. $\mathrm{v}_i < \mathrm{v}_k$, 2. $\mathrm{v}_k < \mathrm{v}_i$.

5.3.3 Regular Tree Graphs

In Fig. 5.7b the labeling of vertices and arcs and the sense of directions of the arcs were intentionally unsystematic in order to show that (5.12) relating the matrix $\underline{S}_\mathrm{t}$ and the path matrix $\underline{T}$ is universally valid for directed tree-structured graphs. In what follows a regular labeling and regular arc directions are defined.

A tree-structured graph is regularly directed if all arcs $1, \ldots, n$ are pointing toward vertex 0. A labeling is called regular if the following two conditions are satisfied

- along every branch starting from vertex 0 the sequence of vertex numbers is monotonically increasing
- the inboard arc of vertex j $(j = 1, \ldots, n)$ is arc j.

In general, there is more than one way in which numbers can be assigned satisfying these conditions. Any such labeling is called regular. A directed tree graph is called regular if it has regular labeling and regular arc directions. Then, the two functions $i^+(a)$ and $i^-(a)$ have the properties

$$i^+(a) = a \,, \qquad i^-(a) < a \qquad (a = 1, \ldots, n) \,. \tag{5.27}$$

Under these conditions the function $i^-(a)$ alone suffices for defining the interconnection structure. Furthermore, the matrices $\underline{S}_\mathrm{t}$ and $\underline{T}$ are both upper triangular matrices. Both matrices have elements +1 along the diagonal, and all nonzero elements of $\underline{T}$ are +1.

In what follows an algorithm is explained which starts out from given functions $i^+(a)$ and $i^-(a)$ of an unsystematically labeled and unsystematically directed graph. The vertices and arcs are regularly relabeled and the arcs are regularly redirected. Between the original and the new labeling a one-to-one relationship is established. The method is explained by taking as example

Table 5.2. Conversion from old labeling to new regular labeling

1 arc a; old labeling	1	2	3	4	5	6	7
2 vertex $i^+(a)$; old labeling	1	1	2	7	3	4	1
3 vertex $i^-(a)$; old labeling	0	7	5	6	7	1	5
4 vertex; old labeling	1	7	2	6	3	4	5
5 vertex and inboard arc j; new labeling	1	2	7	4	3	5	6
6 inboard vertex $i^-(j)$; new labeling	0	1	6	2	2	1	1

the graph in Fig. 5.7b. Its functions $i^+(a)$ and $i^-(a)$ are copied from Table 5.1 into rows 1, 2 and 3 of Table 5.2. Imagine that rows 4, 5 and 6 of the table are still empty. Following Table 5.1 it has been explained that a terminal vertex together with its inboard arc and inboard vertex is identified by the fact that the vertex number occurs only once in rows 2 and 3. Furthermore, if this vertex number occurs in row 2 (in row 3) then the inboard arc a is pointing toward vertex 0 (away from vertex 0). The numbers occurring only once are 2, 3, 4 and 6. Arbitrarily, the number 2 in column $a = 3$ is chosen. Arc 3 is pointing toward vertex 0. To the vertex 2 the new number 7 is given ($n = 7$ is the highest number available). In column 3 the numbers 2 and 7 are repeated in rows 4 and 5, respectively. In row 6 no entry is made at this point. Following this procedure column 3 in rows 1, 2 and 3 is deleted. This means that one terminal body and its inboard arc are removed from the graph. For the resulting smaller graph the same procedure is repeated. The vertex numbers occurring only once are 3, 4, 5 and 6. Arbitrarily, the number 5 in column $a = 7$ is chosen. To the vertex with the old number 5 the new number 6 is given (the highest number still available). The numbers 5 (old) and 6 (new) are filled into rows 4 and 5, respectively, of column 7. This procedure is repeated until only the vertex number 1 is left. By the same procedure it is given the new number 1. As final step row 6 is filled in. Consider column 3 again. The vertex labeled 2 (old) and 7 (new) is connected by its inboard arc $a = 3$ (old) to its inboard vertex 5 (old). According to rows 4 and 5 this inboard vertex has the new number 6. This number 6 is the entry in row 6. The same procedure is repeated in every column.

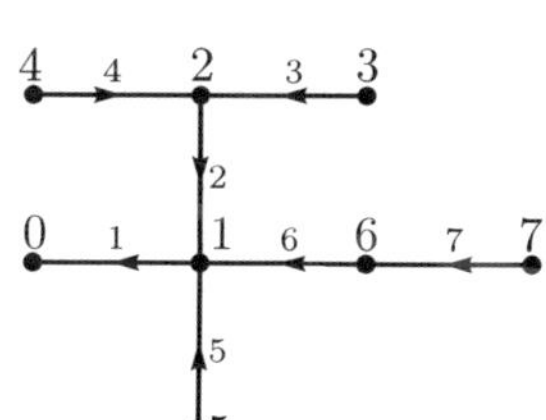

Fig. 5.8. Directed graph of Fig. 5.7b regularly relabeled and regularly redirected

Rows 4 and 5 of the table relate old to new vertex numbers and vice versa. Rows 1 and 5 relate old to new arc numbers and vice versa. Rows 5 and 6 together define the function $i^-(j)$ of the regularly labeled and regularly directed graph. This graph is shown in Fig. 5.8. The matrix $\underline{S}_\mathrm{t}$ and the path matrix $\underline{T}$ of this graph are

$$\underline{S}_\mathrm{t} = \begin{bmatrix} +1 & -1 & 0 & 0 & -1 & -1 & 0 \\ 0 & +1 & -1 & -1 & 0 & 0 & 0 \\ 0 & 0 & +1 & 0 & 0 & 0 & 0 \\ 0 & 0 & 0 & +1 & 0 & 0 & 0 \\ 0 & 0 & 0 & 0 & +1 & 0 & 0 \\ 0 & 0 & 0 & 0 & 0 & +1 & -1 \\ 0 & 0 & 0 & 0 & 0 & 0 & +1 \end{bmatrix}, \qquad \underline{T} = \begin{bmatrix} +1 & +1 & +1 & +1 & +1 & +1 & +1 \\ 0 & +1 & +1 & +1 & 0 & 0 & 0 \\ 0 & 0 & +1 & 0 & 0 & 0 & 0 \\ 0 & 0 & 0 & +1 & 0 & 0 & 0 \\ 0 & 0 & 0 & 0 & +1 & 0 & 0 \\ 0 & 0 & 0 & 0 & 0 & +1 & +1 \\ 0 & 0 & 0 & 0 & 0 & 0 & +1 \end{bmatrix}. \tag{5.28}$$

The matrix $\underline{T}$ is constructed from rows 5 and 6 of Table 5.2 as follows. For arbitrary j $(j = 1, \ldots, n)$ the arcs on the path from s_j to s_0 have the numbers j, $i^-(j)$, $i^-(i^-(j))$, $i^-(i^-(i^-(j)))$, $\ldots, 1$. In column j of $\underline{T}$ the elements in these rows are $+1$. All other elements are zero. Example: $j = 7$ yields the sequence 7, $i^-(7) = 6$ and $i^-(6) = 1$. Hence, $T_{77} = T_{67} = T_{17} = +1$. This is in accordance with (5.28).

Problem 5.6. For a system of bodies $i = 0, \ldots, n$ interconnected by joints $a = 1, \ldots, m$ the following quantities are defined

- absolute angular velocity $\boldsymbol{\omega}_i$ of body i $(i = 1, \ldots, n)$; $\boldsymbol{\omega}_0 = \mathbf{0}$
- angular velocity $\boldsymbol{\Omega}_a$ of body $i^-(a)$ relative to body $i^+(a)$ in joint a $(a = 1, \ldots, m)$
- internal forces $+\mathbf{F}_a$ and $-\mathbf{F}_a$ produced by a spring in joint a $(a = 1, \ldots, m)$; $+\mathbf{F}_a$ is applied to body $i^+(a)$ and $-\mathbf{F}_a$ to body $i^-(a)$
- resultant force $\mathbf{F}_{i_\mathrm{res}}$ on body i produced by the springs in all joints on body i $(i = 1, \ldots, n)$.

Use the incidence matrix for expressing $\boldsymbol{\Omega}_a$ $(a = 1, \ldots, m)$ in terms of $\boldsymbol{\omega}_i$ $(i = 1, \ldots, n)$ and $\mathbf{F}_{i_\mathrm{res}}$ $(i = 1, \ldots, n)$ in terms of $\mathbf{F}_a$ $(a = 1, \ldots, m)$.

Consider the special case of a system with tree structure with bodies $i = 0, \ldots, n$ and joints $a = 1, \ldots, n$. Define the column matrices $\underline{\boldsymbol{\omega}} = [\boldsymbol{\omega}_1 \; \ldots \; \boldsymbol{\omega}_n]^T$, $\underline{\boldsymbol{\Omega}} = [\boldsymbol{\Omega}_1 \; \ldots \; \boldsymbol{\Omega}_n]^T$, $\underline{\mathbf{F}} = [\mathbf{F}_1 \; \ldots \; \mathbf{F}_n]^T$ and $\underline{\mathbf{F}}_\mathrm{res} = [\mathbf{F}_{1_\mathrm{res}} \; \cdots \; \mathbf{F}_{n_\mathrm{res}}]^T$ and write the two sets of n equations each in matrix form. Use the path matrix for resolving these equations for $\underline{\boldsymbol{\omega}}$ and for $\underline{\mathbf{F}}$.

Denote by $\mathbf{c}_{i^+(a),a}$ the vector from the center of mass of body $i^+(a)$ to the point of application of the spring force $+\mathbf{F}_a$ on this body and, likewise, by $\mathbf{c}_{i^-(a),a}$ the vector from the center of mass of body $i^-(a)$ to the point of application of the spring force $-\mathbf{F}_a$ on this body. Define, furthermore, the vectors $\mathbf{C}_{ia} = S_{ia}\mathbf{c}_{ia}$ $(i, a = 1, \ldots, n)$ and the $(n \times n)$-matrix $\underline{\mathbf{C}}$ with these vectors as elements. It is a weighted incidence matrix. Express with this matrix the column matrix $\underline{\mathbf{M}}_\mathrm{res} = [\mathbf{M}_{1_\mathrm{res}} \; \cdots \; \mathbf{M}_{n_\mathrm{res}}]^T$ of resultant spring torques on the bodies $i = 1, \ldots, n$.

5.4 Principle of Virtual Power for Multibody Systems

Dynamics equations of motion can be obtained either by analytical methods or by the synthetical method starting from Newton's and Euler's equations for isolated bodies. In this section the analytical method based on the principle of virtual power is chosen. In Sect. 3.5 the principle of virtual power has been formulated for a single rigid body (see (3.44)). For an arbitrary system of n rigid bodies the principle has the form

$$\sum_{i=1}^{n} [\delta\dot{\mathbf{r}}_i \cdot (m_i\ddot{\mathbf{r}}_i - \mathbf{F}_i) + \delta\boldsymbol{\omega}_i \cdot (\mathsf{J}_i \cdot \dot{\boldsymbol{\omega}}_i + \boldsymbol{\omega}_i \times \mathsf{J}_i \cdot \boldsymbol{\omega}_i - \mathbf{M}_i)] = 0\,. \tag{5.29}$$

Each quantity carries the index i of the individual body. The carrier body 0 is excluded from the sum because its virtual velocity change is zero. The index C referring to the body center of mass has been omitted. So, $\mathbf{r}_i$ is the position vector of the body i center of mass, $\mathbf{F}_i$ is the resultant force acting on body i and applied at the body i center of mass, and $\mathbf{M}_i$ is the resultant torque about the body i center of mass. Contributions to $\mathbf{F}_i$ and to $\mathbf{M}_i$ are made by gravity, by force elements, by sliding friction and by other forces which contribute to virtual power. Constraint forces caused by ideal kinematical constraints in joints do not contribute because for any pair of constraint forces, say $\mathbf{F}_1 = +\mathbf{F}$ and $\mathbf{F}_2 = -\mathbf{F}$ (actio = reactio) the term $\delta\dot{\mathbf{r}}_1 \cdot \mathbf{F}_1 + \delta\dot{\mathbf{r}}_2 \cdot \mathbf{F}_2$ is equal to zero.

In (5.29) the quantities $\delta\dot{\mathbf{r}}_i$, $\delta\boldsymbol{\omega}_i$, $\ddot{\mathbf{r}}_i$ and $\dot{\boldsymbol{\omega}}_i$ appear in linear form. For this reason the following matrix formulation of the equation is possible:

$$\delta\underline{\dot{\mathbf{r}}}^T \cdot (\underline{m}\,\underline{\ddot{\mathbf{r}}} - \underline{\mathbf{F}}) + \delta\underline{\boldsymbol{\omega}}^T \cdot (\underline{\mathsf{J}} \cdot \underline{\dot{\boldsymbol{\omega}}} - \underline{\mathbf{M}}^*) = 0 \tag{5.30}$$

(column matrices $\underline{\mathbf{r}} = [\mathbf{r}_1 \ldots \mathbf{r}_n]^T$, $\underline{\boldsymbol{\omega}} = [\boldsymbol{\omega}_1 \ldots \boldsymbol{\omega}_n]^T$, $\underline{\mathbf{F}} = [\mathbf{F}_1 \ldots \mathbf{F}_n]^T$, diagonal mass matrix $\underline{m}$, diagonal matrix $\underline{\mathsf{J}}$ of inertia tensors). The column matrix $\underline{\mathbf{M}}^*$ is introduced for abbreviation. It has the elements

$$\mathbf{M}_i^* = \mathbf{M}_i - \boldsymbol{\omega}_i \times \mathsf{J}_i \cdot \boldsymbol{\omega}_i \qquad (i = 1, \ldots, n)\,. \tag{5.31}$$

5.4.1 Systems Without Constraints to Inertial Space

If a system is free of constraints to inertial space such as an orbiting spacecraft or a flying or freely falling system then Newton's equation of motion for the composite system center of mass C can be decoupled from the remaining equations. This is done as follows. The radius vector of the composite system center of mass C is called $\mathbf{r}_C$ and the vector from C to the body i center of mass is called $\mathbf{R}_i$. Thus, by definition

$$\mathbf{r}_i = \mathbf{r}_C + \mathbf{R}_i \quad (i = 1, \ldots, n)\,, \qquad \sum_{i=1}^{n} \mathbf{R}_i m_i = \mathbf{0}\,. \tag{5.32}$$

The expression for $\mathbf{r}_i$ is substituted into (5.29). Multiplying out and using the second Eq. (5.32) one obtains the equation

$$\delta\dot{\mathbf{r}}_{\mathrm{C}} \cdot \left(M\ddot{\mathbf{r}}_{\mathrm{C}} - \sum_{i=1}^{n} \mathbf{F}_i\right) + \sum_{i=1}^{n} \left[\delta\dot{\mathbf{R}}_i \cdot (m_i\ddot{\mathbf{R}}_i - \mathbf{F}_i) + \delta\boldsymbol{\omega}_i \cdot (\mathsf{J}_i \cdot \dot{\boldsymbol{\omega}}_i - \mathbf{M}_i^*)\right] = 0 \tag{5.33}$$

with M being the total system mass. The special property of a system without constraints to inertial space is the independence of $\delta\dot{\mathbf{r}}_{\mathrm{C}}$. From this follows the equation

$$M\ddot{\mathbf{r}}_{\mathrm{C}} = \sum_{i=1}^{n} \mathbf{F}_i \,. \tag{5.34}$$

This is Newton's law for the composite system center of mass. The rest of the equation is written in the matrix form

$$\delta\dot{\underline{\mathbf{R}}}^T \cdot (\underline{m}\,\ddot{\underline{\mathbf{R}}} - \underline{\mathbf{F}}) + \delta\underline{\boldsymbol{\omega}}^T \cdot (\underline{\mathsf{J}} \cdot \dot{\underline{\boldsymbol{\omega}}} - \underline{\mathbf{M}}^*) = 0 \,. \tag{5.35}$$

The formal difference between this equation and (5.30) for arbitrary systems is that $\underline{\mathbf{R}}$ replaces $\underline{\mathbf{r}}$. Between $\underline{\mathbf{R}}$ and $\underline{\mathbf{r}}$ exists a simple relationship. By definition, the radius vector of the composite system center of mass is

$$\mathbf{r}_{\mathrm{C}} = \frac{1}{M}\sum_{j=1}^{n} m_j \mathbf{r}_j \,. \tag{5.36}$$

Substitution into the first Eq. (5.32) yields

$$\mathbf{R}_i = \sum_{j=1}^{n} \left(\delta_{ij} - \frac{m_j}{M}\right)\mathbf{r}_j \qquad (i = 1, \ldots, n) \,. \tag{5.37}$$

Let $\underline{\mu}$ be the dimensionless constant matrix with elements

$$\mu_{ij} = \delta_{ij} - \frac{m_i}{M} \qquad (i, j = 1, \ldots, n) \,. \tag{5.38}$$

Then, the matrix form of all n Eqs. (5.37) is

$$\underline{\mathbf{R}} = \underline{\mu}^T \underline{\mathbf{r}} \,. \tag{5.39}$$

The matrix $\underline{\mu}$ has remarkable properties. It satisfies the three equations

$$\underline{\mu}^T \underline{1} = \underline{0} \,, \qquad \underline{\mu}\,\underline{\mu} = \underline{\mu} \,, \qquad \underline{\mu}\,\underline{m} = \underline{m}\,\underline{\mu}^T = \underline{\mu}\,\underline{m}\,\underline{\mu}^T \,. \tag{5.40}$$

The first equation states that the sum of all rows is a row of zeros which means that $\underline{\mu}$ is singular. Hence, (5.39) cannot be resolved for $\underline{\mathbf{r}}$. This is obvious for physical reasons. The positions $\underline{\mathbf{r}}$ of the body centers of mass in inertial space cannot be determined if only the positions relative to the composite system center of mass are known. For a proof of the other two equations[1] one must show that $(\underline{\mu}\,\underline{\mu})_{ij} = \mu_{ij}$ and that $(\underline{\mu}\,\underline{m}\,\underline{\mu}^T)_{ij} = (\underline{\mu}\,\underline{m})_{ij} = \mu_{ij}m_j$. For this purpose one calculates

$$(\underline{\mu}\,\underline{\mu})_{ij} = \sum_{k=1}^{n} \mu_{ik}\mu_{kj} = \sum_{k=1}^{n} \mu_{ik}\Big(\delta_{kj} - \frac{m_k}{M}\Big) = \mu_{ij} - \frac{1}{M}\sum_{k=1}^{n} \mu_{ik}m_k\ ,$$

$$\begin{aligned}(\underline{\mu}\,\underline{m}\,\underline{\mu}^T)_{ij} &= \sum_{k=1}^{n} \mu_{ik}m_k\mu_{jk} = \sum_{k=1}^{n} \mu_{ik}m_k\Big(\delta_{jk} - \frac{m_j}{M}\Big)\\ &= \mu_{ij}m_j - \frac{m_j}{M}\sum_{k=1}^{n} \mu_{ik}m_k\ .\end{aligned}$$

The sum $\sum_{k=1}^{n} \mu_{ik}m_k$ appearing in both equations equals zero. End of proof.

Note: Newton's law (5.34) for the composite system center of mass is valid not only for rigid-body systems but for arbitrary systems. So are the definition (5.36) of the composite system center of mass and the relationship (5.39).

5.4.2 Generalized Coordinates

Let $\underline{q} = [q_1, \ldots, q_N]^T$ be an arbitrary set of generalized coordinates which are suitable for specifying the location and the orientation of a multibody system. The coordinates may be either joint variables or variables of position relative to inertial space or a combination of the two. At this point it is also not necessary to know whether $\underline{q}$ represents a minimal set of variables equal in number to the degree of freedom of the entire system or whether N exceeds the degree of freedom so that there exist constraint equations for the variables.

The radius vectors $\mathbf{r}_i$ of the body i centers of mass ($i = 1, \ldots, n$) can be expressed as some more or less complicated nonlinear functions $\mathbf{r}_i(q_1, \ldots, q_N, t)$ of the chosen variables and of time t. Time t appears explicitly because the position of the carrier body 0 is prescribed as function of time. Differentiation with respect to time produces equations of the general matrix forms

$$\underline{\dot{\mathbf{r}}} = \underline{\mathbf{a}}_1\underline{\dot{q}} + \underline{\mathbf{a}}_{10}\ , \qquad \delta\underline{\dot{\mathbf{r}}} = \underline{\mathbf{a}}_1\delta\underline{\dot{q}}\ , \qquad \underline{\ddot{\mathbf{r}}} = \underline{\mathbf{a}}_1\underline{\ddot{q}} + \underline{\mathbf{b}}_1\ . \tag{5.41}$$

Here, $\underline{\mathbf{a}}_1$ is an $(n \times N)$-matrix of as yet unknown vectors which depend on $q_1, \ldots, q_N$. The elements of the column matrix $\underline{\mathbf{a}}_{10}$ are the partial derivatives $\partial\mathbf{r}_i(q_1, \ldots, q_N, t)/\partial t$. They depend on $q_1, \ldots, q_N$ and on t. The column matrix $\underline{\mathbf{b}}_1$ depends on $q_1, \ldots, q_N$, on t and, in addition, on $\dot{q}_1, \ldots, \dot{q}_N$. If the carrier

[1] A matrix having the property $\underline{\mu}\,\underline{\mu} = \underline{\mu}$ is said to be idempotent (Gantmacher [19]). Every idempotent matrix can be expressed in the form $\underline{A}\,\underline{\Delta}\,\underline{A}^{-1}$ where $\underline{\Delta}$ is a diagonal matrix with elements 0 and 1 along the diagonal. From this it follows that the unit matrix is the only nonsingular idempotent matrix.

body is inertial space then time t does not appear explicitly, whence follows, in particular, that $\underline{\mathbf{a}}_{10} = \underline{\mathbf{0}}$. The formula given for $\delta\dot{\underline{\mathbf{r}}}$ is explained by the fact, that time t as well as $q_1, \ldots, q_N$ are held fixed.

For the vectors from the composite system center of mass to the body centers of mass the kinematical relationship (5.39), $\underline{\mathbf{R}} = \underline{\mu}^T\underline{\mathbf{r}}$, has been established. From this it follows that

$$\dot{\underline{\mathbf{R}}} = \underline{\mu}^T(\underline{\mathbf{a}}_1\dot{\underline{q}} + \underline{\mathbf{a}}_{10}) , \qquad \delta\dot{\underline{\mathbf{R}}} = \underline{\mu}^T\underline{\mathbf{a}}_1\delta\dot{\underline{q}} , \qquad \ddot{\underline{\mathbf{R}}} = \underline{\mu}^T(\underline{\mathbf{a}}_1\ddot{\underline{q}} + \underline{\mathbf{b}}_1) . \tag{5.42}$$

In analogy to (5.41) there exist relationships of the forms

$$\underline{\boldsymbol{\omega}} = \underline{\mathbf{a}}_2\dot{\underline{q}} + \underline{\mathbf{a}}_{20} , \qquad \delta\underline{\boldsymbol{\omega}} = \underline{\mathbf{a}}_2\delta\dot{\underline{q}} , \qquad \dot{\underline{\boldsymbol{\omega}}} = \underline{\mathbf{a}}_2\ddot{\underline{q}} + \underline{\mathbf{b}}_2 \tag{5.43}$$

with other matrices $\underline{\mathbf{a}}_2$, $\underline{\mathbf{a}}_{20}$ and $\underline{\mathbf{b}}_2$.

The expressions (5.43) and (5.41) are substituted into (5.30) of the principle of virtual power. The terms $\delta\dot{\underline{\mathbf{r}}}^T = \delta\dot{\underline{q}}^T\underline{\mathbf{a}}_1^T$ and $\delta\underline{\boldsymbol{\omega}}^T = \delta\dot{\underline{q}}^T\underline{\mathbf{a}}_2^T$ allow factoring out $\delta\dot{\underline{q}}^T$. Ordering of terms results in the equation

$$\delta\dot{\underline{q}}^T\Big\{\big(\underline{\mathbf{a}}_1^T\cdot\underline{m}\,\underline{\mathbf{a}}_1 + \underline{\mathbf{a}}_2^T\cdot\underline{\mathbf{J}}\cdot\underline{\mathbf{a}}_2\big)\,\ddot{\underline{q}} - \big[\underline{\mathbf{a}}_1^T\cdot(\underline{\mathbf{F}} - \underline{m}\,\underline{\mathbf{b}}_1) + \underline{\mathbf{a}}_2^T\cdot(\underline{\mathbf{M}}^* - \underline{\mathbf{J}}\cdot\underline{\mathbf{b}}_2)\big]\Big\} = 0 \tag{5.44}$$

or abbreviated

$$\delta\dot{\underline{q}}^T(\underline{A}\ddot{\underline{q}} - \underline{B}) = 0 \tag{5.45}$$

with the matrices

$$\left.\begin{aligned} \underline{A} &= \underline{\mathbf{a}}_1^T\cdot\underline{m}\,\underline{\mathbf{a}}_1 + \underline{\mathbf{a}}_2^T\cdot\underline{\mathbf{J}}\cdot\underline{\mathbf{a}}_2 , \\ \underline{B} &= \underline{\mathbf{a}}_1^T\cdot(\underline{\mathbf{F}} - \underline{m}\,\underline{\mathbf{b}}_1) + \underline{\mathbf{a}}_2^T\cdot(\underline{\mathbf{M}}^* - \underline{\mathbf{J}}\cdot\underline{\mathbf{b}}_2) . \end{aligned}\right\} \tag{5.46}$$

For systems without constraints to inertial space such as orbiting spacecraft and freely falling systems the principle of virtual power has the form (5.35). Substitution of the expressions (5.42) and (5.43) into this equation results in the equation

$$\delta\dot{\underline{q}}^T(\hat{\underline{A}}\ddot{\underline{q}} - \hat{\underline{B}}) = 0 \tag{5.47}$$

with matrices (replace in (5.46) $\underline{\mathbf{a}}_1$ by $\underline{\mu}^T\underline{\mathbf{a}}_1$ and $\underline{\mathbf{b}}_1$ by $\underline{\mu}^T\underline{\mathbf{b}}_1$)

$$\left.\begin{aligned} \hat{\underline{A}} &= \underline{\mathbf{a}}_1^T\underline{\mu}\cdot\underline{m}\,\underline{\mu}^T\underline{\mathbf{a}}_1 + \underline{\mathbf{a}}_2^T\cdot\underline{\mathbf{J}}\cdot\underline{\mathbf{a}}_2 , \\ \hat{\underline{B}} &= \underline{\mathbf{a}}_1^T\underline{\mu}\cdot(\underline{\mathbf{F}} - \underline{m}\,\underline{\mu}^T\underline{\mathbf{b}}_1) + \underline{\mathbf{a}}_2^T\cdot(\underline{\mathbf{M}}^* - \underline{\mathbf{J}}\cdot\underline{\mathbf{b}}_2) . \end{aligned}\right\} \tag{5.48}$$

In what follows it is assumed that the variables $q_1, \ldots, q_N$ represent a minimal set of independent variables equal in number to the degree of freedom of the system under consideration. Then, the variations $\delta\dot{\underline{q}}$ in (5.45) and (5.47) are independent. Hence, two minimal sets of differential equations are obtained in the forms

$$\underline{A}\,\ddot{\underline{q}} = \underline{B} , \tag{5.49}$$

$$\hat{\underline{A}}\ddot{\underline{q}} = \hat{\underline{B}} . \tag{5.50}$$

The second set of equations is valid only for systems without constraints to inertial space, while the first set is valid for arbitrary systems. The column matrices $\underline{B}$ and $\hat{\underline{B}}$ depend explicitly on $q_1, \ldots, q_N$, on t and, in addition, on $\dot{q}_1, \ldots, \dot{q}_N$. The matrices $\underline{A}$ and $\hat{\underline{A}}$ depend explicitly on $q_1, \ldots, q_N$ only. They are symmetric. They are also positive definite. For $\underline{A}$ this is shown in the case when the carrier body is inertial space ($\underline{\mathbf{a}}_{10} = \underline{\mathbf{0}}$, $\underline{\mathbf{a}}_{20} = \underline{\mathbf{0}}$). The total kinetic energy T of the system is

$$\begin{aligned} 2T &= \sum_{i=1}^{n} \left(m_i \dot{\mathbf{r}}_i^2 + \boldsymbol{\omega}_i \cdot \mathsf{J}_i \cdot \boldsymbol{\omega}_i \right) = \underline{\dot{\mathbf{r}}}^T \cdot \underline{m}\, \underline{\dot{\mathbf{r}}} + \underline{\boldsymbol{\omega}}^T \cdot \underline{\mathsf{J}} \cdot \underline{\boldsymbol{\omega}} \\ &= \underline{\dot{q}}^T \left(\underline{\mathbf{a}}_1^T \cdot \underline{m}\, \underline{\mathbf{a}}_1 + \underline{\mathbf{a}}_2^T \cdot \underline{\mathsf{J}} \cdot \underline{\mathbf{a}}_2 \right) \underline{\dot{q}} = \underline{\dot{q}}^T \underline{A}\, \underline{\dot{q}} \, . \end{aligned} \tag{5.51}$$

Thus, the matrix $\underline{A}$ is the coefficient matrix of the total kinetic energy. Since the kinetic energy is positive definite also the matrix $\underline{A}$ is.

As a conclusion of this section it can be stated that a minimal set of equations of motion for a multibody system is explicitly available as soon as the kinematical matrices $\underline{\mathbf{a}}_1$, $\underline{\mathbf{a}}_{10}$, $\underline{\mathbf{b}}_1$ and $\underline{\mathbf{a}}_2$, $\underline{\mathbf{a}}_{20}$, $\underline{\mathbf{b}}_2$ in (5.41) and (5.43) are known in terms of a minimal set of independent variables. Compact expressions can be formulated most easily for multibody systems with tree structure. The next section is devoted to such systems. Based on formulations for tree-structured systems also systems without tree structure can be handled. This is shown in Sect. 5.6.

5.5 Systems with Tree Structure

Systems with tree structure are simple for two reasons which have already been explained. The first reason is the kinematical independence of joint variables. For this reason equations of motion are formulated for joint variables. The second reason is the existence of the path matrix as inverse of the incidence matrix. The kinematics of individual joints in terms of joint variables is the subject of Sect. 5.5.1. In Sect. 5.5.2 the kinematics of entire systems is formulated with the help of path matrix and incidence matrix. Sects. 5.5.3–5.5.6 are devoted to various aspects of the resulting equations of motion.

5.5.1 Kinematics of Individual Joints

This section focuses on a single joint of a multibody system. In the directed system graph arc a is pointing from vertex $i^+(a)$ toward vertex $i^-(a)$. In Fig. 5.9 a single joint a is shown schematically without indication of its nature. The joint connects the bodies $i^+(a)$ and $i^-(a)$. The arrow indicates the sense of direction of arc a in the directed graph. Body-fixed reference bases $\underline{\mathbf{e}}^{i^+(a)}$ and $\underline{\mathbf{e}}^{i^-(a)}$ are attached to the two bodies at the body centers of mass.

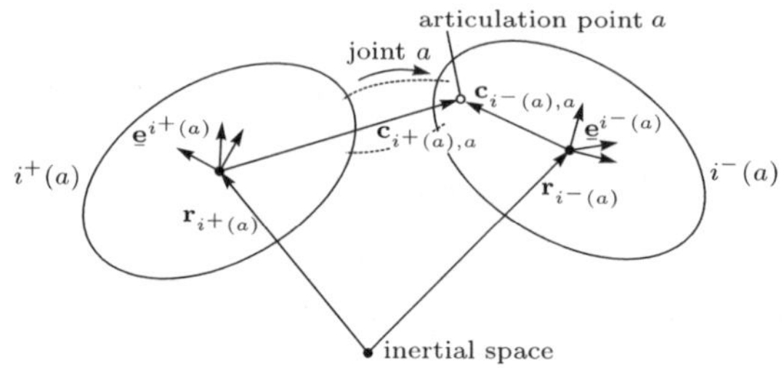

Fig. 5.9. Vectors on two bodies coupled by joint a of unspecified nature

Exception: The center of mass of the carrier body 0 is of no interest. The base $\underline{\mathbf{e}}^0$ on this body is attached at some conveniently chosen point. Joint a has a degree of freedom in the range $1 \le f_a \le 6$ (see Figs. 5.3a–f). An equal number of joint variables $q_{a\ell}$ ($\ell = 1, \dots, f_a$) is chosen. For the majority of joints the variables are rotation angles about certain axes or cartesian coordinates. Other types of variables are not ruled out, however. The following sign convention is adopted. The joint variables describe the position of body $i^-(a)$ relative to body $i^+(a)$.

Altogether six kinematical quantities are formulated for joint a as functions of joint variables. Three of them are the angular orientation, the angular velocity and the angular acceleration of body $i^-(a)$ relative to body $i^+(a)$. First, the other three quantities are explained. These are the position, the velocity and the acceleration relative to body $i^+(a)$ of a single point fixed on body $i^-(a)$. How to choose this point will be explained later. The chosen point is referred to as *articulation point* a. Its constant position on body $i^-(a)$ is specified by the vector $\mathbf{c}_{i^-(a),a}$ in base $\underline{\mathbf{e}}^{i^-(a)}$ (see Fig. 5.9). The constant coordinates in this base are system parameters. The position vector of the articulation point in base $\underline{\mathbf{e}}^{i^-(a)}$ is denoted $\mathbf{c}_{i^+(a),a}$. It is a known function of some or of all joint variables of joint a. More precisely, the coordinates of the vector in base $\underline{\mathbf{e}}^{i^+(a)}$ are known functions. The articulation point is chosen such that these functions are as simple as possible. Examples: If joint a is a spherical joint then the center of the sphere is chosen. If joint a is a Hooke's joint then the intersection point of the two joint axes on the central cross is chosen. If joint a is a revolute joint then an arbitrary point on the joint axis is chosen. In all three cases the articulation point is fixed on both bodies coupled by joint a, i.e. $\mathbf{c}_{i^+(a),a} = \text{const}$ on body $i^+(a)$.

The next two kinematical quantities are the velocity and the acceleration of the articulation point relative to body $i^+(a)$. They are denoted $\mathbf{v}_a$ and $\mathbf{a}_a$, respectively. They are the first and the second time derivatives of $\mathbf{c}_{i^+(a),a}$ in base $\underline{\mathbf{e}}^{i^+(a)}$. They have the forms

$$\mathbf{v}_a = \sum_{\ell=1}^{f_a} \mathbf{k}_{a\ell} \dot{q}_{a\ell} , \qquad \mathbf{a}_a = \sum_{\ell=1}^{f_a} \mathbf{k}_{a\ell} \ddot{q}_{a\ell} + \mathbf{s}_a . \tag{5.52}$$

A vector $\mathbf{s}_a \neq 0$ exists only if at least one of the vectors $\mathbf{k}_{a\ell}$ is not fixed on body $i^+(a)$.
Examples:

1. Spherical, Hooke's and revolute joints with articulation points chosen as described: $\mathbf{v}_a = \mathbf{0}$, $\mathbf{a}_a = \mathbf{0}$.
2. Cylindrical joint with articulation point on the joint axis, with unit vector $\mathbf{n}$ along the joint axis, with cartesian coordinate q_{a1} along $\mathbf{n}$ and with rotation angle q_{a2} about $\mathbf{n}$: $\mathbf{k}_{a1} = \mathbf{n}$, $\mathbf{k}_{a2} = \mathbf{0}$, $\mathbf{s}_a = \mathbf{0}$.

Next, the angular orientation, the angular velocity and the angular acceleration of body $i^-(a)$ relative to body $i^+(a)$ are expressed in terms of joint variables. The quantity determining the angular orientation is the direction cosine matrix $\underline{A}_a$ relating the bases $\underline{\mathbf{e}}^{i^+(a)}$ and $\underline{\mathbf{e}}^{i^-(a)}$. The definition is

$$\underline{\mathbf{e}}^{i^-(a)} = \underline{A}_a \underline{\mathbf{e}}^{i^+(a)} \,. \tag{5.53}$$

The matrix is a function of the angular variables among the joint variables of joint a. Example: In a cylindrical joint with variables q_{a1} and q_{a2} as before $\underline{A}_a$ is a function of q_{a2} only.

The angular velocity and the angular acceleration of body $i^-(a)$ relative to body $i^+(a)$ are denoted $\boldsymbol{\Omega}_a$ and $\boldsymbol{\varepsilon}_a$, respectively. They have the forms

$$\boldsymbol{\Omega}_a = \sum_{\ell=1}^{f_a} \mathbf{p}_{a\ell} \dot{q}_{a\ell} \,, \qquad \boldsymbol{\varepsilon}_a = \sum_{\ell=1}^{f_a} \mathbf{p}_{a\ell} \ddot{q}_{a\ell} + \mathbf{w}_a \,. \tag{5.54}$$

Examples:

1. Prismatic joint: $\boldsymbol{\Omega}_a = \mathbf{0}$, $\boldsymbol{\varepsilon}_a = \mathbf{0}$.
2. Cylindrical joint with unit vector $\mathbf{n}$ and with variables q_{a1} and q_{a2} as before: $\mathbf{p}_{a1} = \mathbf{0}$, $\mathbf{p}_{a2} = \mathbf{n}$, $\mathbf{w}_a = \mathbf{0}$.
3. Hooke's joint with axial unit vectors $\mathbf{p}_{a1}$ fixed on body $i^+(a)$ and $\mathbf{p}_{a2}$ fixed on body $i^-(a)$: $\boldsymbol{\Omega}_a = \mathbf{p}_{a1}\dot{q}_{a1} + \mathbf{p}_{a2}\dot{q}_{a2}$, $\boldsymbol{\varepsilon}_a = \mathbf{p}_{a1}\ddot{q}_{a1} + \mathbf{p}_{a2}\ddot{q}_{a2} + \boldsymbol{\Omega}_a \times \mathbf{p}_{a2}\dot{q}_{a2}$, whence follows $\mathbf{w}_a = \mathbf{p}_{a1} \times \mathbf{p}_{a2}\dot{q}_{a1}\dot{q}_{a2}$.

Alternative formulation for spherical joints. It is known that three angular joint variables (Euler angles or Bryan angles) can be inconvenient. More convenient is the following choice of variables. The matrix $\underline{A}_a$ is expressed in the form (2.35) as function of Euler–Rodrigues parameters. Equations (5.54) are replaced by the equations

$$\boldsymbol{\Omega}_a = \sum_{\ell=1}^{3} \mathbf{p}_{a\ell} \Omega_{a\ell} \,, \qquad \boldsymbol{\varepsilon}_a = \sum_{\ell=1}^{3} \mathbf{p}_{a\ell} \dot{\Omega}_{a\ell} \,. \tag{5.55}$$

The scalars $\Omega_{a\ell}$ $(\ell = 1, 2, 3)$ are the coordinates of $\boldsymbol{\Omega}_a$ in base $\underline{\mathbf{e}}^{i^-(a)}$, and the vectors $\mathbf{p}_{a\ell}$ $(\ell = 1, 2, 3)$ are the base vectors themselves. The Euler–Rodrigues parameters and the coordinates $\Omega_{a\ell}$ are related through the kinematical differential equations (2.119).

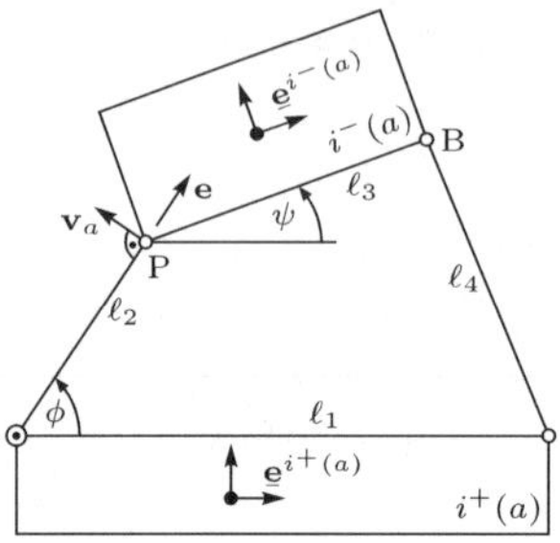

Fig. 5.10. Two massless rods creating a 1-d.o.f. joint

The chosen formulations for the six kinematical quantities of a joint are applicable not only to standard joints but to arbitrarily sophisticated joints. This is demonstrated by the joint shown in Fig. 5.10. The two bodies labeled $i^+(a)$ and $i^-(a)$ are coupled by two rods with revolute joints at both ends. The lengths $\ell_1, \ell_2, \ell_3, \ell_4$ form a planar fourbar. It is assumed that the rods are massless. This has the effect that the two rods together create a 1-d.o.f. joint connecting the bodies $i^+(a)$ and $i^-(a)$. The crank angle ϕ is chosen as joint variable q_{a1} and the point P as articulation point. The figure explains the angle ψ, the unit vector $\mathbf{e}$ along the crank and the unit vector $\mathbf{p}$ normal to the plane. The angle ψ is a function of ϕ. It is left to the reader to show that it is determined by the equation[2] $A\cos\psi + B\sin\psi = C$ with coefficients $A = -2\ell_3(\ell_1 - \ell_2\cos\phi)$, $B = 2\ell_2\ell_3\sin\phi$, $C = 2\ell_1\ell_2\cos\phi - (\ell_1^2 + \ell_2^2 + \ell_3^2 - \ell_4^2)$. The six kinematical quantities are

$$\mathbf{c}_{i^+(a),a} = \ell_2\mathbf{e} + \text{const}\,, \qquad \mathbf{v}_a = \dot{\phi}\mathbf{p}\times\ell_2\mathbf{e}\,, \quad \mathbf{a}_a = \ddot{\phi}\mathbf{p}\times\ell_2\mathbf{e} - \dot{\phi}^2\ell_2\mathbf{e}\,,$$

$$\underline{A}_a = \begin{bmatrix} \cos\psi & \sin\psi & 0 \\ -\sin\psi & \cos\psi & 0 \\ 0 & 0 & 1 \end{bmatrix}, \quad \boldsymbol{\Omega}_a = \mathbf{p}\frac{\mathrm{d}\psi}{\mathrm{d}\phi}\dot{\phi}\,, \quad \boldsymbol{\varepsilon}_a = \mathbf{p}\left(\frac{\mathrm{d}\psi}{\mathrm{d}\phi}\ddot{\phi} + \frac{\mathrm{d}^2\psi}{\mathrm{d}\phi^2}\dot{\phi}^2\right).$$

The vectors $\mathbf{a}_a$ and $\boldsymbol{\varepsilon}_a$ have the forms (5.52) and (5.54), respectively, with vectors $\mathbf{k}_{a1} = \ell_2\mathbf{p}\times\mathbf{e}$, $\mathbf{s}_a = -\dot{\phi}^2\ell_2\mathbf{e}$, $\mathbf{p}_{a1} = \mathbf{p}\,\mathrm{d}\psi/\mathrm{d}\phi$ and $\mathbf{w}_a = \mathbf{p}\dot{\phi}^2\,\mathrm{d}^2\psi/\mathrm{d}\phi^2$. The coordinates of these vectors in base $\underline{\mathbf{e}}^{i^+(a)}$ are functions of ϕ.

In preparation for the following section (5.52) and (5.54) are written in the matrix forms

$$\mathbf{v}_a = \underline{\mathbf{k}}_a^T\underline{\dot{q}}_a\,, \qquad \mathbf{a}_a = \underline{\mathbf{k}}_a^T\underline{\ddot{q}}_a + \mathbf{s}_a\,, \tag{5.56}$$

$$\boldsymbol{\Omega}_a = \underline{\mathbf{p}}_a^T\underline{\dot{q}}_a\,, \qquad \boldsymbol{\varepsilon}_a = \underline{\mathbf{p}}_a^T\underline{\ddot{q}}_a + \mathbf{w}_a \tag{5.57}$$

with row matrices $\underline{\mathbf{k}}_a^T = [\mathbf{k}_{a1} \;\ldots\; \mathbf{k}_{af_a}]$ and $\underline{\mathbf{p}}_a^T = [\mathbf{p}_{a1} \;\ldots\; \mathbf{p}_{af_a}]$. For a multibody system with joints $a = 1, \ldots, n$ column matrices $\underline{\mathbf{v}}$, $\underline{\mathbf{a}}$, $\underline{\mathbf{s}}$, $\underline{\boldsymbol{\Omega}}$,

[2] Cartesian coordinates x_B and y_B of the point B are functions of φ and ψ. These expressions are substituted into the constraint equation $(\ell_1 - x_\mathrm{B})^2 + y_\mathrm{B}^2 = \ell_4^2$. Each angle ϕ is associated with two (not necessarily real) angles ψ.

$\underline{\varepsilon}$ and $\underline{\mathbf{w}}$ of n vectors each are defined, for example $\underline{\mathbf{v}} = [\mathbf{v}_1 \ \ldots \ \mathbf{v}_n]^T$ and $\underline{\mathbf{w}} = [\mathbf{w}_1 \ \ldots \ \mathbf{w}_n]^T$. In terms of these matrices the four sets of n equations each are combined in the matrix forms

$$\underline{\mathbf{v}} = \underline{\mathbf{k}}^T \dot{\underline{q}} \,, \qquad \underline{\mathbf{a}} = \underline{\mathbf{k}}^T \ddot{\underline{q}} + \underline{\mathbf{s}} \,, \tag{5.58}$$

$$\underline{\boldsymbol{\Omega}} = \underline{\mathbf{p}}^T \dot{\underline{q}} \,, \qquad \underline{\varepsilon} = \underline{\mathbf{p}}^T \ddot{\underline{q}} + \underline{\mathbf{w}} \,. \tag{5.59}$$

The matrices $\underline{\mathbf{k}}^T$ and $\underline{\mathbf{p}}^T$ have block-diagonal form with the row matrices $\underline{\mathbf{k}}_a^T$ and $\underline{\mathbf{p}}_a^T$ along the diagonal, and $\ddot{\underline{q}}$ is the column matrix composed of the blocks $\ddot{\underline{q}}_a$ $(a = 1, \ldots, n)$.

5.5.2 Kinematics of Entire Systems

As illustrative example for the general formalism to be developed the tree-structured system shown in Fig. 5.11 is used. It is the system from Fig. 5.6 without force elements. Its directed graph is shown in Fig. 5.7b. For the associated incidence and path matrices see (5.7), (5.8) and (5.11). Each joint is of the general form shown in Fig. 5.9. Dots stand for the articulation points of joints. For joint $a = 2$ the vectors $\mathbf{c}_{72} = \mathbf{c}_{i^+(a),a}$ and $\mathbf{c}_{12} = \mathbf{c}_{i^-(a),a}$ are shown as examples. The carrier body 0 happens to be connected to a single body. The formalism to be developed is not restricted to this special case. Body 0 can be connected to several tree-structured subsystems. As in Fig. 5.7b the labeling of the bodies $1, \ldots, n$ and of the joints $1, \ldots, n$ as well as the directions of arcs in the graph are arbitrary. The following convention is adopted, however. All arcs incident with vertex 0 are directed toward vertex 0. This has the following consequences. The articulation points of all joints located on body 0 are fixed on body 0. The associated vectors $\mathbf{c}_{0a}$ are fixed in base $\underline{\mathbf{e}}^0$ the origin of which has, in inertial space, the prescribed position vector $\mathbf{r}_0(t)$. The position vector $\mathbf{r}_0(t) + \mathbf{c}_{0a}$ of the articulation point in inertial space is a prescribed function of time, too, and so are the absolute velocity

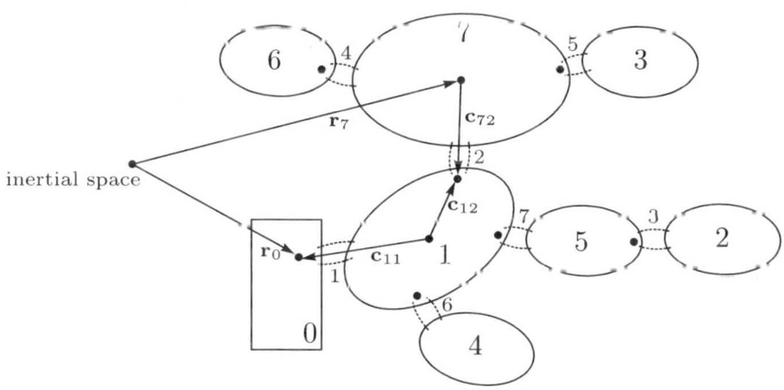

Fig. 5.11. Multibody system with tree structure and with joints of unspecified nature. Dots in joints and vectors are explained in Fig. 5.9

$\dot{\mathbf{r}}_0(t) + \boldsymbol{\omega}_0(t) \times \mathbf{c}_{0a}$ and the absolute acceleration

$$\ddot{\mathbf{r}}_0(t) + \dot{\boldsymbol{\omega}}_0(t) \times \mathbf{c}_{0a} + \boldsymbol{\omega}_0(t) \times [\boldsymbol{\omega}_0(t) \times \mathbf{c}_{0a}] \,. \tag{5.60}$$

Without loss of generality let it be assumed that the origin of the base $\underline{\mathbf{e}}^0$ with the prescribed position vector $\mathbf{r}_0(t)$ is one of the articulation points fixed on body 0. Then, the associated vector $\mathbf{c}_{0a}$ is zero. The advantage of this assumption is that systems with a single joint on body 0 do not have nonzero vectors $\mathbf{c}_{0a}$. This is the situation shown in Fig. 5.11 with the vector $\mathbf{c}_{01} = \mathbf{0}$ in joint 1.

The goal of this section is to express the matrices $\underline{\mathbf{a}}_1$, $\underline{\mathbf{a}}_{10}$, $\underline{\mathbf{b}}_1$, $\underline{\mathbf{a}}_2$, $\underline{\mathbf{a}}_{20}$ and $\underline{\mathbf{b}}_2$ in the relationships $\dot{\underline{\mathbf{r}}} = \underline{\mathbf{a}}_1\dot{\underline{q}} + \underline{\mathbf{a}}_{10}$, $\ddot{\underline{\mathbf{r}}} = \underline{\mathbf{a}}_1\ddot{\underline{q}} + \underline{\mathbf{b}}_1$ and $\underline{\boldsymbol{\omega}} = \underline{\mathbf{a}}_2\dot{\underline{q}} + \underline{\mathbf{a}}_{20}$, $\dot{\underline{\boldsymbol{\omega}}} = \underline{\mathbf{a}}_2\ddot{\underline{q}} + \underline{\mathbf{b}}_2$ (see (5.41) and (5.43)) in terms of joint variables and of time derivatives of joint variables. First, angular velocities are considered. Figure 5.11 shows that the absolute angular velocity $\boldsymbol{\omega}_i$ of an arbitrary body i is the sum of $\boldsymbol{\omega}_0$ and of all vectors $\boldsymbol{\Omega}_a$ (some positive and some negative) along the path from body 0 to body i. The formula for $\boldsymbol{\omega}_i$ is

$$\boldsymbol{\omega}_i = \boldsymbol{\omega}_0 - \sum_{a=1}^{n} T_{ai}\boldsymbol{\Omega}_a \qquad (i = 1, \ldots, n) \,. \tag{5.61}$$

The elements T_{ai} of the path matrix sort out the direct path from body 0 to body i and they provide the correct signs as well. Differentiation with respect to time yields

$$\dot{\boldsymbol{\omega}}_i = \dot{\boldsymbol{\omega}}_0 - \sum_{a=1}^{n} T_{ai}(\boldsymbol{\varepsilon}_a + \underbrace{\boldsymbol{\omega}_{i^-(a)} \times \boldsymbol{\Omega}_a}_{\mathbf{f}_a}) \qquad (i = 1, \ldots, n) \,. \tag{5.62}$$

The matrix forms of these equations are[3]

$$\underline{\boldsymbol{\omega}} = \boldsymbol{\omega}_0\underline{1} - \underline{T}^T\underline{\boldsymbol{\Omega}} \,, \qquad \dot{\underline{\boldsymbol{\omega}}} = \dot{\boldsymbol{\omega}}_0\underline{1} - \underline{T}^T\underline{\boldsymbol{\varepsilon}} - \underline{T}^T\underline{\mathbf{f}} \,. \tag{5.63}$$

Into these equations the expressions (5.59) are substituted. This results in the equations

$$\underline{\boldsymbol{\omega}} = \boldsymbol{\omega}_0\underline{1} - \underline{T}^T\underline{\mathbf{p}}^T\dot{\underline{q}} \,, \qquad \dot{\underline{\boldsymbol{\omega}}} = \dot{\boldsymbol{\omega}}_0\underline{1} - \underline{T}^T\underline{\mathbf{p}}^T\ddot{\underline{q}} - \underline{T}^T(\underline{\mathbf{w}} + \underline{\mathbf{f}}) \tag{5.64}$$

or finally

$$\underline{\boldsymbol{\omega}} = -(\underline{\mathbf{p}}\,\underline{T})^T\dot{\underline{q}} + \boldsymbol{\omega}_0\underline{1} \,, \qquad \dot{\underline{\boldsymbol{\omega}}} = -(\underline{\mathbf{p}}\,\underline{T})^T\ddot{\underline{q}} + \dot{\boldsymbol{\omega}}_0\underline{1} - \underline{T}^T(\underline{\mathbf{w}} + \underline{\mathbf{f}}) \,. \tag{5.65}$$

[3] The definition $\mathbf{f}_a = \boldsymbol{\omega}_{i^-(a)} \times \boldsymbol{\Omega}_a$ is equivalent to $\mathbf{f}_a = \boldsymbol{\omega}_{i^+(a)} \times \boldsymbol{\Omega}_a$.
The expression for $\underline{\boldsymbol{\omega}}$ can be obtained without making use of Fig. 5.11. Figure 5.9 yields $\boldsymbol{\Omega}_a = \boldsymbol{\omega}_{i^-(a)} - \boldsymbol{\omega}_{i^+(a)}$ or, with the definition of the incidence matrix, $\boldsymbol{\Omega}_a = -\sum_{i=0}^{n} S_{ia}\boldsymbol{\omega}_i$ $(a = 1, \ldots, n)$. This is written in the matrix form $\underline{\boldsymbol{\Omega}} = -\boldsymbol{\omega}_0\underline{S}_0^T - \underline{S}^T\underline{\boldsymbol{\omega}}$. Multiplication from the left by $\underline{T}^T$ and application of (5.12) and (5.13) result in (5.63).

These equations have the desired forms $\underline{\omega} = \underline{\mathbf{a}}_2\dot{\underline{q}} + \underline{\mathbf{a}}_{20}$ and $\dot{\underline{\omega}} = \underline{\mathbf{a}}_2\ddot{\underline{q}} + \underline{\mathbf{b}}_2$. The matrices are

$$\underline{\mathbf{a}}_2 = -(\underline{\mathbf{p}}\,\underline{T})^T\,, \qquad \underline{\mathbf{a}}_{20} = \boldsymbol{\omega}_0\underline{1}\,, \qquad \underline{\mathbf{b}}_2 = \dot{\boldsymbol{\omega}}_0\underline{1} - \underline{T}^T(\underline{\mathbf{w}} + \underline{\mathbf{f}})\,. \tag{5.66}$$

Next, the matrices $\underline{\mathbf{a}}_1$, $\underline{\mathbf{a}}_{10}$ and $\underline{\mathbf{b}}_1$ are formulated. Figure 5.11 shows that the position vector $\mathbf{r}_i$ of body i is

$$\mathbf{r}_i = \mathbf{r}_0 - \sum_{a=1}^{n} T_{ai}(\mathbf{c}_{i^+(a),a} - \mathbf{c}_{i^-(a),a}) \qquad (i = 1, \ldots, n)\,. \tag{5.67}$$

Of the two vectors $\mathbf{c}_{i^+(a),a}$ and $\mathbf{c}_{i^-(a),a}$ the latter one is fixed on body $i^-(a)$. This explains the expressions for the first and for the second time derivative of the difference vector:

$$\left.\begin{aligned}\dot{\mathbf{c}}_{i^+(a),a} - \dot{\mathbf{c}}_{i^-(a),a} &= -\mathbf{c}_{i^+(a),a} \times \boldsymbol{\omega}_{i^+(a)} + \mathbf{v}_a\\ &\quad + \mathbf{c}_{i^-(a),a} \times \boldsymbol{\omega}_{i^-(a)}\,,\\ \ddot{\mathbf{c}}_{i^+(a),a} - \ddot{\mathbf{c}}_{i^-(a),a} &= -\mathbf{c}_{i^+(a),a} \times \dot{\boldsymbol{\omega}}_{i^+(a)} + \mathbf{a}_a + \mathbf{h}_a\\ &\quad + \mathbf{c}_{i^-(a),a} \times \dot{\boldsymbol{\omega}}_{i^-(a)}\end{aligned}\right\} (a = 1, \ldots, n)\,. \tag{5.68}$$

The vectors $\mathbf{v}_a$ and $\mathbf{a}_a$ are known from (5.52), and $\mathbf{h}_a$ is

$$\begin{aligned}\mathbf{h}_a &= \boldsymbol{\omega}_{i^+(a)} \times (\boldsymbol{\omega}_{i^+(a)} \times \mathbf{c}_{i^+(a),a}) - \boldsymbol{\omega}_{i^-(a)} \times (\boldsymbol{\omega}_{i^-(a)} \times \mathbf{c}_{i^-(a),a})\\ &\quad + 2\boldsymbol{\omega}_{i^+(a)} \times \mathbf{v}_a \qquad (a = 1, \ldots, n)\,.\end{aligned} \tag{5.69}$$

From the definition (5.4) of the incidence matrix it follows that the vector differences in (5.67) and (5.68) can be written in the forms

$$\mathbf{c}_{i^+(a),a} - \mathbf{c}_{i^-(a),a} = \sum_{i=0}^{n} S_{ia}\mathbf{c}_{ia}\,, \tag{5.70}$$

$$\dot{\mathbf{c}}_{i^+(a),a} - \dot{\mathbf{c}}_{i^-(a),a} = -\sum_{i=0}^{n} S_{ia}\mathbf{c}_{ia} \times \boldsymbol{\omega}_i + \mathbf{v}_a\,, \tag{5.71}$$

$$\ddot{\mathbf{c}}_{i^+(a),a} - \ddot{\mathbf{c}}_{i^-(a),a} = -\sum_{i=0}^{n} S_{ia}\mathbf{c}_{ia} \times \dot{\boldsymbol{\omega}}_i + \mathbf{a}_a + \mathbf{h}_a \tag{5.72}$$

$(a = 1, \ldots, n)$. These formulations suggest to define the vectors

$$\mathbf{C}_{ia} = S_{ia}\mathbf{c}_{ia} \qquad (i = 0, \ldots, n;\ a = 1, \ldots, n) \tag{5.73}$$

and to construct a matrix with these vectors as elements[4]. This matrix represents a weighted incidence matrix. Like the incidence matrix it is partitioned

[4] To be precise one must define that $\mathbf{c}_{ia} = \mathbf{0}$ if $S_{ia} = 0$ $(i = 0, \ldots, n;\ a = 1, \ldots, n)$.

into the row matrix $\underline{\mathbf{C}}_0$ with elements $\mathbf{C}_{0a}$ $(a = 1,\ldots,n)$ and the $(n \times n)$-matrix $\underline{\mathbf{C}}$ with elements $\mathbf{C}_{ia}$ $(i, a = 1,\ldots,n)$. With these matrices the n Eqs. (5.67) are written in the matrix form $\underline{\mathbf{r}} = \mathbf{r}_0\underline{1} - \underline{T}^T\underline{\mathbf{C}}_0^T - \underline{T}^T\underline{\mathbf{C}}^T\underline{1}$ or[5]

$$\underline{\mathbf{r}} = \mathbf{r}_0\underline{1} - (\underline{\mathbf{C}}_0\underline{T})^T - (\underline{\mathbf{C}}\,\underline{T})^T\underline{1}\,. \tag{5.74}$$

For the first and for the second time derivative (5.71) and (5.72) yield the expressions

$$\dot{\underline{\mathbf{r}}} = \dot{\mathbf{r}}_0\underline{1} - \boldsymbol{\omega}_0 \times (\underline{\mathbf{C}}_0\underline{T})^T + (\underline{\mathbf{C}}\,\underline{T})^T \times \underline{\boldsymbol{\omega}} - \underline{T}^T\underline{\mathbf{v}}\,, \tag{5.75}$$

$$\ddot{\underline{\mathbf{r}}} = \ddot{\mathbf{r}}_0\underline{1} - \dot{\boldsymbol{\omega}}_0 \times (\underline{\mathbf{C}}_0\underline{T})^T + (\underline{\mathbf{C}}\,\underline{T})^T \times \dot{\underline{\boldsymbol{\omega}}} - \underline{T}^T(\underline{\mathbf{a}} + \underline{\mathbf{h}}) \tag{5.76}$$

with column matrices $\underline{\mathbf{v}} = [\mathbf{v}_1 \ldots \mathbf{v}_n]^T$, $\underline{\mathbf{a}} = [\mathbf{a}_1 \ldots \mathbf{a}_n]^T$ and $\underline{\mathbf{h}} = [\mathbf{h}_1 \ldots \mathbf{h}_n]^T$. For $\underline{\mathbf{v}}$ and $\underline{\mathbf{a}}$ the expressions (5.58) are substituted and for $\underline{\boldsymbol{\omega}}$ and $\dot{\underline{\boldsymbol{\omega}}}$ the expressions $\underline{\boldsymbol{\omega}} = \underline{\mathbf{a}}_2\dot{\underline{q}} + \boldsymbol{\omega}_0\underline{1}$ and $\dot{\underline{\boldsymbol{\omega}}} = \underline{\mathbf{a}}_2\ddot{\underline{q}} + \underline{\mathbf{b}}_2$ (see (5.65)). This yields

$$\begin{aligned}
\dot{\underline{\mathbf{r}}} &= [(\underline{\mathbf{C}}\,\underline{T})^T \times \underline{\mathbf{a}}_2 - (\underline{\mathbf{k}}\,\underline{T})^T]\dot{\underline{q}} \\
&\quad + \dot{\mathbf{r}}_0\underline{1} - \boldsymbol{\omega}_0 \times [(\underline{\mathbf{C}}_0\underline{T})^T + (\underline{\mathbf{C}}\,\underline{T})^T\underline{1}]\,,
\end{aligned} \tag{5.77}$$

$$\begin{aligned}
\ddot{\underline{\mathbf{r}}} &= [(\underline{\mathbf{C}}\,\underline{T})^T \times \underline{\mathbf{a}}_2 - (\underline{\mathbf{k}}\,\underline{T})^T]\ddot{\underline{q}} \\
&\quad + \ddot{\mathbf{r}}_0\underline{1} - \dot{\boldsymbol{\omega}}_0 \times (\underline{\mathbf{C}}_0\underline{T})^T + (\underline{\mathbf{C}}\,\underline{T})^T \times \underline{\mathbf{b}}_2 - \underline{T}^T(\underline{\mathbf{s}} + \underline{\mathbf{h}})\,.
\end{aligned} \tag{5.78}$$

These equations have the desired forms $\dot{\underline{\mathbf{r}}} = \underline{\mathbf{a}}_1\dot{\underline{q}} + \underline{\mathbf{a}}_{10}$ and $\ddot{\underline{\mathbf{r}}} = \underline{\mathbf{a}}_1\ddot{\underline{q}} + \underline{\mathbf{b}}_1$. The matrices are

$$\underline{\mathbf{a}}_1 = (\underline{\mathbf{C}}\,\underline{T})^T \times \underline{\mathbf{a}}_2 - (\underline{\mathbf{k}}\,\underline{T})^T\,, \tag{5.79}$$

$$\underline{\mathbf{a}}_{10} = \dot{\mathbf{r}}_0\underline{1} - \boldsymbol{\omega}_0 \times [(\underline{\mathbf{C}}_0\underline{T})^T + (\underline{\mathbf{C}}\,\underline{T})^T\underline{1}]\,, \tag{5.80}$$

$$\underline{\mathbf{b}}_1 = \ddot{\mathbf{r}}_0\underline{1} - \dot{\boldsymbol{\omega}}_0 \times (\underline{\mathbf{C}}_0\underline{T})^T + (\underline{\mathbf{C}}\,\underline{T})^T \times \underline{\mathbf{b}}_2 - \underline{T}^T(\underline{\mathbf{s}} + \underline{\mathbf{h}})\,. \tag{5.81}$$

With these matrices and with the matrices in (5.66),

$$\underline{\mathbf{a}}_2 = -(\underline{\mathbf{p}}\,\underline{T})^T\,, \qquad \underline{\mathbf{a}}_{20} = \boldsymbol{\omega}_0\underline{1}\,, \qquad \underline{\mathbf{b}}_2 = \dot{\boldsymbol{\omega}}_0\underline{1} - \underline{T}^T(\underline{\mathbf{w}} + \underline{\mathbf{f}})\,, \tag{5.82}$$

the final goal of the analysis has been achieved.

5.5.3 Equations of Motion

The matrices (5.79)–(5.82) determine the matrices (cf. (5.46))

$$\left.\begin{aligned}
\underline{A} &= \underline{\mathbf{a}}_1^T \cdot \underline{m}\,\underline{\mathbf{a}}_1 + \underline{\mathbf{a}}_2^T \cdot \underline{\mathbf{J}} \cdot \underline{\mathbf{a}}_2\,, \\
\underline{B} &= \underline{\mathbf{a}}_1^T \cdot (\underline{\mathbf{F}} - \underline{m}\,\underline{\mathbf{b}}_1) + \underline{\mathbf{a}}_2^T \cdot (\underline{\mathbf{M}}^* - \underline{\mathbf{J}} \cdot \underline{\mathbf{b}}_2)
\end{aligned}\right\} \tag{5.83}$$

[5] The expression for $\underline{\mathbf{r}}$ can be obtained without making use of Fig. 5.11. Figure 5.9 yields $\mathbf{r}_{i^+(a)} - \mathbf{r}_{i^-(a)} = \mathbf{c}_{i^-(a),a} - \mathbf{c}_{i^+(a),a}$ $(a = 1,\ldots,n)$ or $\sum_{i=0}^{n} S_{ia}\mathbf{r}_i = -\sum_{i=0}^{n} \mathbf{C}_{ia}$ $(a = 1,\ldots,n)$. This is written in the matrix form $\mathbf{r}_0\underline{S}_0^T + \underline{S}^T\underline{\mathbf{r}} = -\underline{\mathbf{C}}_0^T - \underline{\mathbf{C}}^T\underline{1}$. Multiplication from the left by $\underline{T}^T$ yields (5.74).

in the equations of motion (5.49) of tree-structured systems:

$$\underline{A}\,\underline{\ddot{q}} = \underline{B}\,. \tag{5.84}$$

The prescribed motion of the carrier body is represented by the vectors $\dot{\mathbf{r}}_0(t)$, $\ddot{\mathbf{r}}_0(t)$, $\boldsymbol{\omega}_0(t)$ and $\dot{\boldsymbol{\omega}}_0(t)$ which enter the right-hand side $\underline{B}$ of the equations. These terms are zero if the carrier body is inertial space. The term $\ddot{\mathbf{r}}_0\underline{1} - \dot{\boldsymbol{\omega}}_0 \times (\underline{\mathbf{C}}_0\underline{T})^T$ in (5.81) accounts for accelerations of articulation points fixed on the carrier body. Following (5.60) it has been said that $\underline{\mathbf{C}}_0$ equals zero if the carrier body 0 is connected to the system via a single joint and if $\mathbf{r}_0$ is defined as position vector of the articulation point chosen for this joint.

The equations of motion are particularly simple for systems which have revolute joints only and which are mounted on a stationary body 0. Many robots are systems of this kind. Each joint a has a single axial unit vector $\mathbf{p}_a$ and a single rotation angle q_a around this vector. The matrix $\underline{\mathbf{p}}$ is the diagonal matrix of the vectors $\mathbf{p}_1, \ldots, \mathbf{p}_n$. As articulation points points on the joint axes are chosen. This has the consequence that not only the vectors $\mathbf{c}_{i^-(a),a}$ but also the vectors $\mathbf{c}_{i^+(a),a}$ and, hence, all vectors in the matrix $\underline{\mathbf{C}}$ are body-fixed vectors. Furthermore, $\underline{\mathbf{k}} = \underline{\mathbf{0}}$, $\underline{\mathbf{w}} = \underline{\mathbf{0}}$ and $\underline{\mathbf{s}} = \underline{\mathbf{0}}$. The essential kinematics equations have the special forms

$$\left.\begin{aligned} \underline{\dot{\mathbf{r}}} &= \underline{\mathbf{a}}_1\underline{\dot{q}}\,, & \underline{\ddot{\mathbf{r}}} &= \underline{\mathbf{a}}_1\underline{\ddot{q}} + \underline{\mathbf{b}}_1\,,\\ \underline{\boldsymbol{\omega}} &= \underline{\mathbf{a}}_2\underline{\dot{q}}\,, & \underline{\dot{\boldsymbol{\omega}}} &= \underline{\mathbf{a}}_2\underline{\ddot{q}} + \underline{\mathbf{b}}_2 \end{aligned}\right\} \tag{5.85}$$

with matrices

$$\left.\begin{aligned} \underline{\mathbf{a}}_1 &= (\underline{\mathbf{C}}\,\underline{T})^T \times \underline{\mathbf{a}}_2\,, & \underline{\mathbf{b}}_1 &= (\underline{\mathbf{C}}\,\underline{T})^T \times \underline{\mathbf{b}}_2 - \underline{T}^T\underline{\mathbf{h}}\,,\\ \underline{\mathbf{a}}_2 &= -(\underline{\mathbf{p}}\,\underline{T})^T\,, & \underline{\mathbf{b}}_2 &= -\underline{T}^T\,\underline{\mathbf{f}}\,. \end{aligned}\right\} \tag{5.86}$$

In Sect. 5.7 systems are investigated in which all joints are spherical joints, and in Sect. 5.8 the special case of planar motions of systems with revolute joints is considered.

For systems without kinematical constraints to inertial space equations of motion have the special form (5.50):

$$\underline{\hat{A}}\underline{\ddot{q}} = \underline{\hat{B}}\,. \tag{5.87}$$

According to (5.48) the matrices $\underline{\hat{A}}$ and $\underline{\hat{B}}$ are obtained from (5.83) if $\underline{\mathbf{a}}_1$ is replaced by $\underline{\mu}^T\underline{\mathbf{a}}_1$ and $\underline{\mathbf{b}}_1$ by $\underline{\mu}^T\underline{\mathbf{b}}_1$. These matrices are

$$\underline{\mu}^T\underline{\mathbf{a}}_1 = (\underline{\mathbf{C}}\,\underline{T}\,\underline{\mu})^T \times \underline{\mathbf{a}}_2 - (\underline{\mathbf{k}}\,\underline{T}\,\underline{\mu})^T\,, \tag{5.88}$$

$$\underline{\mu}^T\underline{\mathbf{b}}_1 = (\underline{\mathbf{C}}\,\underline{T}\,\underline{\mu})^T \times \underline{\mathbf{b}}_2 - (\underline{T}\,\underline{\mu})^T(\underline{\mathbf{s}} + \underline{\mathbf{h}})\,, \quad \underline{\mathbf{b}}_2 = -\underline{T}^T(\underline{\mathbf{w}} + \underline{\mathbf{f}})\,. \tag{5.89}$$

The terms representing the motion of body 0 are eliminated because of the first Eq. (5.40).

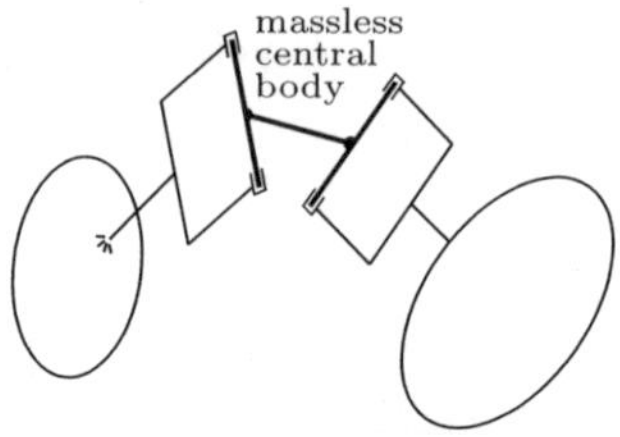

Fig. 5.12. System with a massless central body

Problem 5.7. Formulate the matrices $\underline{A}$ and $\underline{B}$ in (5.83) for a system in which the joints allow only translational motions of bodies relative to one another and in which, furthermore, body 0 does not rotate relative to inertial space.

Problem 5.8. The system shown in Fig. 5.12 with a massless central body can be modeled either as a system of two bodies coupled by a two-degree-of-freedom joint or as system of three bodies coupled by two revolute joints. Does the masslessness of the central body have the effect that the matrix $\underline{A}$ in the equations of motion $\underline{A}\,\ddot{\underline{q}} = \underline{B}$ is singular?

5.5.4 Augmented Bodies

In the matrices $\underline{A}$ and $\underline{B}$ of the equations of motion (5.84) a dominant role is played by the matrix product $\underline{\mathbf{C}}\,\underline{T}$. It originated from (5.74): $\underline{\mathbf{r}} = \mathbf{r}_0\underline{1} - (\underline{\mathbf{C}}_0\underline{T})^T - (\underline{\mathbf{C}}\,\underline{T})^T\underline{1}$. As a consequence of (5.39), $\underline{\mathbf{R}} = \underline{\mu}^T\underline{\mathbf{r}}$, the matrix product $\underline{\mathbf{C}}\,\underline{T}\,\underline{\mu}$ is equally dominant in the equations of motion (5.87) for systems without constraints to inertial space. In what follows geometrical interpretations are given for the elements of these two matrix products.

For simplicity, it is assumed that body 0 is connected to the system by a single joint, and that the vector $\mathbf{r}_0$ fixed on body 0 is the articulation point of this joint. This has the consequence that $\underline{\mathbf{C}}_0 = \underline{\mathbf{0}}$. Hence,

$$\underline{\mathbf{r}} = \mathbf{r}_0\underline{1} - (\underline{\mathbf{C}}\,\underline{T})^T\underline{1} \tag{5.90}$$

and, because of (5.40), $\underline{\mu}^T\underline{1} = \underline{0}$,

$$\underline{\mathbf{R}} = -(\underline{\mathbf{C}}\,\underline{T}\,\underline{\mu})^T\underline{1}\,. \tag{5.91}$$

First, the matrix $\underline{\mathbf{C}}\,\underline{T}$ is investigated. Its elements are abbreviated $\mathbf{d}_{ij}$. Equation (5.90) then yields for a single vector $\mathbf{r}_j$ the formula

$$\mathbf{r}_j = \mathbf{r}_0 - \sum_{i=1}^{n} \mathbf{d}_{ij} \qquad (j = 1,\ldots,n)\,. \tag{5.92}$$

With (5.73) the vector $\mathbf{d}_{ij}$ is

$$\mathbf{d}_{ij} = (\underline{\mathbf{C}}\,\underline{T})_{ij} = \sum_{a=1}^{n} S_{ia}\mathbf{c}_{ia}T_{aj} \qquad (i,j = 1,\ldots,n)\,. \tag{5.93}$$

Both indices of $\mathbf{d}_{ij}$ are associated with bodies since the first index of $\underline{\mathbf{C}}$ as well as the second index of $\underline{T}$ is associated with a body. Since all vectors $\mathbf{c}_{ia}$ are located on body i also $\mathbf{d}_{ij}$ is located on body i. For (5.92) all vectors $\mathbf{d}_{ij}$ $(i = 1, \ldots, n)$ are required for a fixed value of j. The products $S_{ia}T_{aj}$ are different from zero only for those arcs a which are incident with vertex i $(S_{ia} \neq 0)$ and which are, furthermore, on the path from vertex 0 to vertex j $(T_{aj} \neq 0)$. There are two such arcs if vertex $i <$ vertex j, a single arc if $i = j$ and no arc at all otherwise. Hence,

$$\mathbf{d}_{ij} = \mathbf{0} \quad \text{if neither } i = j \text{ nor vertex } i < \text{vertex } j \qquad (i, j = 1, \ldots, n) \,. \tag{5.94}$$

The case vertex $i <$ vertex j: Let the two arcs be arc b and arc c with arc b being the inboard arc of vertex i. Independent of the sense of direction of these arcs $S_{ib}T_{bj} = +1$ and $S_{ic}T_{cj} = -1$. From this it follows that $\mathbf{d}_{ij} = \mathbf{c}_{ib} - \mathbf{c}_{ic}$. The case $i = j$: The only arc is the inboard arc b of vertex j. Hence, $\mathbf{d}_{jj} = \mathbf{c}_{jb}$. For illustration the system shown in Fig. 5.11 is used. Let body j be body 7. The only nonzero vectors $\mathbf{d}_{i7}$ are $\mathbf{d}_{17} = \mathbf{c}_{11} - \mathbf{c}_{12}$ and $\mathbf{d}_{77} = \mathbf{c}_{72}$. Equation (5.92) reads $\mathbf{r}_7 = \mathbf{r}_0 - \mathbf{d}_{17} - \mathbf{d}_{77}$. These vectors are shown in Fig. 5.13. This concludes the geometrical interpretation of the elements of the matrix $\underline{\mathbf{C}}\,\underline{T}$.

Next, the elements of the matrix $\underline{\mathbf{C}}\,\underline{T}\,\underline{\mu}$ are investigated. In order to do so the important concept of augmented bodies is introduced. For each of the bodies $i = 1, \ldots, n$ an augmented body is constructed is follows. At the tip of each vector $\mathbf{c}_{ia}$ on body i a point mass is attached which is equal to the sum of the masses of all bodies except body 0 which are connected with body i either directly or indirectly via the respective joint a. To give an example, in the system of Fig. 5.13 the augmented body 7 is obtained from the original body 7 by attaching three point masses, namely the point mass $m_1 + m_4 + m_5 + m_2$ at the tip of the vector $\mathbf{c}_{72}$, the point mass m_3 at the tip of $\mathbf{c}_{75}$ and the point mass m_6 at the tip of $\mathbf{c}_{74}$. The augmented body 2 is obtained by attaching to body 2 the point mass $m_1 + m_3 + m_4 + m_5 + m_6 + m_7$ at the tip of $\mathbf{c}_{23}$. Each augmented body has the mass M of the total system.

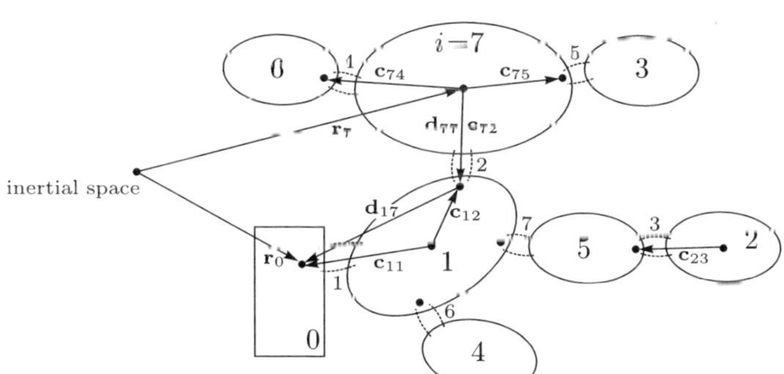

Fig. 5.13. Multibody system with chain of vectors $\mathbf{r}_7 = \mathbf{r}_0 - \mathbf{d}_{17} - \mathbf{d}_{77}$

Augmented bodies are of particular interest in the case of multibody systems all joints of which are either spherical or revolute or Hooke's joints. Then, the articulation point of each joint a is chosen such that it is fixed not only on body $i^-(a)$ but also on body $i^+(a)$. This has the consequence that each augmented body is a rigid body which does not change when the system is moving. In what follows arbitrary joints are admitted. In this general case the augmented body i has an instantaneous mass distribution. It has – instantaneously – a center of mass which, in general, does not coincide with the center of mass C_i of the original body i. The center of mass of the augmented body i is called the barycenter B_i of the body. In Fig. 5.14 a body i is depicted with its center of mass C_i and its barycenter B_i. Also shown are body 0 and another body j. Both bodies are connected with body i either directly or indirectly. Broken lines indicate the paths between bodies. On the augmented body i vectors $\mathbf{b}_{ij}$ $(j = 0, \ldots, n)$ are defined. The vector $\mathbf{b}_{ii}$ points from the barycenter B_i to the center of mass C_i. The vector $\mathbf{b}_{ij}$ with $j \neq i$ points from the barycenter B_i to the tip of the vector $\mathbf{c}_{ia}$ which leads either directly or indirectly to body j. In Fig. 5.14 vectors $\mathbf{b}_{ii}$, $\mathbf{b}_{i0}$ and $\mathbf{b}_{ij}$ are shown. The vectors $\mathbf{b}_{ij}$ with $j \neq i$ play for the augmented bodies the role that is played by the vectors $\mathbf{c}_{ia}$ for the original bodies. Notice, however, the following differences. The second index of $\mathbf{c}_{ia}$ corresponds to a joint whereas the second index of $\mathbf{b}_{ij}$ corresponds to a body. Vectors $\mathbf{b}_{ij}$ exist for all index combinations $i = 1, \ldots, n;\ j = 0, \ldots, n$. The number of different vectors $\mathbf{b}_{ij}$ is smaller than the number of different combinations of indices i, j. In the system of Fig. 5.13, for instance, the identities $\mathbf{b}_{17} = \mathbf{b}_{13} = \mathbf{b}_{16}$ hold. From the definition of the vectors it follows that

$$\sum_{k=1}^{n} m_k \mathbf{b}_{ik} = \mathbf{0} \qquad (i = 1, \ldots, n)\,. \tag{5.95}$$

Proposition: The vectors $\mathbf{d}_{ij}$ and $\mathbf{b}_{ij}$ are related through the equation

$$\mathbf{d}_{ij} = \mathbf{b}_{i0} - \mathbf{b}_{ij} \qquad (i, j = 1, \ldots, n)\,. \tag{5.96}$$

For a proof four cases have to be distinguished: (i) $i = j$, (ii) vertex $i <$ vertex j, (iii) vertex $j <$ vertex i and (iv) otherwise. Consider case (iv).

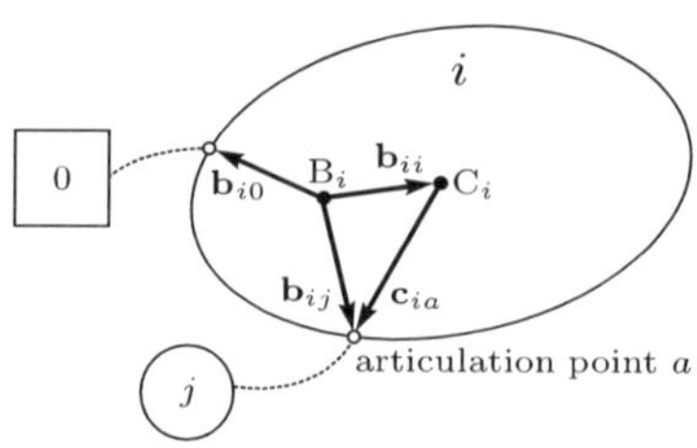

Fig. 5.14. Body i with center of mass C_i, barycenter B_i and vectors $\mathbf{b}_{ii}$, $\mathbf{b}_{i0}$, $\mathbf{b}_{ij}$

According to (5.94) $\mathbf{d}_{ij} = \mathbf{0}$. The vector $\mathbf{b}_{ij}$ is, indeed, identical with $\mathbf{b}_{i0}$. The other cases are equally elementary.

Equation (5.96) together with (5.95) is the key to the geometrical interpretation of the matrix $\underline{\mathbf{C}}\,\underline{T}\,\underline{\mu}$. The elements of the matrix are

$$\begin{aligned}(\underline{\mathbf{C}}\,\underline{T}\,\underline{\mu})_{ij} &= \sum_{k=1}^{n} (\underline{\mathbf{C}}\,\underline{T})_{ik}\,\mu_{kj} = \sum_{k=1}^{n} \mathbf{d}_{ik}\,\mu_{kj} \\ &= \sum_{k=1}^{n} (\mathbf{b}_{i0} - \mathbf{b}_{ik})\Big(\delta_{kj} - \frac{m_k}{M}\Big) = -\mathbf{b}_{ij} + \frac{m_k}{M}\sum_{k=1}^{n} \mathbf{b}_{ik} m_k \\ &= -\mathbf{b}_{ij} \qquad (i,j = 1,\ldots,n)\,. \end{aligned} \tag{5.97}$$

Through this equation the augmented body vectors $\mathbf{b}_{ij}$ $(i, j = 1, \ldots, n)$ are shown to be dominant system parameters. The equation provides a simple algorithm for calculating the coordinates of the vectors. The combination of (5.97) with (5.91) yields for a single vector $\mathbf{R}_j$ the remarkable relationship

$$\mathbf{R}_j = \sum_{i=1}^{n} \mathbf{b}_{ij} \qquad (j = 1,\ldots,n)\,. \tag{5.98}$$

Every vector $\mathbf{b}_{ij}$ in this sum is located on another body i. As has already been pointed out this equation is valid even in the case when the augmented bodies are not rigid bodies, i.e. for systems with arbitrary joints. Rigidity is important only when the vectors $\mathbf{d}_{ij}$ and $\mathbf{b}_{ij}$ are differentiated with respect to time. Only then $\dot{\mathbf{d}}_{ij} = \boldsymbol{\omega}_i \times \mathbf{d}_{ij}$ and $\dot{\mathbf{b}}_{ij} = \boldsymbol{\omega}_i \times \mathbf{b}_{ij}$. This will be the case in Sect. 5.7 which is devoted to systems with spherical joints.

Problem 5.9. Identify in Fig. 5.13 the vectors $\mathbf{d}_{ij}$, $\mathbf{b}_{ij}$ and $\mathbf{b}_{i0}$ in (5.96) for the following sets of indices (i, j): $(1,2)$, $(1,3)$, $(2,1)$, $(2,2)$ and $(2,6)$.

5.5.5 Force Elements

Subject of this section are force elements in a system governed by (5.84) or by (5.87). Force elements are either passive (linear or nonlinear springs and dampers) or active (actuators). They exert forces or torques of equal magnitude and of opposite direction on the two bodies connected by the force element. These forces and torques appear in the matrix $\underline{B}$ and in this matrix in the term

$$\underline{\mathbf{a}}_1^T \cdot \underline{\mathbf{F}} + \underline{\mathbf{a}}_2^T \cdot \underline{\mathbf{M}}\,. \tag{5.99}$$

By definition (see (5.29)) $\mathbf{F}_i$ is the resultant force on body i with its line of action passing through the body center of mass, and $\mathbf{M}_i$ is the resultant torque about the center of mass. The mathematical formulation is simplest for force elements which are torsional springs or dampers resisting the relative rotation in joint axes. This case is treated first. It is assumed, again, that each joint $a = 1, \ldots, n$ of the system is a revolute joint with an angular variable q_a and with unit vector $\mathbf{p}_a$ along the joint axis. In each joint a torsional spring with spring constant k_a is located which is unstressed in the position $q_a = q_{a0}$. The first term in (5.99) is zero. With (5.86) the second term is $-\underline{\mathbf{p}}\underline{T} \cdot \underline{\mathbf{M}}$ where $\underline{\mathbf{p}}$ is the diagonal matrix of the altogether n unit vectors. The torque exerted by the spring in joint a is $-\mathbf{p}_a\, k_a(q_a - q_{a0})$ on body $i^-(a)$ and $+\mathbf{p}_a\, k_a(q_a - q_{a0})$ on body $i^+(a)$. The incidence matrix allows the statement: The torque exerted on an arbitrary body i is $S_{ia}\mathbf{p}_a\, k_a(q_a - q_{a0})$. The resultant of all torques exerted on body i is $\mathbf{M}_i = \sum_{a=1}^{n} S_{ia}\mathbf{p}_a\, k_a(q_a - q_{a0})$. This yields for the column matrix $\underline{\mathbf{M}}$ the expression $\underline{\mathbf{M}} = \underline{S}\,\underline{\mathbf{p}}\,\underline{K}(\underline{q} - \underline{q}_0)$ where $\underline{K}$ is the diagonal matrix of spring constants. Thus,

$$-\underline{\mathbf{p}}\underline{T} \cdot \underline{\mathbf{M}} = -\underline{\mathbf{p}}\underline{T} \cdot \underline{S}\,\underline{\mathbf{p}}\,\underline{K}(\underline{q} - \underline{q}_0) = -\underline{K}(\underline{q} - \underline{q}_0) \tag{5.100}$$

because $\underline{T}\,\underline{S} = \underline{I}$ and $\underline{\mathbf{p}} \cdot \underline{\mathbf{p}} = \underline{I}$. If, in addition to torsional springs, torsional dampers with damper constants d_a are mounted in the joints then the term in the equations of motion is $-\underline{D}\,\underline{\dot{q}} - \underline{K}(\underline{q} - \underline{q}_0)$ with a diagonal damping matrix $\underline{D}$.

Next, a single force element is considered which is attached to points fixed on two bodies as is shown in Fig. 5.15. The force exerted on the two bodies has the direction of the line connecting the attachment points. If the force element is a spring or damper then the magnitude of the force is a given function of the distance L and of the rate of change of the distance of the attachment points. In the case of an active element the magnitude of the force is determined by a control law from kinematical data. In the course of numerical integration of the equations of motion the positions and the velocities of the attachment points as well as other kinematical data are available at every time t. Hence, also direction and magnitude of the force are known. A simple algorithm requires the following calculations. Let k and $\ell \neq k$ (arbitrary) be the labels of the two bodies and let, furthermore, $\mathbf{c}_k$ and $\mathbf{c}_\ell$ be the body-fixed vectors from the respective body centers of mass to the attachment points. The position vectors and the velocities of the attachment points are

$$\left.\begin{aligned} \mathbf{u}_k &= \mathbf{r}_k + \mathbf{c}_k\,, & \mathbf{v}_k &= \dot{\mathbf{r}}_k + \boldsymbol{\omega}_k \times \mathbf{c}_k\,,\\ \mathbf{u}_\ell &= \mathbf{r}_\ell + \mathbf{c}_\ell\,, & \mathbf{v}_\ell &= \dot{\mathbf{r}}_\ell + \boldsymbol{\omega}_\ell \times \mathbf{c}_\ell\,. \end{aligned}\right\} \tag{5.101}$$

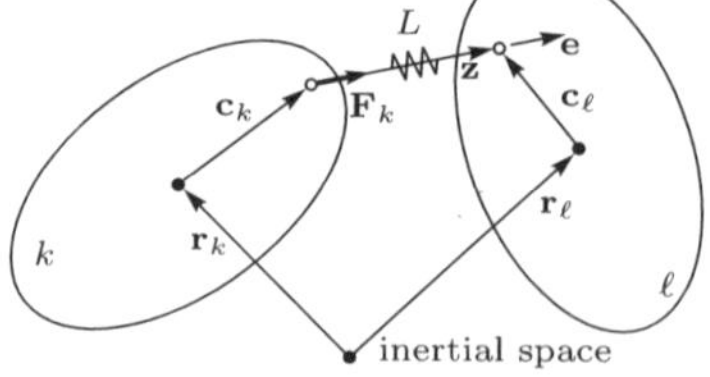

Fig. 5.15. Force element connecting two bodies k and ℓ

These quantities determine the vector $\mathbf{z} = \mathbf{u}_\ell - \mathbf{u}_k$, the distance $L = |\mathbf{z}|$, the unit vector $\mathbf{e} = \mathbf{z}/L$ and the rate of change $\dot{L} = (\mathbf{v}_\ell - \mathbf{v}_k) \cdot \mathbf{e}$. The magnitude of the force is a known function $F(L, \dot{L})$. Let $\mathbf{F}_k = +F\mathbf{e}$ be the force acting on body k. Then, $\mathbf{F}_\ell = -\mathbf{F}_k$ is acting on body ℓ. The torques exerted on the two bodies are $\mathbf{M}_k = \mathbf{c}_k \times \mathbf{F}_k$ and $\mathbf{M}_\ell = \mathbf{c}_\ell \times \mathbf{F}_\ell$. The four vectors $\mathbf{F}_k$, $\mathbf{F}_\ell$, $\mathbf{M}_k$ and $\mathbf{M}_\ell$ have to be substituted into (5.99).

If a system has more than a single force element then the same calculations have to be made for each element separately with individual parameters k, ℓ, $\mathbf{c}_k$ and $\mathbf{c}_\ell$. For each force element the respective vectors $\mathbf{F}_k$, $\mathbf{F}_\ell$, $\mathbf{M}_k$ and $\mathbf{M}_\ell$ have to be substituted into (5.99).

In what follows a different formulation with new notations is presented in order to demonstrate applications of system graphs without tree structure. The force elements are labeled $a = n+1, \ldots, m$. As is shown in Fig. 5.16 the labels of the two bodies coupled by the force element a are denoted $i^+(a)$ and $i^-(a)$, respectively. The vectors from the body centers of mass to the attachment points are denoted $\mathbf{c}_{i^+(a),a}$ and $\mathbf{c}_{i^-(a),a}$, respectively (for the vectors $\mathbf{c}_{0a}$ on body 0 see the text following Fig. 5.11). The vector from the attachment point on body $i^+(a)$ to the attachment point on body $i^-(a)$ is called $\mathbf{z}_a$.

To the tree-structured directed graph representing the interconnection by joints $1, \ldots, n$ directed arcs $n+1, \ldots, m$ representing the interconnection by the force elements are added. This graph is called the complete graph. The directed graph displaying the interconnections by joints alone is a spanning tree of the complete graph, and the added arcs $a = n+1, \ldots, m$ are chords of the complete graph. As illustrative example see the two graphs in Figs. 5.7a and b for the system of Fig. 5.6. The complete graph has an $(n \times m)$ incidence matrix defined in (5.4) with elements S_{ia} $(i = 0, \ldots, n;\ a = 1, \ldots, m)$. Through (5.73) the vectors $\mathbf{C}_{ia} = S_{ia}\mathbf{c}_{ia}$ were defined. This definition is now generalized to include the chords:

$$\mathbf{C}_{ia} = S_{ia}\mathbf{c}_{ia} \qquad (i = 0, \ldots, n;\ a = 1, \ldots, m)\,. \tag{5.102}$$

Equation (5.6) showed the partitioning $\begin{bmatrix} \underline{S}_{0\mathrm{t}} & \underline{S}_{0\mathrm{c}} \\ \underline{S}_{\mathrm{t}} & \underline{S}_{\mathrm{c}} \end{bmatrix}$ of the incidence matrix. With the vectors $\mathbf{C}_{ia} = S_{ia}\mathbf{c}_{ia}$ the partitioned weighted incidence matrix $\begin{bmatrix} \underline{\mathbf{C}}_{0\mathrm{t}} & \underline{\mathbf{C}}_{0\mathrm{c}} \\ \underline{\mathbf{C}}_{\mathrm{t}} & \underline{\mathbf{C}}_{\mathrm{c}} \end{bmatrix}$ is formed. In the preceding sections only the submatrices

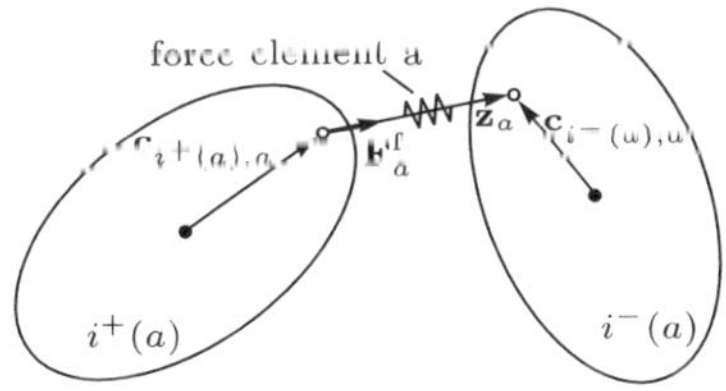

Fig. 5.16. Force element a connecting bodies $i^+(a)$ and $i^-(a)$

$\underline{S}_{0t}$, $\underline{S}_{t}$, $\underline{\mathbf{C}}_{0t}$ and $\underline{\mathbf{C}}_{t}$ were used. They were called $\underline{S}_0$, $\underline{S}$, $\underline{\mathbf{C}}_0$ and $\underline{\mathbf{C}}$, respectively.

After this introduction the forces exerted by the force elements are considered. Notation and sign convention: The force element a $(a = n+1, \ldots, m)$ produces the force $+\mathbf{F}^{\mathrm{f}}_a$ and the torque $+\mathbf{c}_{i^+(a),a} \times \mathbf{F}^{\mathrm{f}}_a$ on body $i^+(a)$ and the force $-\mathbf{F}^{\mathrm{f}}_a$ and the torque $-\mathbf{c}_{i^-(a),a} \times \mathbf{F}^{\mathrm{f}}_a$ on body $i^-(a)$. The upper index f (force element) is necessary because $\mathbf{F}_a$ is, by definition, the resultant force on body a. With the definition of the incidence matrix the force exerted on body i $(i = 1, \ldots, n)$ by the force element a is $S_{ia}\mathbf{F}^{\mathrm{f}}_a$ and the torque is $S_{ia}\mathbf{c}_{ia} \times \mathbf{F}_a = \mathbf{C}_{ia} \times \mathbf{F}^{\mathrm{f}}_a$. The resultant force on body i $(i = 1, \ldots, n)$ by all force elements is $\mathbf{F}_i = \sum_{a=n+1}^{m} S_{ia}\mathbf{F}^{\mathrm{f}}_a$ and the resultant of all torques is $\mathbf{M}_i = \sum_{a=n+1}^{m} \mathbf{C}_{ia} \times \mathbf{F}^{\mathrm{f}}_a$. Define $\underline{\mathbf{F}}^{\mathrm{f}} = [\mathbf{F}^{\mathrm{f}}_{n+1} \; \cdots \; \mathbf{F}^{\mathrm{f}}_m]^T$. With this matrix the column matrices of resultants needed in (5.99) are $\underline{\mathbf{F}} = \underline{S}_{\mathrm{c}}\underline{\mathbf{F}}^{\mathrm{f}}$ and $\underline{\mathbf{M}} = \underline{\mathbf{C}}_{\mathrm{c}} \times \underline{\mathbf{F}}^{\mathrm{f}}$. Thus, the expression (5.99) becomes

$$\left(\underline{\mathbf{a}}_1^T \underline{S}_{\mathrm{c}} + \underline{\mathbf{a}}_2^T \times \underline{\mathbf{C}}_{\mathrm{c}}\right) \cdot \underline{\mathbf{F}}^{\mathrm{f}} . \tag{5.103}$$

For expressing the force $\mathbf{F}^{\mathrm{f}}_a$ the vector $\mathbf{z}_a$ is required. The appropriate tool for this purpose is the $(n \times m)$ circuit matrix $\underline{U}$ defined in (5.15). Each chord $a = n+1, \ldots, m$ creates a circuit (see Fig. 5.7a). The closure of the circuit formed in Fig. 5.16 by the vector $\mathbf{z}_a$ and by a chain of vectors located on bodies is expressed in the form

$$\mathbf{z}_a + \sum_{b=1}^{m} U_{ab}\left(\mathbf{c}_{i^+(b),b} - \mathbf{c}_{i^-(b),b}\right) = \mathbf{0} \qquad (a = n+1, \ldots, m) . \tag{5.104}$$

Let $\underline{\mathbf{D}}$ be the column matrix with elements $\mathbf{D}_b = \mathbf{c}_{i^+(b),b} - \mathbf{c}_{i^-(b),b}$ $(b = 1, \ldots, m)$. Then, the column matrix $\underline{\mathbf{z}} = [\mathbf{z}_{n+1} \; \cdots \; \mathbf{z}_m]^T$ is the product

$$\underline{\mathbf{z}} = -\underline{U}\,\underline{\mathbf{D}} . \tag{5.105}$$

The column matrix $\underline{\mathbf{D}}$ is the transpose of the sum of all rows of the weighted incidence matrix $\begin{bmatrix} \underline{\mathbf{C}}_{0t} & \underline{\mathbf{C}}_{0c} \\ \underline{\mathbf{C}}_{t} & \underline{\mathbf{C}}_{c} \end{bmatrix}$.

Problem 5.10. Modify (5.100) to be valid for systems with Hooke's joints and with torsional springs in both axes of the central cross-shaped body. Consider the case of nonorthogonal axes.

Problem 5.11. Develop (5.105) for $\underline{\mathbf{z}}$ from the equation $\mathbf{z}_a = (\mathbf{c}_{i^-(a),a} - \mathbf{c}_{i^+(a),a}) - (\mathbf{r}_{i^+(a)} - \mathbf{r}_{i^-(a)})$ $(a = n+1, \ldots, m)$ in combination with the expression (5.74) for $\underline{\mathbf{r}}$.

5.5.6 Constraint Forces and Torques in Joints

Kinematical constraints in joints are caused by rigid-body contacts either at individual points or along lines or surfaces. Constraint forces are, therefore, distributed forces. Since the contacting lines and surfaces are assumed rigid, it is impossible to determine the distribution of constraint forces. All one can do is to define for each joint an equivalent force system which consists of a single force and a single torque. The torque depends upon the choice of the point at which the single force is thought to be acting. It is natural to choose for each joint a the articulation point fixed on body $i^-(a)$. Let $\mathbf{X}_a$ and $\mathbf{Y}_a$ be the constraint force and the constraint torque, respectively, thus defined for joint a. More precisely, $+\mathbf{X}_a$ and $+\mathbf{Y}_a$ are acting on body $i^+(a)$ and $-\mathbf{X}_a$ and $-\mathbf{Y}_a$ on body $i^-(a)$. The incidence matrix allows the statement: Body i (arbitrary) is subject to $S_{ia}\mathbf{X}_a$ and to $S_{ia}\mathbf{Y}_a$. The resultant of all constraint forces and the resultant of all constraint torques acting on body i are

$$\sum_{a=1}^{n} S_{ia}\mathbf{X}_a \qquad \text{and} \qquad \sum_{a=1}^{n} S_{ia}(\mathbf{c}_{ia}\times\mathbf{X}_a+\mathbf{Y}_a) \qquad (i=1,\ldots,n)\,, \tag{5.106}$$

respectively. With these expressions Newtons's and Euler's equations for isolated bodies have the forms

$$\left.\begin{aligned} m_i\ddot{\mathbf{r}}_i &= \mathbf{F}_i+\sum_{a=1}^{m} S_{ia}\mathbf{X}_a\,,\\ \mathsf{J}_i\cdot\dot{\boldsymbol{\omega}}_i+\boldsymbol{\omega}_i\times\mathsf{J}_i\cdot\boldsymbol{\omega}_i &= \mathbf{M}_i+\sum_{a=1}^{m}\mathbf{C}_{ia}\times\mathbf{X}_a+\sum_{a=1}^{m} S_{ia}\mathbf{Y}_a \end{aligned}\right\} \tag{5.107}$$

$(i=1,\ldots,n)$. In matrix form these equations read

$$\underline{m}\,\underline{\ddot{\mathbf{r}}}=\underline{\mathbf{F}}+\underline{S}\,\underline{\mathbf{X}}\,,\qquad \underline{\mathsf{J}}\cdot\underline{\dot{\boldsymbol{\omega}}}=\underline{\mathbf{M}}^*+\underline{\mathbf{C}}\times\underline{\mathbf{X}}+\underline{S}\,\underline{\mathbf{Y}} \tag{5.108}$$

with column matrices $\underline{\mathbf{X}}=[\mathbf{X}_1\ \ldots\ \mathbf{X}_n]^T$ and $\underline{\mathbf{Y}}=[\mathbf{Y}_1\ \ldots\ \mathbf{Y}_n]^T$. The other quantities are defined as in (5.30). Premultiplication with the path matrix $\underline{T}$ yields for the constraint forces and torques the explicit expressions

$$\underline{\mathbf{X}}=\underline{T}(\underline{m}\,\underline{\ddot{\mathbf{r}}}-\underline{\mathbf{F}})\,,\qquad \underline{\mathbf{Y}}=\underline{T}(\underline{\mathsf{J}}\cdot\underline{\dot{\boldsymbol{\omega}}}-\underline{\mathbf{M}}^*-\underline{\mathbf{C}}\times\underline{\mathbf{X}})\,. \tag{5.109}$$

The expression for $\underline{\mathbf{X}}$ is substituted into the equation for $\underline{\mathbf{Y}}$. In the course of numerical integration of the equations of motion the terms on the right-hand side of these equations are known as functions of time.

The constraint force $\mathbf{X}_a$ in joint a satisfies the orthogonality conditions

$$\mathbf{X}_a\cdot\mathbf{k}_{a\ell}=0 \qquad (\ell=1,\ldots,f_a)\,. \tag{5.110}$$

In revolute joints and in Hooke's joints the constraint torque $\mathbf{Y}_a$ satisfies the analogous orthogonality conditions

$$\mathbf{Y}_a\cdot\mathbf{p}_{a\ell}=0 \qquad (\ell=1,\ldots,f_a) \tag{5.111}$$

if the articulation point is located on the axis of the revolute joint (on both axes of the Hooke's joint). For the definitions of $\mathbf{k}_{a\ell}$ and $\mathbf{p}_{a\ell}$ see (5.52), (5.54).

5.5.7 Software Tools

The compact expressions for the matrices $\underline{\mathbf{a}}_1$, $\underline{\mathbf{b}}_1$, $\underline{\mathbf{a}}_2$ and $\underline{\mathbf{b}}_2$ and for the matrices $\underline{A}$ and $\underline{B}$ in the equations of motion are easily programmable. The change from one multibody system to another is possible by a simple change of input data. Input data are

- number n of bodies and of joints
- integer functions $i^+(a)$ and $i^-(a)$ $(a = 1, \ldots, n)$
- body masses m_i and inertia tensors J_i $(i = 1, \ldots, n)$
- forces $\mathbf{F}_i$ and torques $\mathbf{M}_i$ $(i = 1, \ldots, n)$ acting on bodies
- for each joint $a = 1, \ldots, n$
  - matrix $\underline{A}_a$
  - vectors $\mathbf{c}_{i^-(a),a}$ and $\mathbf{c}_{i^+(a),a}$
  - vectors in the matrices $\underline{\mathbf{k}}_a$ and $\underline{\mathbf{p}}_a$
  - vectors $\mathbf{s}_a$ and $\mathbf{w}_a$.

Up to this point not a single vector or tensor has been decomposed in any reference frame. The execution of vector products in terms of scalar coordinates is left to the computer. The only scalar coordinates visible to the user of a program are the coordinates of the vectors and tensors listed as input data. For each of these vectors the coordinates are known in a reference base fixed on a particular body. To give an example, the vector $\mathbf{c}_{i^-(a),a}$ has constant coordinates in base $\underline{\mathbf{e}}^{i^-(a)}$ fixed on body $i^-(a)$. Input data are these coordinates and, in addition, the label $i^-(a)$ of the base. In this way the input data of every vector and of every tensor consists of coordinates together with the label of the base in which the coordinates are given. The functions $i^+(a)$ and $i^-(a)$ contain the information how to transform vector and tensor coordinates via a chain of joints and with the help of the matrices $\underline{A}_a$ from given reference frames into other reference frames. Transformations are carried out automatically.

The matrix notation of mathematical expressions has the advantage of being compact and easily interpretable. When programming such expressions a different strategy should be followed. In many expressions the path matrix $\underline{T}$ appears in the form $\underline{T}^T$ multiplied by some column matrix. Examples are (5.63) and (5.74):

$$\underline{\boldsymbol{\omega}} = \boldsymbol{\omega}_0 \underline{1} - \underline{T}^T \underline{\boldsymbol{\Omega}}\,, \qquad \underline{\mathbf{r}} = \mathbf{r}_0 \underline{1} - (\underline{\mathbf{C}}_0 \underline{T})^T - (\underline{\mathbf{C}}\,\underline{T})^T \underline{1}\,. \tag{5.112}$$

The factor $\underline{T}^T$ is an indication that the elements of the product are calculated recursively starting with body 0 and terminating at terminal bodies of the system. For the elements of $\underline{\boldsymbol{\omega}}$ and $\underline{\mathbf{r}}$ the recursive equations are (see Fig. 5.9): $\boldsymbol{\omega}_{i^+(a)} = \boldsymbol{\omega}_{i^-(a)} - \boldsymbol{\Omega}_a$ $(a = 1, \ldots, n)$, $\mathbf{r}_{i^+(a)} = \mathbf{r}_{i^-(a)} + \mathbf{c}_{i^-(a),a} - \mathbf{c}_{i^+(a),a}$ $(a = 1, \ldots, n)$. For an efficient evaluation it is essential that the graph is regularly labeled and regularly directed. Then, the recursion formulas are $\boldsymbol{\omega}_i = \boldsymbol{\omega}_{i^-(i)} - \boldsymbol{\Omega}_i$, $\mathbf{r}_i = \mathbf{r}_{i^-(i)} + \mathbf{c}_{i^-(i),i} - \mathbf{c}_{ii}$ $(i = 1, \ldots, n)$. The recursion starts with $i = 1$, $\boldsymbol{\omega}_{i^-(1)} = \boldsymbol{\omega}_0(t)$ and $\mathbf{r}_{i^-(1)} = \mathbf{r}_0(t)$. Note: The labeling of bodies and of joints chosen by the user of the program need not be regular. The conversion to regular labeling is done by the program with the algorithm described in Sect. 5.3.3. Provided efficient recursive programming techniques are used the computation time required for calculating the matrices $\underline{A}$ and $\underline{B}$ in the equations of motion $\underline{A}\,\ddot{\underline{q}} = \underline{B}$ for an n-body system is of the order O(n) (Keppler [35]).

The first software tool based on the formalism was a FORTRAN program written by the author in 1975 for Daimler-Benz AG for simulating the dynamics of a human dummy in car accidents (passenger inside the car or pedestrian outside). The dummy was modeled as system of eleven bodies. A six-degree-of-freedom joint connected one body of the trunk with the car. The neck had three degrees of freedom of rotation and two of translation. The other joints were spherical joints or revolute joints (knees and elbows). The three-dimensional motion of the car (carrier body 0) was prescribed by arbitrary functions of time. Seat and safety belts were treated as force elements with arbitrary geometry and arbitrary material laws. Interaction forces between head and windshield were modeled as functions of windshield deformation. The program calculated as functions of time positions, velocities and accelerations as well as forces acting on the dummy. For many years the program was intensively used for the development of safety belts, airbags and other measures in car safety research.

Based on the same formalism Liu, author of [48], developed in 1985 for the Shanghai Sport Committee a computer program simulating high-jump and other maneuvers of the human body without contact to the ground. The human body was modeled as a system of fifteen rigid bodies. Conservation of total angular momentum was achieved by Baumgarte's method of stabilizing integrals of motion [4]. One of the purposes of the program was to study the effect of mid-air maneuvers by arms and legs on the performance in high-jump.

The formalism is the basis of the software tool **MESA VERDE** (**ME**chanism, **SA**tellite, **VE**hicle, **R**obot **D**ynamics **E**quations) originally created by Wolz [106] in 1985. MESA VERDE is not restricted to systems with tree structure. Its characteristic feature is the generation of symbolic expressions. Input data as well as output data are symbolic expressions (Wittenburg, Wolz, Schmidt [104]). To give an example, the input data for body 2 might have the form

$$\underline{\text{body}}\ 2\ \underline{\text{alias}}\ \text{frame}\,.$$

Here, <u>body</u> and <u>alias</u> are key words, and "frame" is the name given to body 2 by the program user. This input has the effect that for body 2 symbolic expressions for mass, for moments of inertia and for products of inertia are automatically defined (symbols for products of inertia are not defined if an additional key word indicates that principal axes of inertia are used). Input data for joints is equally simple. Suppose, for example, that joint 3 is a revolute joint connecting body 2 ("frame") to body 0 referred to as "inertial space" and that, furthermore, $i^-(3) = 0$. Then, the input data might have the form

$$\underline{\text{joint}}\ 3\ \underline{\text{type}}\ \text{revolute}\ \underline{\text{from}}\ \text{frame}\ \underline{\text{to}}\ \text{inertial space}\ .$$

The key words from to indicate the sense of direction of the arc. The input "revolute" has the effect that symbolic expressions are defined for all quantities associated with a revolute joint. These are the elements of the matrix $\underline{A}_3$, coordinates of the vectors $\mathbf{c}_{03}$ and $\mathbf{c}_{23}$ and the coordinates of the single vector $\mathbf{p}_3$ along the joint axis. These expressions are provided by an expandable *joint library* which is part of MESA VERDE. For each type of joint such as "revolute" the associated symbolic expressions are automatically provided.

In terms of the symbolic expressions representing input data the program produces symbolic expressions for the elements of the matrices $\underline{A}$ and $\underline{B}$ in the equations of motion $\underline{A}\,\ddot{\underline{q}} = \underline{B}$ together with symbolic expressions for various kinematical quantities and for forces of interest. These expressions are stored for subsequent numerical evaluations in other software packages. The availability of symbolic expressions is important for several reasons. The first reason is that only nonzero expressions have to be evaluated. If products of inertia are known to be zero then no symbolic expressions are defined. If a joint is known to be a revolute joint then no symbolic expressions are defined for quantities which occur only in more sophisticated joints. These aspects of saving computation time played a key-role in the 1980s and in the early 1990s, when computers were still comparatively slow and expensive. Today, the availability of symbolic expressions in C/C++ code is of prime importance for other reasons. First, symbolic expressions are open in the sense that they can be integrated into other software tools with graphical user interfaces, with 3D-visualization and with other post-processing capabilities. Second, symbolic expressions can be compiled in real-time operating systems such as REAL-TIME Linux, for example. The technology called *Hardware-in-the-Loop* (HiL) allows real-time simulations of systems one component of which is the mathematical model of a multibody system while other components are physical hardware.

Collaboration with *IPG Automotive, Karlsruhe* opened the way to the development of professional engineering software tools for a large range of applications in the automotive industry. MESA VERDE-generated kinematics and dynamics equations for vehicles form the backbone of IPG's *CarMaker®* product range which has become a powerful tool for vehicle dynamics analysis and HiL-testing of vehicle electronic control systems. CarMaker is, in turn, the basis of *AVL InMotion®* which is used for developing and testing engines and powertrains. In these HiL applications engines or entire powertrains (consisting of engine, gearbox, differential and halfshafts) are the hardware part while the rest of the vehicle is modelled mathematically. MESA VERDE-generated equations are used in the software tool *FADYNA* developed by IPG for Daimler-Chrysler [107]. It is a comprehensive library of vehicle simulation models covering the entire range of Daimler-Chrysler commercial vehicles ranging from vans to 40-ton semi-trailer trains. MESA VERDE is also used by Renault, PSA Peugeot Citroen and Opel.

5.6 Systems with Closed Kinematic Chains

A system with closed kinematic chains has a graph of joint connections which does not have tree structure. From Sect. 5.2 it is known that the joint variables in the joints forming a closed kinematic chain are subject to constraint equations. This has the effect, that the number of independent joint variables is smaller than the total number of joint variables. The goal of this section is the formulation of equations of motion for a set of independent joint variables. This is achieved in two steps. In the first step the system is converted into a system with tree structure. For this system equations of motion are formulated by the method described in the previous sections. In the second step constraint equations for joint variables are formulated. These constraint equations are then incorporated into the formalism for the system with tree structure. The conversion of a system with closed kinematic chains into a system with tree structure can be achieved in different ways. In what follows two methods referred to as *removal of joints* and *duplication of bodies* are explained.

5.6.1 Removal of Joints. Holonomic Constraints

The system under consideration is converted into a system with tree structure by removing suitably selected joints. The resulting system is referred to as spanning tree of the original system. For the spanning tree joint variables $\underline{q}$ are chosen as usual. In terms of these variables the matrices $\underline{A}$ and $\underline{B}$ in (5.83) are formulated. The original system does not have the equations of motion $\underline{A}\ddot{\underline{q}} = \underline{B}$. However, the principle of virtual power in the form (5.45) is valid:

$$\delta\dot{\underline{q}}^T(\underline{A}\ddot{\underline{q}} - \underline{B}) = 0\,. \tag{5.113}$$

The variations $\delta\dot{\underline{q}}$ are subject to the kinematical constraints caused by the removed joints. Depending on the nature of these joints kinematical constraints are either holonomic or nonholonomic. The present section is devoted to holonomic constraints. Nonholonomic constraints are the subject of Sect. 5.6.4. Let N be the number of variables represented in $\underline{q}$. Then, a holonomic constraint equation has the general form $f(q_1, \ldots, q_N, t) = 0$ where f is some function. In general, more than a single constraint equation exists. Let ν be the total number of *independent* constraint equations:

$$f_i(q_1, \ldots, q_N, t) = 0 \qquad (i = 1, \ldots, \nu)\,. \tag{5.114}$$

If time t appears explicitly then the constraint is called holonomic-rheonomic, otherwise holonomic-skleronomic. Among the N variables $N-\nu$ are independent and the remaining ν are dependent. Which ones are chosen as independent is, in some cases, dictated by the equations. Often, it is a matter of taste. In a 1-d.o.f. mechanism, for example, one might choose the input variable as independent variable. In many cases the equations are too complex for an explicit solution for the dependent variables in terms of the independent variables.

Implicit total time differentiation of the constraint equations produces the equations $\dot{f}_i = 0$ and $\ddot{f}_i = 0$ $(i = 1, \ldots, \nu)$. These are the equations

$$\sum_{j=1}^{N} \frac{\partial f_i}{\partial q_j} \dot{q}_j + \frac{\partial f_i}{\partial t} = 0 \qquad (i = 1, \ldots, \nu)\,, \tag{5.115}$$

$$\sum_{j=1}^{N} \frac{\partial f_i}{\partial q_j} \ddot{q}_j + \sum_{j=1}^{N} \sum_{k=1}^{N} \frac{\partial^2 f_i}{\partial q_j \partial q_k} \dot{q}_j \dot{q}_k + 2 \sum_{j=1}^{N} \frac{\partial^2 f_i}{\partial q_j \partial t} \dot{q}_j + \frac{\partial^2 f_i}{\partial t^2} = 0 \tag{5.116}$$

$(i = 1, \ldots, \nu)$. The various partial derivatives are themselves functions of $q_1, \ldots, q_N$ and of t. In the case of skleronomic constraints the second term in (5.115) and the last two terms in (5.116) are missing.

The first set of equations is linear with respect to $\dot{q}_j$ and the second is linear with respect to $\ddot{q}_j$ $(j = 1, \ldots, N)$. Both sets are solved for the ν dependent quantities in terms of the $N - \nu$ independent quantities. The result of this procedure are expressions of the form

$$\underline{\dot{q}} = \underline{G}\,\underline{\dot{q}}^* + \underline{Q}\,, \qquad \delta\underline{\dot{q}} = \underline{G}\,\delta\underline{\dot{q}}^*\,, \qquad \underline{\ddot{q}} = \underline{G}\,\underline{\ddot{q}}^* + \underline{H}\,. \tag{5.117}$$

Here, $\underline{q}$ is the column matrix of all N variables, and $\underline{q}^*$ is the column matrix of the $N - \nu$ independent variables. $\underline{G}$ is an $[N \times (N - \nu)]$-matrix which depends explicitly on $q_1, \ldots, q_N$ and on t. Also the column matrix $\underline{Q}$ depends explicitly on $q_1, \ldots, q_N$ and on t. The column matrix $\underline{H}$ depends, in addition, also on $\dot{q}_1, \ldots, \dot{q}_N$. In the case of skleronomic constraints $\underline{Q}$ is zero. Since each element of $\underline{q}^*$ is identical with one element of $\underline{q}$ $N-\nu$ of the N equations are identities. Suppose, for example, that $q^*_\ell = q_k$. Then $Q_k = 0$, $H_k = 0$, and the kth row of $\underline{G}$ is $\underline{1}_\ell = [0\;0\;\ldots\;1\;\ldots\;0]$ with the element 1 in the position ℓ.

Substitution of (5.117) into (5.113) results in the equation

$$\delta\underline{\dot{q}}^{*T}(\underline{A}^*\underline{\ddot{q}}^* - \underline{B}^*) = 0 \tag{5.118}$$

with

$$\underline{A}^* = \underline{G}^T \underline{A}\,\underline{G}\,, \qquad \underline{B}^* = \underline{G}^T(\underline{B} - \underline{A}\,\underline{H})\,. \tag{5.119}$$

The independence of the variations $\delta\underline{\dot{q}}^*$ has the consequence that

$$\underline{A}^*\underline{\ddot{q}}^* = \underline{B}^*. \tag{5.120}$$

This is the desired minimal set of equations of motion. The matrices $\underline{A}^*$ and $\underline{B}^*$ depend on all variables $q_1, \ldots, q_N$. In the course of numerical integration the matrices $\underline{A}^*$ and $\underline{B}^*$ must be evaluated at every time step. If the constraint Eqs. (5.114) cannot be solved explicitly for the dependent variables then they must be solved numerically. For two reasons this is not a very time-consuming task. First, the Jacobian $\underline{G}$ is available so that a Newton–Raphson method can be applied. Second, the solution for the previous time step is always a good approximation for the actual solution. Only the determination of initial values for $t = 0$ is more difficult.

In Sects. 5.6.6.1–5.6.6.3 the matrices $\underline{G}$ and $\underline{H}$ are formulated for a planar system and for two spatial systems. The spatial systems demonstrate that the formulation of constraint equations can be a difficult task[6]. Even more difficult is sometimes the proof that constraint equations are independent. If no analytical proof is available then the number of independent equations must be determined numerically. It equals the rank of the Jacobian with elements $\partial f_i/\partial q_j$ $(i = 1, \ldots, \nu;\ j = 1, \ldots, N)$.

General-purpose program packages for multibody system dynamics are not accepted by the engineering community if the user is required to formulate the matrices $\underline{G}$, $\underline{Q}$ and $\underline{H}$ himself. What is needed is an algorithm which generates the matrices automatically from kinematical expressions for a tree-structured system. The algorithm described in the following section meets these requirements. It was first published by Lilov/Chirikov [46].

5.6.2 Duplication of Bodies

The method is explained for a system with a single closed kinematical chain. The generalization to several closed kinematical chains is obvious. The first step is the creation of a tree-structured system. In the previous section this has been done by the removal of a single joint. Now it is done by the duplication of a single body. Let this be the triangular body in Fig. 5.17a. As is shown in Fig. 5.17b it is replaced by two identically shaped twin-bodies labeled, say, k and ℓ. The masses and the inertia tensors of the two bodies

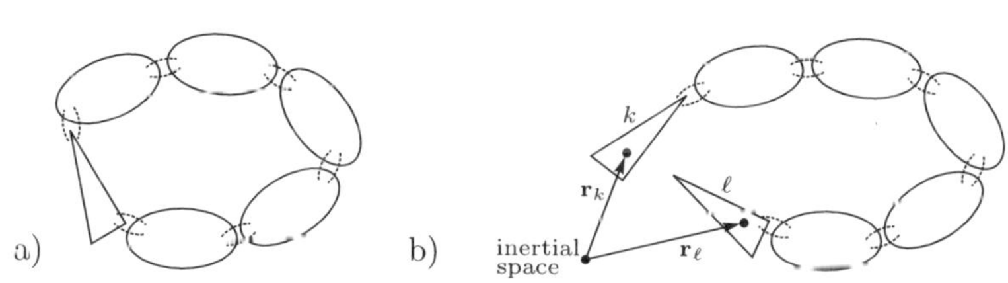

Fig. 5.17. Duplication of the triangular body in system **(a)** results in system **(b)**

[6] For a systematic formulation of constraint equations for closed kinematic chains with seven variables in revolute, prismatic and cylindrical joints see Woernle [105], Li [42], Wittenburg [100, 102].

must add up to the mass and to the inertia tensor, respectively, of the original body. The original body has two joints which belong to the closed kinematic chain under consideration. One of these joints is eliminated on body k and the other is eliminated on body ℓ. The bodies k and ℓ are allowed to drift apart (see Fig. 5.17b). This results in a tree-structured system with one additional body and with the original set of joints. Constraint equations express conditions that the bodies do not drift apart. These conditions are formulated on the level of joint variables and on the level of accelerations. Level of joint variables: The two body centers of mass must coincide. This is the condition $\mathbf{r}_k = \mathbf{r}_\ell$. Both vectors are elements of the matrix $\underline{\mathbf{r}}$ in (5.74):

$$\underline{\mathbf{r}} = \mathbf{r}_0\underline{1} - (\underline{\mathbf{C}}_0\underline{T})^T - (\underline{\mathbf{C}}\,\underline{T})^T\underline{1}\,. \tag{5.121}$$

Both vectors are functions of joint variables. The vector equation yields three scalar equations.

Not only the two body centers of mass must coincide. In addition, the angular orientation of the bodies k and ℓ must be identical. This is a condition on the direction cosine matrices $\underline{A}_a$ defined in (5.53). More precisely speaking, the ordered product of the matrices $\underline{A}_a$ (or $\underline{A}_a^T$ depending on the sense of direction of the joints) of all joints in the closed chain must be the unit matrix. Of the nine equations represented by this condition only three need be considered. Level of accelerations: The two bodies must satisfy the conditions $\ddot{\mathbf{r}}_k = \ddot{\mathbf{r}}_\ell$ and $\dot{\boldsymbol{\omega}}_k = \dot{\boldsymbol{\omega}}_\ell$. The vectors in these equations are elements of column matrices in (5.41) and (5.43):

$$\underline{\ddot{\mathbf{r}}} = \underline{\mathbf{a}}_1\underline{\ddot{q}} + \underline{\mathbf{b}}_1\,, \qquad \underline{\dot{\boldsymbol{\omega}}} = \underline{\mathbf{a}}_2\underline{\ddot{q}} + \underline{\mathbf{b}}_2\,. \tag{5.122}$$

The two vector equations establish six inhomogeneous linear equations for $\underline{\ddot{q}}$. In the course of numerical integration of the equations of motion these equations are generated numerically. The rank of the coefficient matrix is determined. This rank determines the number of independent acceleration variables. Once these variables have been identified numerical values for the matrices $\underline{G}$ and $\underline{H}$ are obtained. For more details see Wolz [106] and Weber [89].

5.6.3 Controlled Joint Variables

This section deals with a particularly simple, yet important rheonomic constraint. A single angle of rotation q_k in an individual revolute joint of a system is prescribed as function of time $q_k(t)$. Then, also $\dot{q}_k$ and $\ddot{q}_k$ are prescribed functions of time. Typical examples of interest are the functions $\dot{q}_k =$ const and $\dot{q}_k = A\cos\omega t$ with $\omega =$ const. For producing the function a motor mounted on the axis of the joint must produce an appropriate torque M_k which acts with opposite signs on the two bodies coupled by the joint. With q_k being prescribed the degree of freedom of the entire system is reduced by one. To be formulated is a system of differential equations of motion for the reduced set of $N-1$ variables. In addition, an expression is to be formulated for the motor torque M_k. In what follows this problem is, first, solved for systems with tree structure. The equations of motion for the complete set of variables $q_1, \ldots, q_N$ including q_k have the form $\underline{A}\,\ddot{\underline{q}} = \underline{B}$. The unknown motor torque M_k appears in the matrix $\underline{B}$ and in this matrix in the expression $-\underline{\mathbf{p}} \cdot \underline{T}\,\underline{\mathbf{M}}$. Let b be the label of the joint with the variable q_k and with the axial unit vector $\mathbf{p}_k$. Let, furthermore, $+M_k\mathbf{p}_k$ be the torque on body $i^+(b)$ and $-M_k\mathbf{p}_k$ the torque on body $i^-(b)$. In the column matrix $\underline{T}\,\underline{\mathbf{M}}$ the element b is the only nonzero element, and this element is $+M_k\mathbf{p}_k$. In column b of the matrix $\underline{\mathbf{p}}$ the kth element is the only nonzero element, and this element is $\mathbf{p}_k$. From this it follows that the kth element of $-\underline{\mathbf{p}} \cdot \underline{T}\,\underline{\mathbf{M}}$ is $-M_k$ and that all other elements are free of M_k. Writing the kth element of the matrix $\underline{B}$ in the form $B_k = B'_k - M_k$ the equations of motion are

$$\begin{bmatrix} A_{11} & \ldots & A_{1k} & \ldots & A_{1N} \\ \vdots & & & & \\ A_{k1} & \ldots & A_{kk} & \ldots & A_{kN} \\ \vdots & & & & \\ A_{N1} & \ldots & A_{Nk} & \ldots & A_{NN} \end{bmatrix} \begin{bmatrix} \ddot{q}_1 \\ \vdots \\ \ddot{q}_k(t) \\ \vdots \\ \ddot{q}_N \end{bmatrix} = \begin{bmatrix} B_1 \\ \vdots \\ B'_k - M_k \\ \vdots \\ B_N \end{bmatrix} . \tag{5.123}$$

The kth equation is solved for the unknown motor torque M_k:

$$M_k = B'_k - [\, A_{k1} \ldots A_{kk} \ldots A_{kN} \,]\ddot{\underline{q}} \,. \tag{5.124}$$

In the remaining $N-1$ equations the terms involving $\ddot{q}_k(t)$ are shifted to the right-hand side. The result is the desired reduced set of differential equations:

$$\begin{bmatrix} A_{11} & \ldots A_{1,k-1} & A_{1,k+1} \ldots & A_{1N} \\ \vdots & & & \\ A_{k-1,1} & & & \\ A_{k+1,1} & & & \\ \vdots & & & \\ A_{N1} & & & A_{NN} \end{bmatrix} \begin{bmatrix} \ddot{q}_1 \\ \vdots \\ \ddot{q}_{k-1} \\ \ddot{q}_{k+1} \\ \vdots \\ \ddot{q}_N \end{bmatrix} = \begin{bmatrix} B_1 \\ \vdots \\ B_{k-1} \\ B_{k+1} \\ \vdots \\ B_N \end{bmatrix} - \ddot{q}_k(t) \begin{bmatrix} A_{1k} \\ \vdots \\ A_{k-1,k} \\ A_{k+1,k} \\ \vdots \\ A_{Nk} \end{bmatrix} . \tag{5.125}$$

The coefficient matrix is symmetric, again. This matrix is now an explicit function of time since it depends on the variable $q_k(t)$. Once the solution is obtained by numerical integration the motor torque $M_k(t)$ is calculated from (5.124). The method just described can be generalized to the case with more than one variable prescribed as function of time. In the extreme all N variables are prescribed. Then, the left-hand side of (5.123) is prescribed and the right hand side has the form $\underline{B}' - \underline{M}$ with the column matrix $\underline{M}$ of motor torques.

In what follows the same problem is treated when the equations of motion have the form (5.120):

$$(\underline{G}^T \underline{A}\,\underline{G})\ddot{\underline{q}}^{\,*} = \underline{G}^T(\underline{B} - \underline{A}\,\underline{H}) \,. \tag{5.126}$$

The elements of $\underline{q}^*$ are the independent variables of a larger set $\underline{q}$. The two sets are related through (5.117): $\ddot{\underline{q}} = \underline{G}\,\ddot{\underline{q}}^* + \underline{H}$. As before, the kth variable q_k of the set $\underline{q}$ is prescribed as function of time and the associated torque M_k is unknown. It is assumed that q_k is one of the variables $\underline{q}^*$ and that it is the ℓth of these variables. Following (5.117) it has been explained that then the kth row of $\underline{G}$ is $\underline{1}_\ell = [0\ 0\ \ldots\ 1\ \ldots\ 0]$ with the element 1 in the position ℓ. The kth element of $\underline{B}$ is $B'_k - M_k$ as before. Hence, the matrix $\underline{B}$ is identical with the one on the right-hand side of (5.123). It is now written in the form $\underline{B} = \underline{B}' - M_k \underline{1}_k^T$. It follows that $\underline{G}^T\underline{B} = \underline{G}^T\underline{B}' - M_k\underline{1}_\ell^T$. Thus, (5.126) becomes

$$(\underline{G}^T\underline{A}\,\underline{G})\ddot{\underline{q}}^* = \underline{G}^T(\underline{B}' - \underline{A}\,\underline{H}) - M_k\underline{1}_\ell^T\,. \tag{5.127}$$

Compare this with (5.123). On the left-hand side $\ddot{q}_\ell^*(t)$ is a prescribed function of time. On the right-hand side the unknown M_k occurs in the ℓth equation only. This ℓth equation is solved for M_k. The result is an equation of the form (5.124). In the remaining equations the terms involving $\ddot{q}_\ell^*(t)$ are shifted to the right-hand side. This results in a reduced set of differential equations for the remaining variables which has the structure of (5.125).

In what follows the transition from (5.123) to (5.125) is made in a different way. Equation (5.123) resulted originally from the principle of virtual power (5.29) in combination with the kinematical relationships (5.41) and (5.43). The condition that an individual variable q_k is a prescribed function of time is now incorporated into (5.41) and (5.43). In (5.41) $\dot{q}_k(t)$ and $\ddot{q}_k(t)$ are multiplied by the kth column of $\underline{\mathbf{a}}_1$. The three equations are written in the form (note that $\delta\dot{q}_k = 0$)

$$\dot{\underline{\mathbf{r}}} = \underline{\mathbf{a}}_{1\mathrm{r}}\dot{\underline{q}}_\mathrm{r} + \dot{q}_k(t)\underline{\mathbf{a}}_1^k + \underline{\mathbf{a}}_{10}\,,\quad \delta\dot{\underline{\mathbf{r}}} = \underline{\mathbf{a}}_{1\mathrm{r}}\delta\dot{\underline{q}}_\mathrm{r}\,,\quad \ddot{\underline{\mathbf{r}}} = \underline{\mathbf{a}}_{1\mathrm{r}}\ddot{\underline{q}}_\mathrm{r} + \ddot{q}_k(t)\underline{\mathbf{a}}_1^k + \underline{\mathbf{b}}_1 \tag{5.128}$$

where $\underline{\mathbf{a}}_1^k$ denotes the kth column of $\underline{\mathbf{a}}_1$, $\underline{\mathbf{a}}_{1\mathrm{r}}$ the reduced matrix $\underline{\mathbf{a}}_1$ (reduced by its kth column) and $\underline{q}_\mathrm{r}$ the reduced set of variables (reduced by q_k). Similarly, (5.43) is written in the form

$$\underline{\boldsymbol{\omega}} = \underline{\mathbf{a}}_{2\mathrm{r}}\dot{\underline{q}}_\mathrm{r} + \dot{q}_k(t)\underline{\mathbf{a}}_2^k + \underline{\mathbf{a}}_{20}\,,\quad \delta\underline{\boldsymbol{\omega}} = \underline{\mathbf{a}}_{2\mathrm{r}}\delta\dot{\underline{q}}_\mathrm{r}\,,\quad \dot{\underline{\boldsymbol{\omega}}} = \underline{\mathbf{a}}_{2\mathrm{r}}\ddot{\underline{q}}_\mathrm{r} + \ddot{q}_k(t)\underline{\mathbf{a}}_2^k + \underline{\mathbf{b}}_2\,. \tag{5.129}$$

Substitution of these expressions into (5.29) yields instead of (5.44) the equation

$$\begin{aligned}\delta\dot{\underline{q}}_\mathrm{r}^T\Big\{ &\big(\underline{\mathbf{a}}_{1\mathrm{r}}^T\cdot\underline{m}\,\underline{\mathbf{a}}_{1\mathrm{r}} + \underline{\mathbf{a}}_{2\mathrm{r}}^T\cdot\underline{\mathbf{J}}\cdot\underline{\mathbf{a}}_{2\mathrm{r}}\big)\ddot{\underline{q}}_\mathrm{r} - \big[\underline{\mathbf{a}}_{1\mathrm{r}}^T\cdot(\underline{\mathbf{F}} - \underline{m}\,\underline{\mathbf{b}}_1) + \underline{\mathbf{a}}_{2\mathrm{r}}^T\cdot(\underline{\mathbf{M}}^* - \underline{\mathbf{J}}\cdot\underline{\mathbf{b}}_2)\\ &-\ddot{q}_k(t)\big(\underline{\mathbf{a}}_{1\mathrm{r}}^T\cdot\underline{m}\,\underline{\mathbf{a}}_1^k + \underline{\mathbf{a}}_{2\mathrm{r}}^T\cdot\underline{\mathbf{J}}\cdot\underline{\mathbf{a}}_2^k\big)\big]\Big\} = 0\,.\end{aligned} \tag{5.130}$$

The expression in curled brackets is zero. This is seen to be the reduced set of Eqs. (5.125).

5.6.4 Nonholonomic Constraints

Nonholonomic constraint equations for a set of variables $q_1, \ldots, q_N$ have the Pfaffian form

$$g_i(\underline{q}, \dot{\underline{q}}, t) = a_{i1}\dot{q}_1 + \ldots + a_{iN}\dot{q}_N + a_{i0} = 0 \qquad (i = 1, \ldots, \mu) . \qquad (5.131)$$

The coefficients $a_{i1}, \ldots, a_{iN}, a_{i0}$ are functions of $q_1, \ldots, q_N$ and, possibly, of time t. If t does not appear explicitly, then the constraints are called nonholonomic-skleronomic, otherwise nonholonomic-rheonomic. In the skleronomic case the term a_{i0} is missing. The coefficients are such that it is impossible to reduce the constraint equations by integration to the form $f_i(q_1, \ldots, q_N, t) = 0$ $(i = 1, \ldots, \mu)$ of holonomic constraint equations. Nonholonomic constraint equations do not reduce the degree of freedom of a system. The constraints are placed on velocities only. In what follows the general case is investigated that a tree-structured system governed by the equation

$$\delta\dot{\underline{q}}^T(\underline{A}\ddot{\underline{q}} - \underline{B}) = 0 \qquad (5.132)$$

is subject to μ nonholonomic constraint Eqs. (5.131) and, in addition, to ν holonomic constraint equations $f_i(q_1, \ldots, q_N, t) = 0$ $(i = 1, \ldots, \nu)$. These holonomic constrained equations are treated as before. Their first and second time-derivatives resulted in two sets of ν linear equations for the velocities $\dot{q}_j$ and the accelerations $\ddot{q}_j$ (see (5.115) and (5.116)). The nonholonomic constraint equations have already the form of linear equations for velocities. They are differentiated only once in order to produce linear equations for accelerations. Combining both types of constraints, one has $\nu + \mu$ linear equations for velocities and $\nu + \mu$ linear equations for accelerations. These equations lead again to equations of the forms (see (5.117))

$$\dot{\underline{q}} = \underline{G}\,\dot{\underline{q}}^* + \underline{Q} , \qquad \delta\dot{\underline{q}} = \underline{G}\,\delta\dot{\underline{q}}^* , \qquad \ddot{\underline{q}} = \underline{G}\,\ddot{\underline{q}}^* + \underline{H} . \qquad (5.133)$$

As before, $\dot{\underline{q}}$ is the column matrix of all N generalized velocities, and $\dot{\underline{q}}^*$ is the column matrix of $N-(\nu+\mu)$ independent generalized velocities. Substitution into (5.132) results again in an equation of the form (5.120):

$$(\underline{G}^T\underline{A}\,\underline{G})\ddot{\underline{q}}^* = \underline{G}^T(\underline{B} - \underline{A}\,\underline{H}) . \qquad (5.134)$$

This is a set of $N - (\nu + \mu)$ second-order differential equations. However, the total degree of freedom of the system is $N - \nu$. Hence, another μ first-order differential equations are required. These are the nonholonomic constraint Eqs. (5.131). The two sets of equations are integrated simultaneously in order to determine all position variables. In Sect. 5.6.6.4 the matrices $\underline{G}$, $\underline{Q}$ and $\underline{H}$ are formulated for a system with six nonholonomic constraint equations.

5.6.5 Constraint Forces and Torques in Joints

In a system with closed kinematic chains with bodies $i = 0, \ldots, n$ and with joints $a = 1, \ldots, m$ $(m > n)$ each constraint force $\mathbf{X}_a$ and each constraint torque $\mathbf{Y}_a$ $(a = 1, \dots, m)$ satisfies the orthogonality conditions (5.110) and (5.111), respectively (the latter ones under the conditions specified there). For the constraint forces and torques of a spanning tree constructed either by the removal of joints or by duplication of bodies Eqs. (5.109) are valid. The entire system of constraint forces and torques is indeterminate if the entire system of equations does not have full rank. This is illustrated by the planar fourbar shown in Fig. 5.18. Constraint torques in three out of the four revolute joints suffice to keep the system in plane motion. The constraint torque in the fourth revolute joint is unnecessary and, therefore, indeterminate. In a good engineering design the bearing in one joint is self-aligning so that no constraint torque can be produced. The indeterminacy of internal reactions in systems with closed kinematic chains is familiar from statics. The only difference in dynamics is the presence of inertia forces in addition to external forces.

5.6.6 Illustrative Examples

In what follows constraint equations and the associated matrices $\underline{G}$, $\underline{Q}$ and $\underline{H}$ in (5.117) and (5.133) are formulated for four multibody systems with closed kinematic chains.

5.6.6.1 Planar Fourbar

The closed kinematic chain to be analyzed is the planar fourbar shown in Fig. 5.18. Its link lengths are ℓ, r, a and b. Removal of the joint B creates a spanning tree having the form of a triple pendulum. As joint variables the angles $q_1 = \varphi$, $q_2 = \alpha$ and $q_3 = \beta$ are chosen. The set of constraint Eqs. (5.114) expressing the restitution of the removed joint reads

$$\left.\begin{aligned} r\sin\varphi + a\sin(\varphi+\alpha) + b\sin(\varphi+\alpha+\beta) - \ell &= 0\,,\\ r\cos\varphi + a\cos(\varphi+\alpha) + b\cos(\varphi+\alpha+\beta) \quad &= 0\,. \end{aligned}\right\} \tag{5.135}$$

These equations are independent. Hence, there are two dependent variables. As independent variable the input crank angle φ is chosen. Differentiation

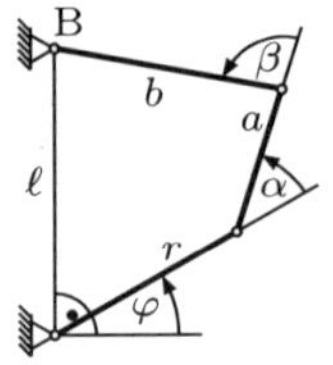

Fig. 5.18. Planar fourbar

with respect to time produces the set of Eqs. (5.115):

$$\left.\begin{aligned}\dot\alpha[a\cos(\varphi+\alpha)+b\cos(\varphi+\alpha+\beta)]+\dot\beta b\cos(\varphi+\alpha+\beta)&=0\,,\\ \dot\alpha[a\sin(\varphi+\alpha)+b\sin(\varphi+\alpha+\beta)]+\dot\beta b\sin(\varphi+\alpha+\beta)&=-\ell\dot\varphi\,.\end{aligned}\right\}\quad(5.136)$$

The coefficient determinant is $ab\sin\beta$. The solutions are

$$\dot\alpha=\frac{\ell\cos(\varphi+\alpha+\beta)}{a\sin\beta}\,\dot\varphi\,,\qquad \dot\beta=\frac{-\ell[a\cos(\varphi+\alpha)+b\cos(\varphi+\alpha+\beta)]}{ab\sin\beta}\,\dot\varphi\,.\quad(5.137)$$

In positions with $\sin\beta=0$ (lengths a and b collinear) both angular velocities are indeterminate. The first Eq. (5.117) has the form

$$\begin{bmatrix}\dot\varphi\\ \dot\alpha\\ \dot\beta\end{bmatrix}=\begin{bmatrix}1\\ \dfrac{\ell\cos(\varphi+\alpha+\beta)}{a\sin\beta}\\ \dfrac{-\ell[a\cos(\varphi+\alpha)+b\cos(\varphi+\alpha+\beta)]}{ab\sin\beta}\end{bmatrix}\dot\varphi\,.\quad(5.138)$$

The formulation of the matrix $\underline{H}$ in (5.117) is left to the reader.

The nonlinear Eqs. (5.135) can be resolved explicitly for α and β. For this purpose they are rewritten in the form

$$\left.\begin{aligned}b\sin(\varphi+\alpha)\cos\beta+b\cos(\varphi+\alpha)\sin\beta&=\ell-r\sin\varphi-a\sin(\varphi+\alpha)\,,\\ b\cos(\varphi+\alpha)\cos\beta-b\sin(\varphi+\alpha)\sin\beta&=-r\cos\varphi-a\cos(\varphi+\alpha)\,.\end{aligned}\right\}\quad(5.139)$$

This yields

$$\cos\beta=\frac{1}{b}\,[\ell\sin(\varphi+\alpha)-r\cos\alpha-a]\,,\quad \sin\beta=\frac{1}{b}\,[\ell\cos(\varphi+\alpha)+r\sin\alpha]\,.\quad(5.140)$$

The condition $\cos^2\beta+\sin^2\beta=1$ is the equation $A\cos\alpha+B\sin\alpha=C$ with $A=2a(\ell\sin\varphi-r)$, $B=2a\ell\cos\varphi$, $C=\ell^2+r^2+a^2-b^2-2r\ell\sin\varphi$. This equation determines for each value of φ two (not necessarily real) solutions for $\cos\alpha$ and for $\sin\alpha$:

$$\left.\begin{aligned}\cos\alpha_k&=\left[AC+(-1)^kB\sqrt{A^2+B^2-C^2}\right]/(A^2+B^2)\,,\\ \sin\alpha_k&=\left[BC-(-1)^kA\sqrt{A^2+B^2-C^2}\right]/(A^2+B^2)\end{aligned}\right\}\ (k=1,2).\quad(5.141)$$

With each solution α_k (5.140) determines the associated values $\cos\beta$ and $\sin\beta$.

5.6.6.2 Orthogonal Bricard Mechanism

The spatial mechanism shown in Fig. 5.19 is a special case of a family known as orthogonal Bricard mechanisms (see Bricard [10]). It is composed of six

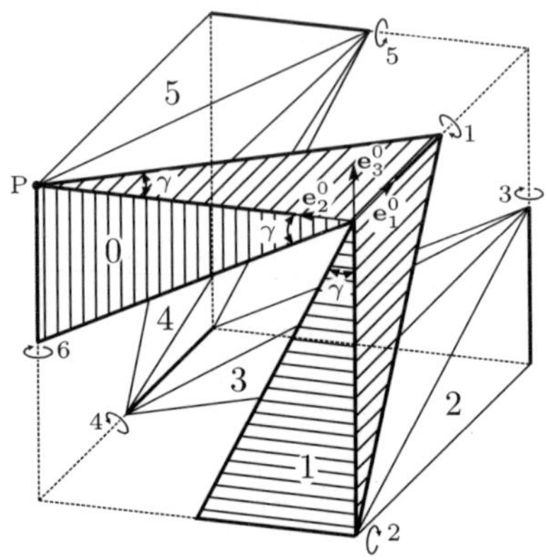

Fig. 5.19. Orthogonal Bricard mechanism

bodies labeled $0, \dots, 5$. Body 0 is fixed in inertial space. The bodies are connected by six revolute joints labeled $1, \dots, 6$. The two joint axes of each body are mutually orthogonal. Bodies 0, 2 and 4 are identical and bodies 1, 3 and 5 are identical. Furthermore, body 1 is a mirror image of body 0. In the position shown the bodies are inscribed in a cube. This cube configuration is symmetric with respect to the plane spanned by the joint axes 1 and 4. It is also symmetric with respect to the plane spanned by the joint axes 3 and 5. Finally, it is centrally symmetric with respect to the center of the cube. The joint axes 1, 3 and 5 on the one hand and the joint axes 2, 4 and 6, on the other intersect in a single point. On each joint axis a torsional spring and a torsional damper are mounted whose constant coefficients are denoted k_a and d_a, respectively, $(a = 1, \dots, 6)$. In the cube configuration all springs are unstressed. To be determined are the total degree of freedom N of the mechanism, equations of motion $\underline{A}\,\ddot{q} = \underline{B}$ for a spanning tree and the matrices $\underline{G}$ and $\underline{H}$ in the equations of motion for N independent variables.

Solution: From mere inspection it is not clear whether the system has a degree of freedom $N > 0$ at all. Following the procedure described in Sect. 5.6.1 joint 6 is removed. The result is a system with tree structure (a spanning tree) with the degree of freedom $f = 5$, one in each of the joints 1 to 5. Reconstitution of the removed joint introduces five constraint equations. These must be formulated first. Then, the number ν of independent constraint equations must be determined. Only in the case $\nu < 5$ a total degree of freedom $N = f - \nu > 0$ is obtained.

On each body i $(i = 0, \dots, 5)$ a body-fixed base $\underline{\mathbf{e}}^i$ is defined. In Fig. 5.19 only the base $\underline{\mathbf{e}}^0$ is shown. In the cube configuration all body-fixed bases are aligned parallel. The locations of the origins are without interest. Three of the five constraint equations express the fact that, independent of rotation angles in the joints, the chain of vectors leading from the point P on body 0 along body edges to the coincident point P on body 5 is closed. This is the vector equation

$$-\mathbf{e}_2^0 - \mathbf{e}_3^1 + \mathbf{e}_1^2 + \mathbf{e}_2^3 + \mathbf{e}_3^4 - \mathbf{e}_1^5 = \mathbf{0}\,. \tag{5.142}$$

Two more scalar constraint equations express the fact that the vectors $\mathbf{e}_1^5$ and $\mathbf{e}_2^5$ are both orthogonal to $\mathbf{e}_3^0$:

$$\mathbf{e}_1^5 \cdot \mathbf{e}_3^0 = 0\,, \qquad \mathbf{e}_2^5 \cdot \mathbf{e}_3^0 = 0\,. \tag{5.143}$$

The altogether five scalar constraint equations must be expressed in terms of joint variables. In joint i ($i = 1, \ldots, 5$) let q_i be the rotation angle of body i relative to body $i-1$. Sign convention: q_1, q_2, q_3 are positive and q_4, q_5 are negative in the case of a right-handed rotation about the base vector in the respective joint axis. In the cube configuration all angles are zero. Let $\underline{A}_i$ be the transformation matrix which transforms in joint i from $\underline{\mathbf{e}}^{i-1}$ to $\underline{\mathbf{e}}^i$. With the abbreviations $c_i = \cos q_i$, $s_i = \sin q_i$ these matrices are:

$$\underline{A}_1 = \begin{bmatrix} 1 & 0 & 0 \\ 0 & c_1 & s_1 \\ 0 & -s_1 & c_1 \end{bmatrix}, \quad \underline{A}_2 = \begin{bmatrix} c_2 & 0 & -s_2 \\ 0 & 1 & 0 \\ s_2 & 0 & c_2 \end{bmatrix}, \quad \underline{A}_3 = \begin{bmatrix} c_3 & s_3 & 0 \\ s_3 & c_3 & 0 \\ 0 & 0 & 1 \end{bmatrix},$$
$$\underline{A}_4 = \begin{bmatrix} 1 & 0 & 0 \\ 0 & c_4 & -s_4 \\ 0 & s_4 & c_4 \end{bmatrix}, \quad \underline{A}_5 = \begin{bmatrix} c_5 & 0 & s_5 \\ 0 & 1 & 0 \\ -s_5 & 0 & c_5 \end{bmatrix}. \tag{5.144}$$

The coordinate transformations for (5.142) and (5.143) are simplest if all vectors are transformed either into base $\underline{\mathbf{e}}^2$ or into base $\underline{\mathbf{e}}^3$. Using $\underline{\mathbf{e}}^3$ (5.142) takes the form

$$\underline{A}_3\underline{A}_2\underline{A}_1 \begin{bmatrix} 0 \\ -1 \\ 0 \end{bmatrix} + \underline{A}_3\underline{A}_2 \begin{bmatrix} 0 \\ 0 \\ -1 \end{bmatrix} + \underline{A}_3 \begin{bmatrix} 1 \\ 0 \\ 0 \end{bmatrix} + \begin{bmatrix} 0 \\ 1 \\ 0 \end{bmatrix}$$
$$+\underline{A}_4^T \begin{bmatrix} 0 \\ 0 \\ 1 \end{bmatrix} + \underline{A}_4^T\underline{A}_5^T \begin{bmatrix} -1 \\ 0 \\ 0 \end{bmatrix} = \begin{bmatrix} 0 \\ 0 \\ 0 \end{bmatrix}. \tag{5.145}$$

Multiplying out one gets the equations

$$c_3[1 + s_2(1 - s_1)] - s_3c_1 - c_5 = 0\,, \tag{5.146}$$
$$s_3[1 + s_2(1 - s_1)] + c_3c_1 - s_4(1 - s_5) - 1 = 0\,, \tag{5.147}$$
$$c_2(1 - s_1) \qquad - c_4(1 - s_5) = 0\,. \tag{5.148}$$

Also the scalar products in (5.143) are expressed in terms of vector coordinates in base $\underline{\mathbf{e}}^3$. This yields the equations

$$c_5(-c_3s_2c_1 + s_3s_1) + s_5[s_4(s_3s_2c_1 + c_3s_1) + c_4c_2c_1] = 0\,, \tag{5.149}$$
$$c_4(s_3s_2c_1 + c_3s_1) - s_4c_2c_1 = 0\,. \tag{5.150}$$

From the symmetry properties in Fig. 5.19 it can be predicted that in the original mechanism including joint 6, if it is mobile, the joint axes 1, 3 and 5 on the one hand and the joint axes 2, 4 and 6, on the other intersect in a single point in every position of the mechanism. This is expressed by the three independent constraint equations

$$q_3 = q_1 \,, \qquad q_5 = q_1 \,, \qquad q_4 = q_2 \,. \tag{5.151}$$

With these equations, (5.146)–(5.150) get the simple forms

$$\left.\begin{aligned} c_1(s_1 + s_1 s_2 - s_2) &= 0 \,, \\ (s_1 - 1)(s_1 + s_1 s_2 - s_2) &= 0 \,, \\ 0 &= 0 \,, \\ c_1(1 + s_1 + s_1 s_2)(s_1 + s_1 s_2 - s_2) &= 0 \,, \\ c_1 c_2(s_1 + s_1 s_2 - s_2) &= 0 \,. \end{aligned}\right\} \tag{5.152}$$

These equations are satisfied if $s_1 + s_1 s_2 - s_2 = 0$. This is the equation

$$\sin q_2 = \frac{\sin q_1}{1 - \sin q_1} \,. \tag{5.153}$$

In addition to (5.151) this constitutes a fourth independent constraint equation. There are no other independent constraint equations. From this it follows that the mechanism has the degree of freedom $N = 1$. As independent variable the angle q_1 is chosen.

Remark on (5.151): These constraint equations are found without recourse to intuition as follows. Multiply (5.147) by $c_1 c_4$, (5.148) by $-c_1 s_4$, (5.150) by $-(1 - s_1)$ and add. Simple reformulation followed by division through c_4 results in the equation relating q_3 to q_1:

$$(1 - s_1)c_3 = (1 - s_3)c_1 \,. \tag{5.154}$$

The solutions are $q_3 = q_1$ and $q_3 = \pi/2$. Only the first solution is useful. This is the first Eq. (5.151). Because of the equal character of all bodies and of all joints and because of the definitions of the angles this equation holds true if the indices are increased by 1 and by 2. This yields the other two constraint equations $q_4 = q_2$ and $q_5 = q_3$. End of remark on (5.151).

The relationship (5.153) is illustrated in the diagram of Fig. 5.20. Because of the conditions $|\sin q_{1,2}| \le 1$ the angles are restricted to the intervals $-210° \le q_1 \le +30°$ and $-30° \le q_2 \le +210°$. Motion in these intervals is possible without collision of neighboring bodies if the angle γ shown in Fig. 5.19 is $\gamma \le 30°$. The complicated motion of the mechanism is best understood if a model is available. It can be produced by folding cardboard with glued-in strips of plastic as joints.

Differentiation of (5.153) with respect to time produces the relationship between angular velocities:

$$\dot{q}_2 = \dot{q}_1 \,\frac{\cos q_1}{(1 - \sin q_1)^2 \cos q_2} = \dot{q}_1 \,\frac{\cos q_1}{(1 - \sin q_1)\sqrt{1 - 2\sin q_1}} \,. \tag{5.155}$$

Differentiating one more time one gets for the angular acceleration the expression

$$\ddot{q}_2 = \ddot{q}_1\, g(q_1) + \dot{q}_1^2 h(q_1) \tag{5.156}$$

with

$$g(q_1) = \frac{\cos q_1}{(1-\sin q_1)\sqrt{1-2\sin q_1}}\,, \qquad h(q_1) = \frac{2(1-\sin q_1)-\sin^2 q_1}{(1-\sin q_1)(1-2\sin q_1)^{3/2}}\,. \tag{5.157}$$

This equation in combination with the equations $q_5 = q_3 = q_1$ and $q_4 = q_2$ determines the matrices $\underline{G}$ and $\underline{H}$ in the equation $\underline{\ddot{q}} = [\ddot{q}_1 \; \dots \; \ddot{q}_5]^T = \underline{G}\,\ddot{q}_1 + \underline{H}$. Both matrices are column matrices:

$$\left.\begin{aligned} \underline{G} &= [\,1 \quad g(q_1) \quad 1 \quad g(q_1) \quad 1\,]^T\,, \\ \underline{H} &= [\,0 \quad \dot{q}_1^2 h(q_1) \quad 0 \quad \dot{q}_1^2 h(q_1) \quad 0\,]^T\,. \end{aligned}\right\} \tag{5.158}$$

The closed chain is governed by the single scalar differential equation $\underline{G}^T\underline{A}\,\underline{G}\ddot{q}_1 = \underline{G}^T(\underline{B} - \underline{A}\,\underline{H})$ with $\underline{A}$ and $\underline{B}$ being the matrices in the equations of motion $\underline{A}\,\underline{\ddot{q}} = \underline{B}$ for the spanning tree. These matrices are determined from (5.83) and (5.86). The definitions of the variables $q_1, \dots, q_5$ imply that

$$\underline{\mathbf{p}} = \begin{bmatrix} \mathbf{e}_1^1 & & & & \\ & \mathbf{e}_2^2 & & & \\ & & \mathbf{e}_3^3 & & \\ & & & -\mathbf{e}_1^4 & \\ & & & & -\mathbf{e}_2^5 \end{bmatrix}, \qquad \underline{T} = \begin{bmatrix} -1 & -1 & -1 & -1 & -1 \\ 0 & -1 & -1 & -1 & -1 \\ 0 & 0 & -1 & -1 & -1 \\ 0 & 0 & 0 & -1 & -1 \\ 0 & 0 & 0 & 0 & -1 \end{bmatrix}. \tag{5.159}$$

The springs and dampers in joints $1, \dots 5$ contribute to the right-hand side of the equation the term $-\underline{D}\,\underline{\dot{q}} - \underline{K}\,\underline{q}$ with diagonal (5×5)-matrices $\underline{D}$ and $\underline{K}$ of the damper constants and the spring constants, respectively (see (5.100)). The spring-damper torque in joint 6 represents an external torque $\mathbf{M}_5$ on body 5. Let q_6 be the angle of rotation of body 0 relative to body 5 (positive about the axis $-\mathbf{e}_3^5 = -\mathbf{e}_3^0$). This definition is such that $q_6 = q_4 = q_2$. The torque on body 5 is $\mathbf{M}_5 = (d_6\dot{q}_2 + k_6 q_2)\mathbf{e}_3^5$. It enters the right-hand side of

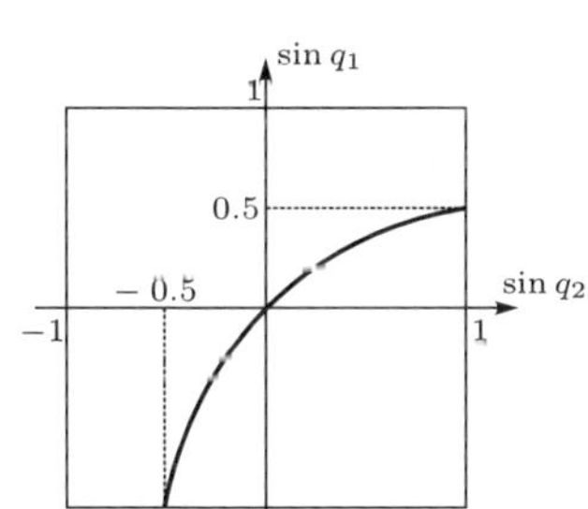

Fig. 5.20. Relationship between q_1 and q_2 in the Bricard mechanism of Fig. 5.19

the equation in the term $-\underline{\mathbf{p}}\,\underline{T}\cdot\underline{\mathbf{M}}$ with $\underline{\mathbf{M}} = [\mathbf{0}\;\;\mathbf{0}\;\;\mathbf{0}\;\;\mathbf{0}\;\;\mathbf{M}_5]^T$. This is the column matrix

$$(d_6\dot{q}_2 + k_6 q_2)\begin{bmatrix}0 & \mathbf{e}_2^2\cdot\mathbf{e}_3^0 & \mathbf{e}_3^3\cdot\mathbf{e}_3^5 & -\mathbf{e}_1^4\cdot\mathbf{e}_3^5 & 0\end{bmatrix}^T \tag{5.160}$$

and with (5.144) and (5.151)

$$(d_6\dot{q}_2 + k_6 q_2)\,[0\quad s_1\quad c_1 c_2\quad s_1\quad 0]^T\,. \tag{5.161}$$

5.6.6.3 Stewart Platform

Figure 5.21 depicts a Stewart platform. The body labeled 1 is the platform proper. Its center of mass C is the origin of the base $\underline{\mathbf{e}}^1$. The platform is supported in inertial space (body 0 with base $\underline{\mathbf{e}}^0$) by six telescopic legs with spherical joints in the articulation points Q_i and P_i ($i = 1,\ldots,6$). The points Q_i have position vectors $\mathbf{R}_i$ in $\underline{\mathbf{e}}^0$, and the points P_i have position vectors $\boldsymbol{\varrho}_i$ in $\underline{\mathbf{e}}^1$. Neither the six points Q_i nor the six points P_i are assumed to be coplanar or otherwise regularly arranged. Due to the speed of operation in combination with compactness and stiffness Stewart platforms find applications whenever objects must be oriented both rapidly and precisely. Examples are grippers of robots, platforms for vehicles in car and flight simulators and for work pieces in heavy-duty machine tools. Mechanical or hydraulic actuators in the legs produce forces which act with equal magnitude and with opposite directions on the two sections of a leg. Within a certain workspace the platform has the degree of freedom six. Each leg is free to rotate about its longitudinal axis. In practice, this freedom is suppressed by replacing in each leg one spherical joint by a Hooke's joint. Here, this problem is eliminated by the assumption that the legs are infinitesimally thin rods with zero moment of inertia about the longitudinal axis and with equal moments of inertia about lateral axes. With this assumption the position of each leg is determined by the position of the platform. Between platform and body 0 a six-degree-of-freedom joint is defined. Joint variables for this joint are chosen as follows. Three of the variables are the coordinates of the position vector $\mathbf{r}$ of the center of mass C in $\underline{\mathbf{e}}^0$. The angular orientation of $\underline{\mathbf{e}}^1$

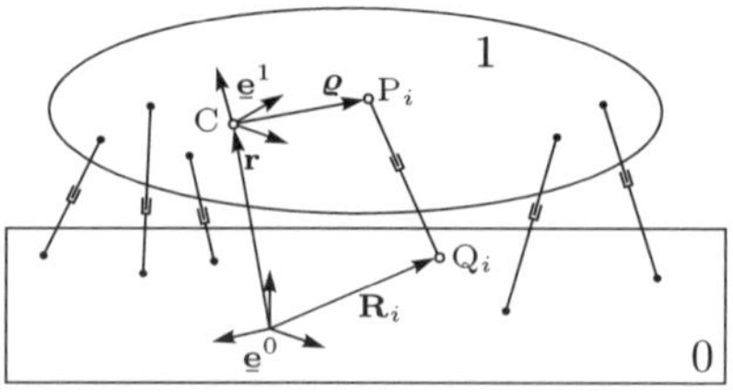

Fig. 5.21. Stewart platform

in $\underline{\mathbf{e}}^0$ is described by Euler–Rodrigues parameters q_0, q_1, q_2, q_3. The direction cosine matrix $\underline{A}^{10}$ relating $\underline{\mathbf{e}}^1$ and $\underline{\mathbf{e}}^0$ is (see (2.35))

$$\underline{A}^{10} = \begin{bmatrix} q_0^2+q_1^2-q_2^2-q_3^2 & 2(q_1q_2+q_0q_3) & 2(q_1q_3-q_0q_2) \\ 2(q_1q_2-q_0q_3) & q_0^2+q_2^2-q_3^2-q_1^2 & 2(q_2q_3+q_0q_1) \\ 2(q_1q_3+q_0q_2) & 2(q_2q_3-q_0q_1) & q_0^2+q_3^2-q_1^2-q_2^2 \end{bmatrix} . \tag{5.162}$$

Let $\boldsymbol{\omega}$ be the angular velocity of the platform and let $\underline{\omega}$ be the column matrix of its coordinates in $\underline{\mathbf{e}}^1$. The parameters q_0, q_1, q_2, q_3 are the solutions of the kinematic differential equations (2.119):

$$\begin{bmatrix} \dot{q}_0 \\ \dot{\underline{q}} \end{bmatrix} = \frac{1}{2} \begin{bmatrix} 0 & -\underline{\omega}^T \\ \underline{\omega} & -\tilde{\underline{\omega}} \end{bmatrix} \begin{bmatrix} q_0 \\ \underline{q} \end{bmatrix} . \tag{5.163}$$

Each leg creates a closed kinematic chain containing the leg, the platform, body 0, the joints of the leg and the six-degree-of-freedom joint between platform and body 0. In what follows the closed kinematic chain created by a single representative leg is investigated. The goal is to express all kinematical quantities of the leg in terms of the variables in the six-degree-of-freedom joint and in terms of derivatives of these variables. In Fig. 5.22 the platform and the single leg are shown together with the bases $\underline{\mathbf{e}}^0$ and $\underline{\mathbf{e}}^1$, with the position vector $\mathbf{r}$ of C and with the position vectors $\mathbf{R}$ and $\boldsymbol{\varrho}$ of the articulation points Q and P, respectively. The position $\mathbf{r}_{\mathrm{P}}$, velocity $\dot{\mathbf{r}}_{\mathrm{P}}$ and acceleration $\ddot{\mathbf{r}}_{\mathrm{P}}$ of P are

$$\mathbf{r}_{\mathrm{P}} = \mathbf{r} + \boldsymbol{\varrho} , \qquad \dot{\mathbf{r}}_{\mathrm{P}} = \dot{\mathbf{r}} + \boldsymbol{\omega} \times \boldsymbol{\varrho} , \qquad \ddot{\mathbf{r}}_{\mathrm{P}} = \ddot{\mathbf{r}} + \dot{\boldsymbol{\omega}} \times \boldsymbol{\varrho} + \boldsymbol{\omega} \times (\boldsymbol{\omega} \times \boldsymbol{\varrho}) . \tag{5.164}$$

The second equation yields

$$\delta\dot{\mathbf{r}}_{\mathrm{P}} = \delta\dot{\mathbf{r}} + \delta\boldsymbol{\omega} \times \boldsymbol{\varrho} . \tag{5.165}$$

The leg is composed of two segments 1 and 2 with centers of mass C_1 and C_2, respectively. The locations of these centers of mass on the two segments are given by the constant lengths ℓ_1 and ℓ_2. The variable leg length is called L, and $\mathbf{e}$ denotes the variable unit vector pointing in the direction from Q to P. Since the leg is treated as infinitesimally thin rod the longitudinal component of its angular velocity can be neglected (if the leg has spherical joints at both ends then this component can actually be zero). Both segments have equal angular velocities perpendicular to the axis $\mathbf{e}$ of the leg. This angular velocity is called $\boldsymbol{\Omega}$. Also the angular accelerations $\dot{\boldsymbol{\Omega}}$ of the two segments are identical and perpendicular to $\mathbf{e}$.

The goal of the kinematics analysis is to express $\boldsymbol{\Omega}$, $\dot{\boldsymbol{\Omega}}$, the position $\mathbf{r}_1$, the velocity $\dot{\mathbf{r}}_1$ and the acceleration $\ddot{\mathbf{r}}_1$ of C_1 in terms of the platform quantities $\boldsymbol{\omega}$, $\dot{\boldsymbol{\omega}}$, $\mathbf{r}$, $\dot{\mathbf{r}}$ and $\ddot{\mathbf{r}}$. The length L and the vector $\mathbf{e}$ depend on the platform position only. Figure 5.22 yields the expressions

$$L = |-\mathbf{R} + \mathbf{r} + \boldsymbol{\varrho}| , \qquad \mathbf{e} = \frac{-\mathbf{R} + \mathbf{r} + \boldsymbol{\varrho}}{L} . \tag{5.166}$$

In terms of these quantities the position and the velocity of P are

$$\mathbf{r}_{\mathrm{P}} = \mathbf{R} + L\mathbf{e}\;, \qquad \dot{\mathbf{r}}_{\mathrm{P}} = \dot{L}\mathbf{e} + L\boldsymbol{\Omega}\times\mathbf{e}\;. \tag{5.167}$$

Dot and cross multiplications of the second equation by $\mathbf{e}$ yield

$$\dot{L} = \mathbf{e}\cdot\dot{\mathbf{r}}_{\mathrm{P}}\;, \qquad \boldsymbol{\Omega} = \frac{1}{L}\mathbf{e}\times\dot{\mathbf{r}}_{\mathrm{P}}\;, \qquad \delta\boldsymbol{\Omega} = \frac{1}{L}\mathbf{e}\times\delta\dot{\mathbf{r}}_{\mathrm{P}} = \frac{1}{L}\mathbf{e}\times(\delta\dot{\mathbf{r}} + \delta\boldsymbol{\omega}\times\boldsymbol{\varrho})\;. \tag{5.168}$$

Time-differentiation of $\boldsymbol{\Omega}$ produces the expression

$$\dot{\boldsymbol{\Omega}} = \frac{1}{L}\left[\mathbf{e}\times\ddot{\mathbf{r}}_{\mathrm{P}} + (\boldsymbol{\Omega}\times\mathbf{e})\times\dot{\mathbf{r}}_{\mathrm{P}}\right] - \frac{\dot{L}}{L^2}\mathbf{e}\times\dot{\mathbf{r}}_{\mathrm{P}} \tag{5.169}$$

and, after substitution of the expressions for $\ddot{\mathbf{r}}_{\mathrm{P}}$, $\dot{L}$ and $\boldsymbol{\Omega}$,

$$\dot{\boldsymbol{\Omega}} = \frac{1}{L}\mathbf{e}\times\left[\ddot{\mathbf{r}} + \dot{\boldsymbol{\omega}}\times\boldsymbol{\varrho} + \boldsymbol{\omega}\times(\boldsymbol{\omega}\times\boldsymbol{\varrho})\right] - \frac{2}{L^2}\mathbf{e}\times\dot{\mathbf{r}}_{\mathrm{P}}(\mathbf{e}\cdot\dot{\mathbf{r}}_{\mathrm{P}})\;. \tag{5.170}$$

From Fig. 5.22 it follows that $\mathbf{r}_1 = \mathbf{r}_{\mathrm{P}} - \ell_1\mathbf{e}$. The second time-derivative is $\ddot{\mathbf{r}}_1 = \ddot{\mathbf{r}}_{\mathrm{P}} - \ell_1(\dot{\boldsymbol{\Omega}}\times\mathbf{e} - \Omega^2\mathbf{e})$ or, with the expressions for $\ddot{\mathbf{r}}_{\mathrm{P}}$, $\dot{\boldsymbol{\Omega}}$ and $\boldsymbol{\Omega}$,

$$\begin{aligned}\ddot{\mathbf{r}}_1 &= \mathsf{T}\cdot\left[\ddot{\mathbf{r}} + \dot{\boldsymbol{\omega}}\times\boldsymbol{\varrho} + \boldsymbol{\omega}\times(\boldsymbol{\omega}\times\boldsymbol{\varrho})\right]\\ &\quad + \frac{\ell_1}{L^2}\left\{2(\mathbf{e}\cdot\dot{\mathbf{r}}_{\mathrm{P}})\dot{\mathbf{r}}_{\mathrm{P}} + \left[\mathbf{v}_{\mathrm{P}}^2 - 3(\mathbf{e}\cdot\dot{\mathbf{r}}_{\mathrm{P}})^2\right]\mathbf{e}\right\}\end{aligned} \tag{5.171}$$

where T is the symmetric tensor

$$\mathsf{T} = (1 - \frac{\ell_1}{L})\mathsf{I} + \frac{\ell_1}{L}\mathbf{e}\mathbf{e}\;. \tag{5.172}$$

From this it follows that

$$\delta\dot{\mathbf{r}}_1 = \mathsf{T}\cdot(\delta\dot{\mathbf{r}} + \delta\boldsymbol{\omega}\times\boldsymbol{\varrho})\;. \tag{5.173}$$

This concludes the kinematics analysis of the leg.

The desired equations of motion are deduced from the principle of virtual power in its original form (5.29):

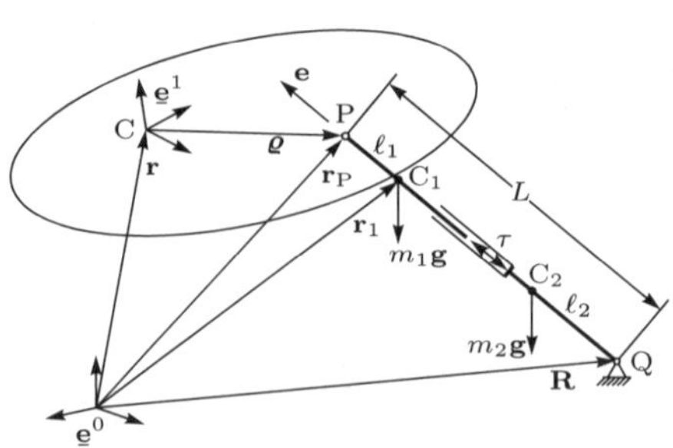

Fig. 5.22. Platform and a single representative leg

$$\begin{aligned}&\delta\dot{\mathbf{r}}\cdot(m\ddot{\mathbf{r}}-\mathbf{F})+\delta\boldsymbol{\omega}\cdot(\mathsf{J}\cdot\dot{\boldsymbol{\omega}}+\boldsymbol{\omega}\times\mathsf{J}\cdot\boldsymbol{\omega}-\mathbf{M})\\&+\sum_{i=1}^{6}\Big[\delta\dot{\mathbf{r}}_{1i}\cdot(m_{1i}\ddot{\mathbf{r}}_{1i}-\mathbf{F}_{1i})+\delta\boldsymbol{\Omega}_i\cdot\Big(\mathsf{J}_{1i}\cdot\dot{\boldsymbol{\Omega}}_i+\boldsymbol{\Omega}_i\times\mathsf{J}_{1i}\cdot\boldsymbol{\Omega}_i-\mathbf{M}_{1i}\Big)\Big]\\&+\sum_{i=1}^{6}\delta\boldsymbol{\Omega}_i\cdot\Big(\mathsf{J}_{2i}\cdot\dot{\boldsymbol{\Omega}}_i+\boldsymbol{\Omega}_i\times\mathsf{J}_{2i}\cdot\boldsymbol{\Omega}_i-\mathbf{M}_{2i}\Big)\ .\end{aligned}\tag{5.174}$$

The first line is the contribution of the platform, the second line the contribution of the segments 1 of the legs and the third the contribution of the segments 2. In the third line use is made of the fact that segment 2 is rotating about a point Q_i fixed in inertial space. Hence, J_{2i} is the inertia tensor with respect to Q_i whereas J_{1i} is the inertia tensor of segment 1 with respect to its center of mass. Both inertia tensors are symmetric with respect to the axis of the leg, and $\boldsymbol{\Omega}_i$ is perpendicular to this axis. From this it follows that $\mathsf{J}_{1i}\cdot\dot{\boldsymbol{\Omega}}_i=J_{1i}\dot{\boldsymbol{\Omega}}_i$ where J_{1i} is the moment of inertia about the lateral axis. Furthermore, $\boldsymbol{\Omega}_i\times\mathsf{J}_{1i}\cdot\boldsymbol{\Omega}_i=0$. The same holds true for segment 2.

Next, the virtual power of forces and torques acting on leg i is formulated. Omitting the index i for a while, this is the expression $\delta\dot{\mathbf{r}}_1\cdot\mathbf{F}+\delta\boldsymbol{\Omega}\cdot\mathbf{M}$. Both segments of the leg are subject to weight and to actuator forces. Let $\tau\mathbf{e}$ be the actuator force acting on segment 1. Its torque about the center of mass C_1 is zero. Its virtual power is $\delta\dot{\mathbf{r}}_1\cdot\tau\mathbf{e}=\tau\delta\dot{\mathbf{r}}_\mathrm{P}\cdot\mathbf{e}$. The virtual power of the force $-\tau\mathbf{e}$ acting on segment 2 is zero. Weight of the two segments is expressed in the forms $-m_1g\mathbf{e}_\mathrm{v}$ and $-m_2g\mathbf{e}_\mathrm{v}$, respectively, where $\mathbf{e}_\mathrm{v}$ is the upward vertical unit vector. The torque of $-m_2g\mathbf{e}_\mathrm{v}$ about Q is $\mathbf{M}=-m_2g\ell_2\mathbf{e}\times\mathbf{e}_\mathrm{v}$. The virtual powers of these forces are $-m_1g\delta\dot{\mathbf{r}}_1\cdot\mathbf{e}_\mathrm{v}=-m_1g\delta\dot{\mathbf{r}}_\mathrm{P}\cdot\mathbf{e}_\mathrm{v}$ and $-m_2g\ell_2\delta\boldsymbol{\Omega}\cdot\mathbf{e}\times\mathbf{e}_\mathrm{v}$, respectively.

The two sums are now combined in a single sum. With the term $-mg\mathbf{e}_\mathrm{v}$ added for weight of the platform (5.174) becomes

$$\begin{aligned}&\delta\dot{\mathbf{r}}\cdot(m\ddot{\mathbf{r}}+mg\mathbf{e}_\mathrm{v}-\mathbf{F})+\delta\boldsymbol{\omega}\cdot(\mathsf{J}\cdot\dot{\boldsymbol{\omega}}+\boldsymbol{\omega}\times\mathsf{J}\cdot\boldsymbol{\omega}-\mathbf{M})\\&+\sum_{i=1}^{6}\Big[\delta\dot{\mathbf{r}}_{1i}\cdot(m_{1i}\ddot{\mathbf{r}}_{1i}+m_{1i}g\mathbf{e}_\mathrm{v})\\&\qquad+\delta\boldsymbol{\Omega}_i\cdot\big[(J_{1i}+J_{2i})\dot{\boldsymbol{\Omega}}_i+m_{2i}g\ell_{2i}\mathbf{e}\times\mathbf{e}_\mathrm{v}\big]-\tau_i\delta\dot{\mathbf{r}}_{\mathrm{P}_i}\cdot\mathbf{e}_i\Big]\ .\end{aligned}\tag{5.175}$$

This is re-arranged such that the terms with highest-order derivatives are separated from the rest:

$$\begin{aligned}&m\,\delta\dot{\mathbf{r}}\cdot\ddot{\mathbf{r}}+\delta\boldsymbol{\omega}\cdot\mathsf{J}\cdot\dot{\boldsymbol{\omega}}+\sum_{i=1}^{6}[m_{1i}\,\delta\dot{\mathbf{r}}_{1i}\cdot\ddot{\mathbf{r}}_{1i}+(J_{1i}+J_{2i})\,\delta\boldsymbol{\Omega}_i\cdot\dot{\boldsymbol{\Omega}}_i]\\&+\delta\dot{\mathbf{r}}\cdot(mg\,\mathbf{e}_\mathrm{v}-\mathbf{F})+\delta\boldsymbol{\omega}\cdot(\boldsymbol{\omega}\times\mathsf{J}\cdot\boldsymbol{\omega}-\mathbf{M})\\&+\sum_{i=1}^{6}[m_{1i}g\,\delta\dot{\mathbf{r}}_{1i}\cdot\mathbf{e}_\mathrm{v}+m_{2i}g\ell_{2i}\,\delta\boldsymbol{\Omega}_i\cdot\mathbf{e}_i\times\mathbf{e}_\mathrm{v}-\tau_i\,\delta\dot{\mathbf{r}}_{\mathrm{P}_i}\cdot\mathbf{e}_i]\ .\end{aligned}\tag{5.176}$$

In what follows only terms with highest-order derivatives and terms with actuator forces are considered. For $\delta\dot{\mathbf{r}}_{\mathrm{P}_i}$, $\delta\dot{\mathbf{r}}_{1i}$, $\ddot{\mathbf{r}}_{1i}$, $\delta\boldsymbol{\Omega}_i$ and $\dot{\boldsymbol{\Omega}}_i$ the expressions in (5.165), (5.173), (5.171), (5.168) and (5.170) are substituted:

$$\left.\begin{aligned}
\delta\dot{\mathbf{r}}_{\mathrm{P}_i} &= \delta\dot{\mathbf{r}} + \delta\boldsymbol{\omega}\times\boldsymbol{\varrho}_i\ , \\
\delta\dot{\mathbf{r}}_{1i} &= \mathsf{T}_i\cdot(\delta\dot{\mathbf{r}} + \delta\boldsymbol{\omega}\times\boldsymbol{\varrho}_i)\ , \qquad & \delta\boldsymbol{\Omega}_i &= \frac{1}{L_i}\,\mathbf{e}_i\times(\delta\dot{\mathbf{r}} + \delta\boldsymbol{\omega}\times\boldsymbol{\varrho}_i)\ , \\
\ddot{\mathbf{r}}_{1i} &= \mathsf{T}_i\cdot(\ddot{\mathbf{r}} + \dot{\boldsymbol{\omega}}\times\boldsymbol{\varrho}_i) + \ldots\ , & \dot{\boldsymbol{\Omega}}_i &= \frac{1}{L_i}\,\mathbf{e}_i\times(\ddot{\mathbf{r}} + \dot{\boldsymbol{\omega}}\times\boldsymbol{\varrho}_i) + \ldots\ .
\end{aligned}\right\}\tag{5.177}$$

Dots indicate terms with lower-order derivatives.

The desired equations of motion are equations for the coordinates of $\mathbf{r}$ in base $\underline{\mathbf{e}}^0$ and for the coordinates of $\boldsymbol{\omega}$ in $\underline{\mathbf{e}}^1$. Let $\underline{r}$ and $\underline{\omega}$ be the column matrices of these coordinates. Since both bases are involved the transformation matrix (5.162) and with it the Euler–Rodrigues parameters q_0, q_1, q_2, q_3 come into play. The vector $\boldsymbol{\varrho}_i$ and the inertia tensor J of the platform have in $\underline{\mathbf{e}}^1$ constant coordinate matrices $\underline{\varrho}_i$ and $\underline{J}$, respectively. The matrix $\underline{J}$ is the diagonal matrix of principal moments of inertia if the base vectors of $\underline{\mathbf{e}}^1$ are principal axes of inertia. The vector $\boldsymbol{\omega}\times\boldsymbol{\varrho}$ has in $\underline{\mathbf{e}}^1$ the coordinate matrix $-\underline{\tilde{\varrho}}_i\,\underline{\omega}$. The coordinate matrix in $\underline{\mathbf{e}}^0$ is $-\underline{A}^{01}\underline{\tilde{\varrho}}_i\,\underline{\omega}$. Let, furthermore, $\underline{e}_i$ and $\underline{T}_i$ be the coordinate matrices of $\mathbf{e}_i$ and of T_i in $\underline{\mathbf{e}}^0$. With this notation the vectors (5.177) have in $\underline{\mathbf{e}}^0$ the coordinate matrices

$$\left.\begin{aligned}
\delta\dot{\underline{r}}_{\mathrm{P}_i} &= \delta\dot{\underline{r}} - \underline{A}^{01}\underline{\tilde{\varrho}}_i\delta\underline{\omega}\ , \\
\delta\dot{\underline{r}}_{1i} &= \underline{T}_i\left(\delta\dot{\underline{r}} - \underline{A}^{01}\underline{\tilde{\varrho}}_i\delta\underline{\omega}\right)\ , \qquad & \delta\underline{\Omega}_i &= \frac{1}{L_i}\,\underline{\tilde{e}}_i\left(\delta\dot{\underline{r}} - \underline{A}^{01}\underline{\tilde{\varrho}}_i\delta\underline{\omega}\right)\ , \\
\ddot{\underline{r}}_{1i} &= \underline{T}_i\left(\ddot{\underline{r}} - \underline{A}^{01}\underline{\tilde{\varrho}}_i\dot{\underline{\omega}}\right) + \ldots\ , & \dot{\underline{\Omega}}_i &= \frac{1}{L_i}\,\underline{\tilde{e}}_i\left(\ddot{\underline{r}} - \underline{A}^{01}\underline{\tilde{\varrho}}_i\dot{\underline{\omega}}\right) + \ldots\ .
\end{aligned}\right\}\tag{5.178}$$

For (5.175) the following products are formed.

$$\left.\begin{aligned}
\delta\dot{\mathbf{r}}\cdot\ddot{\mathbf{r}} &= \delta\dot{\underline{r}}^T\ddot{\underline{r}}\ , \qquad \delta\boldsymbol{\omega}\cdot\mathsf{J}\cdot\dot{\boldsymbol{\omega}} = \delta\underline{\omega}^T\underline{J}\,\dot{\underline{\omega}}\ , \\
\delta\dot{\mathbf{r}}_{1i}\cdot\ddot{\mathbf{r}}_{1i} &= \delta\dot{\underline{r}}_{1i}^T\ddot{\underline{r}}_{1i} = \left(\delta\dot{\underline{r}}^T + \delta\underline{\omega}^T\underline{\tilde{\varrho}}_i\underline{A}^{10}\right)\underline{T}_i\,\underline{T}_i\left(\ddot{\underline{r}} - \underline{A}^{01}\underline{\tilde{\varrho}}_i\dot{\underline{\omega}}\right) + \ldots\ , \\
\delta\boldsymbol{\Omega}_i\cdot\dot{\boldsymbol{\Omega}}_i &= \delta\underline{\Omega}^T\dot{\underline{\Omega}} = -\frac{1}{L_i^2}\left(\delta\dot{\underline{r}}^T + \delta\underline{\omega}^T\underline{\tilde{\varrho}}_i\underline{A}^{10}\right)\underline{\tilde{e}}_i\,\underline{\tilde{e}}_i\left(\ddot{\underline{r}} - \underline{A}^{01}\underline{\tilde{\varrho}}_i\dot{\underline{\omega}}\right) + \ldots\ , \\
\delta\dot{\mathbf{r}}_{\mathrm{P}_i}\cdot\mathbf{e}_i &= \left(\delta\dot{\underline{r}}^T + \delta\underline{\omega}^T\underline{\tilde{\varrho}}_i\underline{A}^{10}\right)\underline{e}_i\ .
\end{aligned}\right\}\tag{5.179}$$

Substitution of these expressions into (5.175) results in an equation of the form $\delta\dot{\underline{r}}^T(\ldots)+\delta\underline{\omega}^T(\ldots)=0$. The expressions in brackets are both zero. This yields the desired equations of motion. They have the form

$$\begin{bmatrix}\underline{A} & \underline{B}\\ \underline{B}^T & \underline{D}\end{bmatrix}\begin{bmatrix}\ddot{\underline{r}}\\ \dot{\underline{\omega}}\end{bmatrix} = \underline{K}\,\underline{\tau} + \ldots \tag{5.180}$$

with a symmetric coefficient matrix. Each of the (3×3)-matrices $\underline{A}$, $\underline{B}$ and $\underline{D}$ contains the matrices

$$\underline{M}_i = m_{1i}\,\underline{T}_i\,\underline{T}_i - \frac{J_{1i}+J_{2i}}{L_i^2}\,\underline{\tilde{e}}_i\,\underline{\tilde{e}}_i \qquad (i=1,\ldots,6)\,. \tag{5.181}$$

With (5.172) for $\underline{T}_i$ and with (1.41) for $\underline{\tilde{e}}_i\,\underline{\tilde{e}}_i$ this is

$$\underline{M}_i = \left[m_{1i}\left(1-\frac{\ell_{1i}}{L_i}\right)+\frac{J_{1i}+J_{2i}}{L_i^2}\right]\underline{I} + \left(m_{1i}\frac{\ell_{1i}}{L_i}-\frac{J_{1i}+J_{2i}}{L_i^2}\right)\underline{e}_i\,\underline{e}_i^T\,. \tag{5.182}$$

The matrices $\underline{A}$, $\underline{B}$ and $\underline{D}$ are

$$\underline{A} = m\underline{I}+\sum_{i=1}^{6}\underline{M}_i\,,\quad \underline{B} = -\sum_{i=1}^{6}\underline{M}_i\underline{A}^{01}\underline{\tilde{\varrho}}_i\,,\quad \underline{D} = \underline{J}-\sum_{i=1}^{6}\underline{\tilde{\varrho}}_i\underline{A}^{10}\underline{M}_i\underline{A}^{01}\underline{\tilde{\varrho}}_i\,. \tag{5.183}$$

In the (6×6) coefficient matrix $\underline{K}$ of $\underline{\tau} = [\tau_1 \ldots \tau_6]^T$ column i is composed of $\underline{e}_i$ placed on top of $\underline{\tilde{\varrho}}_i\underline{A}^{10}\underline{e}_i$ $(i = 1,\ldots,6)$. The matrices $\underline{A}$, $\underline{B}$, $\underline{D}$ and $\underline{K}$ depend on the variables $r_1, r_2, r_3, q_0, q_1, q_2, q_3$. It is left to the reader to formulate from (5.176) the terms indicated by dots. Equations (5.180) are numerically integrated together with (5.163).

5.6.6.4 Table on Wheels

Formulate equations of motion for the three-legged table shown in Fig. 5.23a. At the end of each leg a wheel is held in a vertical position by means of a cage which is free to rotate about a vertical axis fixed in the leg (Fig. 5.23b). It is assumed that the wheels are in contact with ground at all times and that they are rolling without slipping. The inertia of cages and wheels is to be taken into account. The legs are identical, and the main body has its center of mass on the axis of symmetry.

Solution: The rolling condition introduces nonholonomic constraints. First, a system with tree structure is constructed which differs from the given system only in that the nonholonomic constraints are missing. A schematic view is shown in Fig. 5.24 together with its directed graph. Joint 1 between ground (body 0) and the main body 1 is a three-degree-of-freedom joint. The remaining six joints are revolute joints. Body-fixed vector bases and joint variables are explained in Fig. 5.25 which shows, in vertical projection, body 1 with its center of mass C_1 and bodies 2 and 3. On bodies i = 0, 1, 2, 4 and 6 the body-fixed base vector $\mathbf{e}_3^i$ is the upward vertical (in what follows it is denoted $\mathbf{e}_3$). The base vectors $\mathbf{e}_1^2 = \mathbf{e}_1^3$ are directed along the wheel axis. The vector $\mathbf{R}$ from C_1 to the axis of body 2 in the leg lies under a constant angle γ against $\mathbf{e}_1^1$. The angle q_3 is one of the three variables in joint 1. It is the angle of rotation of body 0 relative to body 1 (see the sense of direction of arc 1 in the directed graph of Fig. 5.24). The other two variables in this joint, denoted q_1 and q_2, position the origin C_0 of $\underline{\mathbf{e}}^0$ (not shown) in base $\underline{\mathbf{e}}^1$. This means that C_1 has in $\underline{\mathbf{e}}^0$ the position vector $-q_1\mathbf{e}_1^1 - q_2\mathbf{e}_2^1$. The angular variables in joints 2 and 3 are denoted q_4 (cage 2 relative to table 1) and q_5 (wheel 3 relative to cage 2). For the other two cage-and-wheel units not shown in the figure the angles corresponding to (γ, q_4, q_5) are $(\gamma - 120°, q_6, q_7)$ and $(\gamma + 120°, q_8, q_9)$, respectively. The angles q_4, q_6 and q_8 are zero when the cages are pointing radially away from C_1. For the variables $q_1, \ldots, q_9$ equations of motion $\underline{A}\ddot{\underline{q}} = \underline{B}$ are formulated according to Sect. 5.5.3.

Next, the nonholonomic constraint equations expressing rolling conditions are formulated. Let $\dot{\mathbf{z}}$ be the absolute velocity of that point fixed on a wheel which is, instantaneously, in contact with ground. Rolling requires that $\dot{\mathbf{z}} = \mathbf{0}$. Since the vector $\dot{\mathbf{z}}$ is in the horizontal plane this yields two scalar equations of the form (5.131) for each wheel. The details are worked out for the wheel 3 shown in Fig. 5.25. With the vectors $\mathbf{R}$ and $\mathbf{a}$ of constant lengths and with the wheel radius r

$$\mathbf{z} = -q_1\mathbf{e}_1^1 - q_2\mathbf{e}_2^1 + \mathbf{R} + \mathbf{a} - r\mathbf{e}_3 \, . \tag{5.184}$$

Differentiation with respect to time yields the absolute velocity

$$\dot{\mathbf{z}} = -\dot{q}_1\mathbf{e}_1^1 - \dot{q}_2\mathbf{e}_2^1 + \dot{q}_3\left[q_1\mathbf{e}_2^1 - q_2\mathbf{e}_1^1 - \mathbf{e}_3 \times (\mathbf{R} + \mathbf{a})\right] + \dot{q}_4\mathbf{e}_3 \times \mathbf{a} - \dot{q}_5 r\mathbf{e}_1^2 \times \mathbf{e}_3 \, . \tag{5.185}$$

Rolling requires that $\dot{\mathbf{z}} \cdot \mathbf{e}_1^2 = 0$ and $\dot{\mathbf{z}} \cdot \mathbf{e}_2^2 = 0$. The scalar forms of these conditions represent the nonholonomic constraint equations (5.131):

$$\left.\begin{aligned} g_1 = {} & -\dot{q}_1 \sin(q_4 + \gamma) + \dot{q}_2 \cos(q_4 + \gamma) \\ & +\dot{q}_3[-q_1 \cos(q_4 + \gamma) - q_2 \sin(q_4 + \gamma) + R\cos q_4 + a] - \dot{q}_4 a = 0 \, , \\ g_2 = {} & -\dot{q}_1 \cos(q_4 + \gamma) - \dot{q}_2 \sin(q_4 + \gamma) \\ & +\dot{q}_3[q_1 \sin(q_4 + \gamma) - q_2 \cos(q_4 + \gamma) - R\sin q_4] + \dot{q}_5 r = 0 \, . \end{aligned}\right\} \tag{5.186}$$

Similar equations are valid for the wheels 5 and 7. They are obtained by replacing (γ, q_4, q_5) by $(\gamma - 120°, q_6, q_7)$ and by $(\gamma + 120°, q_8, q_9(t))$, respectively.

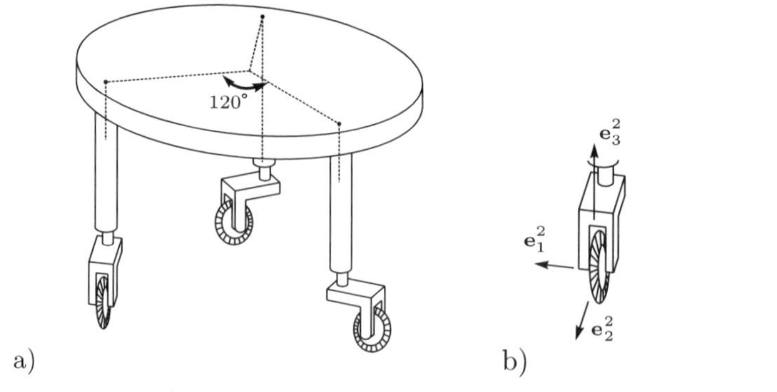

Fig. 5.23. Table on wheels **(a)** and a single cage-and-wheel unit **(b)**

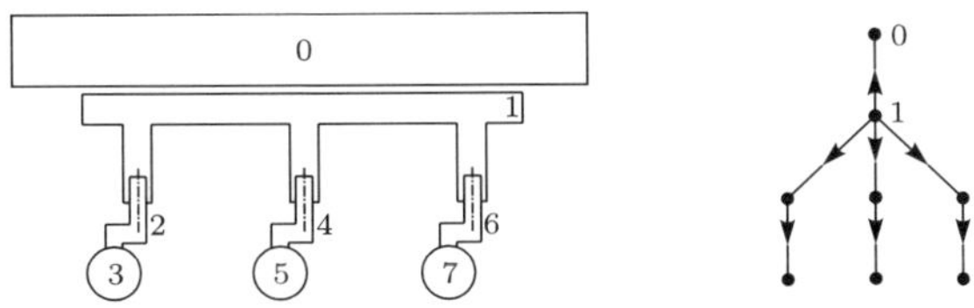

Fig. 5.24. Tree-structured system without nonholonomic constraints and the associated directed graph

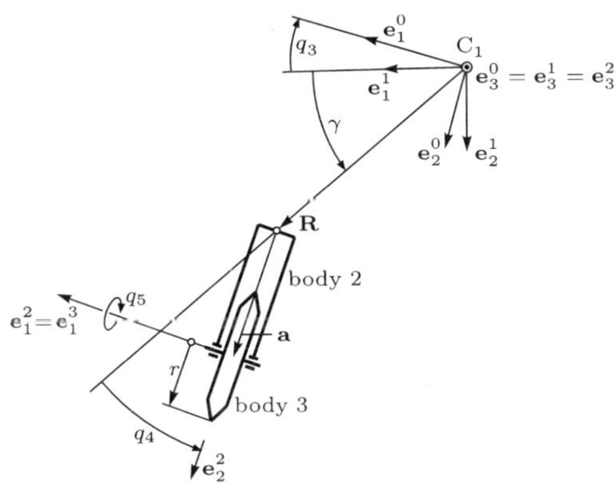

Fig. 5.25. Kinematics of a single cage-and-wheel unit

From the constraint equations the first Eq. (5.133), $\underline{\dot{q}} = \underline{G}\,\underline{\dot{q}}^*$, is formulated. The column matrix on the left-hand side is $\underline{\dot{q}} = [\,\dot{q}_1 \;\; \dot{q}_2 \;\; \dot{q}_3 \;\; \ldots \;\; \dot{q}_9\,]^T$. The constraint Eqs. (5.186) are most easily resolved for $\dot{q}_4$ and $\dot{q}_5$ and, similarly, the other two sets of equations for $\dot{q}_6$, $\dot{q}_7$ and for $\dot{q}_8$, $\dot{q}_9$, respectively. Therefore, it is decided to identify $\underline{\dot{q}}^*$ with $[\,\dot{q}_1 \;\; \dot{q}_2 \;\; \dot{q}_3\,]^T$. The matrix $\underline{G}$ is of size (9×3). The (3×3) submatrix composed of rows 1, 2 and 3 is the unit matrix expressing identities. Rows 4 and 5 are read from (5.186). The elements of these rows are

$$\left.\begin{aligned}
G_{41} &= \tfrac{1}{a}\sin(q_4+\gamma)\,, \qquad G_{42} = -\tfrac{1}{a}\cos(q_4+\gamma)\,,\\
G_{43} &= -\tfrac{1}{a}[q_1\cos(q_4+\gamma) + q_2\sin(q_4+\gamma) - R\cos q_4 - a]\,,\\
G_{51} &= \tfrac{1}{r}\cos(q_4+\gamma)\,, \qquad G_{52} = \tfrac{1}{r}\sin(q_4+\gamma)\,,\\
G_{53} &= -\tfrac{1}{r}[q_1\sin(q_4+\gamma) - q_2\cos(q_4+\gamma) - R\sin q_4]\,.
\end{aligned}\right\} \qquad (5.187)$$

Rows 6 to 9 are obtained by copying rows 4 and 5 with (γ, q_4, q_5) replaced by $(\gamma-120^\circ, q_6, q_7)$ and by $(\gamma+120^\circ, q_8, q_9)$, respectively. It is left to the reader to differentiate the constraint equations with respect to time and to collect the terms forming the column matrix $\underline{H}$. In the final expressions the angle γ can be set equal to zero. With the matrices $\underline{G}$ and $\underline{H}$ thus obtained Eqs. (5.134) are formulated. These equations are integrated numerically together with the six nonholonomic constraint equations.

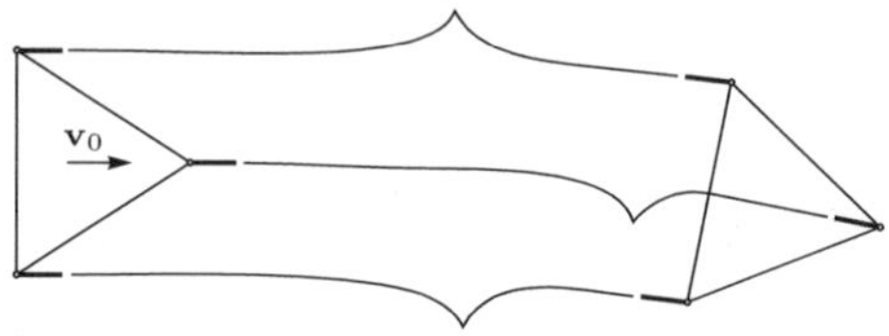

Fig. 5.26. Table in initial position and trajectories of wheel contact points

Suppose that the angular velocity $\dot{q}_9$ of wheel 7 relative to cage 6 is a prescribed function of time $\dot{q}_9 = \Omega(t)$. Then, also q_9 and $\ddot{q}_9$ are prescribed functions of time. These functions of time have to be substituted into the right-hand side of the equations of motion as well as into the nonholonomic constraint equations. In the latter ones the term $r\dot{q}_9 = r\Omega(t)$ represents the term a_{i0} in (5.131).

Figure 5.26 shows results of numerical integration for a system without control on wheel 7. The triangle to the left shows the table in vertical projection in its initial position. The corners of the triangle represent the turning points of the wheel cages (the tip of $\mathbf{R}$ in Fig. 5.23). The tips of the short legs are the wheel contact points on the floor. At time $t = 0$ body 1 is in pure translation with initial velocity $\mathbf{v}_0$. All wheels are pointing forward. Only the front wheel is given an initial angular deviation of 10^{-6} rad. No external forces are acting on the system. For masses and moments of inertia realistic values were chosen. The curved lines represent the trajectories of the wheel contact points. The results are in agreement with experience. The initial angular deviation of the front wheel causes the wheel cages to turn around. In this phase of motion strong nonholonomic constraint forces are acting. The resultant torque causes the table to rotate. Once the wheels are trailing behind the torque is almost zero. Hence, the table continues to rotate.

5.7 Systems with Spherical Joints

Subject of investigation are tree-structured n-body systems all joints of which are spherical joints. The systems are either coupled by a single spherical joint to a moving carrier body 0 or they are in free flight without kinematical constraint to a carrier body. In Fig. 5.27 both cases are illustrated depending on whether body zero and joint 1 exist or not. In engineering, systems with spherical joints are relatively rare. They are studied because equations of motion can be written in such a simple form that it is possible to find analytical solutions for some multibody problems of theoretical interest. In Sects. 5.7.3 and 5.7.4 two such problems will be analyzed in detail.

Equations of motion are developed from the principle of virtual power, again. Starting point are (5.30) and (5.35):

$$\delta\underline{\dot{\mathbf{r}}}^T \cdot (\underline{m}\,\underline{\ddot{\mathbf{r}}} - \underline{\mathbf{F}}) + \delta\underline{\boldsymbol{\omega}}^T \cdot (\underline{\mathsf{J}} \cdot \underline{\dot{\boldsymbol{\omega}}} - \underline{\mathbf{M}}^*) = 0\,, \tag{5.188}$$

$$\delta\underline{\dot{\mathbf{R}}}^T \cdot (\underline{m}\,\underline{\ddot{\mathbf{R}}} - \underline{\mathbf{F}}) + \delta\underline{\boldsymbol{\omega}}^T \cdot (\underline{\mathsf{J}} \cdot \underline{\dot{\boldsymbol{\omega}}} - \underline{\mathbf{M}}^*) = 0\,. \tag{5.189}$$

In both equations $\underline{\mathbf{M}}^*$ is the column matrix with elements

$$\mathbf{M}_i^* = \mathbf{M}_i - \boldsymbol{\omega}_i \times \mathsf{J}_i \cdot \boldsymbol{\omega}_i \qquad (i = 1, \ldots, n)\,. \tag{5.190}$$

The first equation governs motions of systems with or without coupling to inertial space, whereas the second equation governs motions about the composite system center of mass of systems without coupling to inertial space. In the first equation, $\underline{\mathbf{r}}$ is the column matrix of the position vectors $\mathbf{r}_i$ $(i = 1, \ldots, n)$ of the body centers of mass in inertial space. In the second equation, $\underline{\mathbf{R}}$ is the column matrix of the position vectors $\mathbf{R}_i$ $(i = 1, \ldots, n)$ of the body centers of mass measured from the composite system center of mass C (see Fig. 5.27). From (5.90) and (5.91) the expressions are known:

$$\underline{\mathbf{r}} = \mathbf{r}_0\underline{1} - (\underline{\mathbf{C}}\,\underline{T})^T\underline{1}\,, \qquad \underline{\mathbf{R}} = -(\underline{\mathbf{C}}\,\underline{T}\,\underline{\mu})^T\underline{1}\,. \tag{5.191}$$

The elements of the matrix $\underline{\mathbf{C}}\,\underline{T}$ are the vectors $\mathbf{d}_{ij}$ (see (5.93) and Fig. 5.13). The elements of $\underline{\mathbf{C}}\,\underline{T}\,\underline{\mu}$ are the vectors $-\mathbf{b}_{ij}$ on the augmented bodies of the system (see (5.97) and Fig. 5.14). In spherical joints the centers of the joints are chosen as articulation points. This has the consequence that the vectors $\mathbf{d}_{ij}$ and $\mathbf{b}_{ij}$ are fixed on the respective body i (first index). The augmented bodies are rigid bodies. For what follows equations (5.191) are written in the forms

$$\underline{\mathbf{r}} = \mathbf{r}_0\underline{1} - [\mathbf{d}_{ij}]^T\,\underline{1}\,, \tag{5.192}$$

$$\underline{\mathbf{R}} = [\mathbf{b}_{ij}]^T\,\underline{1}\,. \tag{5.193}$$

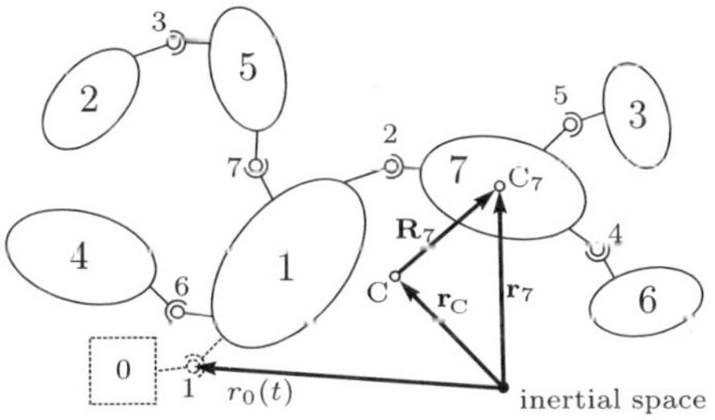

Fig. 5.27. Tree-structured system with spherical joints with or without coupling to a carrier body 0. Composite system center of mass C of the system without body 0

The first time derivatives of these equations are

$$\dot{\underline{\mathbf{r}}} = \dot{\mathbf{r}}_0\underline{1} - [-\mathbf{d}_{ij}\times\boldsymbol{\omega}_i]^T\,\underline{1} = \dot{\mathbf{r}}_0\underline{1} + [\mathbf{d}_{ij}]^T\times\underline{\boldsymbol{\omega}}\,, \tag{5.194}$$

$$\dot{\underline{\mathbf{R}}} = -\,[\mathbf{b}_{ij}]^T\times\underline{\boldsymbol{\omega}}\,. \tag{5.195}$$

From this it follows that

$$\delta\dot{\underline{\mathbf{r}}}^T = -\delta\underline{\boldsymbol{\omega}}^T\times[\mathbf{d}_{ij}]\;, \tag{5.196}$$

$$\delta\dot{\underline{\mathbf{R}}}^T = \quad\delta\underline{\boldsymbol{\omega}}^T\times[\mathbf{b}_{ij}]\;. \tag{5.197}$$

The accelerations are written in the forms

$$\ddot{\underline{\mathbf{r}}} = \ddot{\mathbf{r}}_0\underline{1} - \left[\ddot{\mathbf{d}}_{ij}\right]^T\underline{1}\,, \tag{5.198}$$

$$\ddot{\underline{\mathbf{R}}} = \left[\ddot{\mathbf{b}}_{ij}\right]^T\underline{1}\,. \tag{5.199}$$

Only later the explicit formula $\ddot{\mathbf{d}}_{ij} = \dot{\boldsymbol{\omega}}_i\times\mathbf{d}_{ij} + \boldsymbol{\omega}_i\times(\boldsymbol{\omega}_i\times\mathbf{d}_{ij})$ and a similar formula for $\ddot{\mathbf{b}}_{ij}$ will be substituted.

The expressions (5.196) and (5.198) are substituted into (5.188). This results in the equation

$$\delta\underline{\boldsymbol{\omega}}^T\cdot\left\{-\,[\mathbf{d}_{ij}]\times\left[\underline{m}\left(\ddot{\mathbf{r}}_0\underline{1} - \left[\ddot{\mathbf{d}}_{ij}\right]^T\underline{1}\right) - \underline{\mathbf{F}}\right] + \underline{\mathbf{J}}\cdot\dot{\underline{\boldsymbol{\omega}}} - \underline{\mathbf{M}}^*\right\} = 0 \tag{5.200}$$

(in the first mixed product of vectors the multiplication symbols are interchanged). A system with spherical joints has the property that the angular velocities of its bodies are unconstrained. From this it follows that the expression in curled brackets equals zero. This is the equation

$$[\mathbf{d}_{ij}]\times\underline{m}\left[\ddot{\mathbf{d}}_{ij}\right]^T\underline{1} - [\mathbf{d}_{ij}]\times(\underline{m}\,\ddot{\mathbf{r}}_0\underline{1} - \underline{\mathbf{F}}) + \underline{\mathbf{J}}\cdot\dot{\underline{\boldsymbol{\omega}}} - \underline{\mathbf{M}}^* = 0\,. \tag{5.201}$$

Substitution of the expressions (5.197) and (5.199) into (5.189) results in the similar, yet simpler equation

$$[\mathbf{b}_{ij}]\times\underline{m}\left[\ddot{\mathbf{b}}_{ij}\right]^T\underline{1} - [\mathbf{b}_{ij}]\times\underline{\mathbf{F}} + \underline{\mathbf{J}}\cdot\dot{\underline{\boldsymbol{\omega}}} - \underline{\mathbf{M}}^* = \underline{\mathbf{0}}\,. \tag{5.202}$$

For the further development of this equation see Sect. 5.7.2. Equation (5.201) is the subject of the following section.

5.7.1 Systems Coupled to a Carrier Body

In (5.201) an important role is played by the matrix $[\mathbf{d}_{ij}]\times\underline{m}\left[\ddot{\mathbf{d}}_{ij}\right]^T$. A single element, abbreviated $\mathbf{g}_{ij}$, is the vector

$$\mathbf{g}_{ij} = \sum_{k=1}^{n} m_k\mathbf{d}_{ik}\times\ddot{\mathbf{d}}_{jk} \qquad (i,j = 1,\ldots,n)\,. \tag{5.203}$$

The indices i, j and k refer to vertices in the directed graph. Four cases have to be distinguished: (i) $i = j$, (ii) vertex $i <$ vertex j, (iii) vertex $j <$ vertex i and (iv) otherwise. Because of the properties of the vectors $\mathbf{d}_{ij}$ (see (5.94)) in case (ii) only those vertices k contribute to $\mathbf{g}_{ij}$ for which either $j = k$ or vertex $j <$ vertex k (for all others $\mathbf{d}_{jk}$ equals zero). For these vertices $\mathbf{d}_{ik}$ is, independent of k, identical with $\mathbf{d}_{ij}$. Likewise, in case (iii) only vertices k contribute for which either $i = k$ or vertex $i <$ vertex k, and for them, $\mathbf{d}_{jk}$ is identical with $\mathbf{d}_{ji}$. Finally, in case (iv) at least one of the two vectors $\mathbf{d}_{ik}$ and $\mathbf{d}_{jk}$ is zero. Hence

$$\mathbf{g}_{ij} = \begin{cases} \sum_{k=1}^{n} m_k \mathbf{d}_{ik} \times \ddot{\mathbf{d}}_{ik} & (i = j) \\ \mathbf{d}_{ij} \times \sum_{k=1}^{n} m_k \ddot{\mathbf{d}}_{jk} & (\text{vertex } i < \text{vertex } j) \\ \sum_{k=1}^{n} m_k \mathbf{d}_{ik} \times \ddot{\mathbf{d}}_{ji} & (\text{vertex } j < \text{vertex } i) \\ \mathbf{0} & (\text{otherwise})\,. \end{cases} \tag{5.204}$$

Further simplifications are possible with the help of (5.95) and (5.96). In the case vertex $i <$ vertex j, for example, substitution yields

$$\sum_{k=1}^{n} m_k \ddot{\mathbf{d}}_{jk} = \sum_{k=1}^{n} m_k (\ddot{\mathbf{b}}_{j0} - \ddot{\mathbf{b}}_{jk}) = M \ddot{\mathbf{b}}_{j0} \tag{5.205}$$

(remember that M is the total system mass). An analogous result is obtained for the sum in the case vertex $j <$ vertex i. Together they yield

$$\mathbf{g}_{ij} = \begin{cases} \sum_{k=1}^{n} m_k \mathbf{d}_{ik} \times \ddot{\mathbf{d}}_{ik} & (i = j) \\ M \mathbf{d}_{ij} \times \ddot{\mathbf{b}}_{j0} & (\text{vertex } i < \text{vertex } j) \\ M \mathbf{b}_{i0} \times \ddot{\mathbf{d}}_{ji} & (\text{vertex } j < \text{vertex } i) \\ \mathbf{0} & (\text{otherwise})\,. \end{cases} \tag{5.206}$$

These expressions are substituted into (5.201). At this point the matrix formulation is abandoned replacing the equation again by n individual vector equations. The quantity $\mathbf{M}_i^*$ is replaced by its original expression (5.190). The equations then read

$$\sum_{j=1}^{n} \mathbf{g}_{ij} - \sum_{j=1}^{n} \mathbf{d}_{ij} \times (m_j \ddot{\mathbf{r}}_0 - \mathbf{F}_j) + \mathsf{J}_i \cdot \dot{\boldsymbol{\omega}}_i + \boldsymbol{\omega}_i \times \mathsf{J}_i \cdot \boldsymbol{\omega}_i = \mathbf{M}_i \tag{5.207}$$

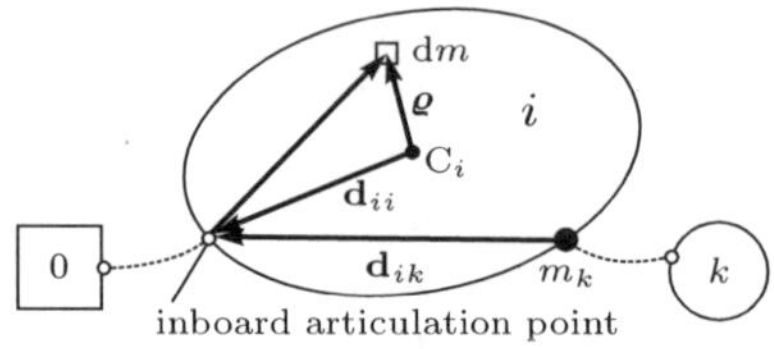

Fig. 5.28. Vectors locating mass particle dm and point mass m_k on body i

$(i = 1, \ldots, n)$ or explicitly

$$\mathsf{J}_i \cdot \dot{\boldsymbol{\omega}}_i + \boldsymbol{\omega}_i \times \mathsf{J}_i \cdot \boldsymbol{\omega}_i + \sum_{k=1}^{n} m_k \mathbf{d}_{ik} \times \ddot{\mathbf{d}}_{ik} - \sum_{j=1}^{n} \mathbf{d}_{ij} \times (m_j \ddot{\mathbf{r}}_0 - \mathbf{F}_j)$$
$$+M \left(\sum_{j:\, \mathrm{v}_i \,<\, \mathrm{v}_j}^{n} \mathbf{d}_{ij} \times \ddot{\mathbf{b}}_{j0} + \mathbf{b}_{i0} \times \sum_{j:\, \mathrm{v}_j \,<\, \mathrm{v}_i}^{n} \ddot{\mathbf{d}}_{ji} \right) = \mathbf{M}_i \tag{5.208}$$

$(i = 1, \ldots, n)$. The symbol $j\colon \mathrm{v}_i < \mathrm{v}_j$ is a short-hand notation indicating that the sum is taken over all vertices j satisfying the condition vertex $i <$ vertex j.

The two leading terms $\mathsf{J}_i \cdot \dot{\boldsymbol{\omega}}_i + \boldsymbol{\omega}_i \times \mathsf{J}_i \cdot \boldsymbol{\omega}_i$ together represent the time derivative of the absolute angular momentum of body i with respect to its center of mass C_i. From (3.14) it is known that

$$\mathsf{J}_i \cdot \dot{\boldsymbol{\omega}}_i + \boldsymbol{\omega}_i \times \mathsf{J}_i \cdot \boldsymbol{\omega}_i = \int_{m_i} \boldsymbol{\varrho} \times \ddot{\boldsymbol{\varrho}} \, \mathrm{d}m \tag{5.209}$$

where $\boldsymbol{\varrho}$ is the vector leading from C_i to the mass element dm of body i. Proposition: The three leading terms in (5.208) together represent the time derivative of the absolute angular momentum of the augmented body i with respect to its inboard articulation point. This is the articulation point which leads toward body 0. It is located by the vector $\mathbf{d}_{ii}$ (see Fig. 5.28).
Proof: The contribution of the distributed mass m_i of body i is the integral (5.209) with $\boldsymbol{\varrho} - \mathbf{d}_{ii}$ instead of $\boldsymbol{\varrho}$. The mass m_k of body $k \neq i$ is attached as point mass at the articulation point which is located by the vector $-\mathbf{d}_{ik}$. Hence, the time derivative of the angular momentum of the augmented body i with respect to its inboard articulation point is

$$\int_m (\boldsymbol{\varrho} - \mathbf{d}_{ii}) \times (\ddot{\boldsymbol{\varrho}} - \ddot{\mathbf{d}}_{ii}) \mathrm{d}m + \sum_{\substack{k=1 \\ \neq i}}^{n} m_k \mathbf{d}_{ik} \times \ddot{\mathbf{d}}_{ik}$$
$$= \int_m \boldsymbol{\varrho} \times \ddot{\boldsymbol{\varrho}} \mathrm{d}m + m_i \mathbf{d}_{ii} \times \ddot{\mathbf{d}}_{ii} + \sum_{\substack{k=1 \\ \neq i}}^{n} m_k \mathbf{d}_{ik} \times \ddot{\mathbf{d}}_{ik}$$
$$= \mathsf{J}_i \cdot \dot{\boldsymbol{\omega}}_i + \boldsymbol{\omega}_i \times \mathsf{J}_i \cdot \boldsymbol{\omega}_i + \sum_{k=1}^{n} m_k \mathbf{d}_{ik} \times \ddot{\mathbf{d}}_{ik} \, . \tag{5.210}$$

End of proof.

Let K_i be the inertia tensor of the augmented body i with respect to its inboard articulation point. It is related to the central inertia tensor J_i of the original body i through the equation

$$\mathsf{K}_i = \mathsf{J}_i + \sum_{k=i}^{n} m_k \left(\mathbf{d}_{ik}^2 \mathsf{I} - \mathbf{d}_{ik}\mathbf{d}_{ik}\right) \qquad (i = 1, \ldots, n) . \tag{5.211}$$

With this tensor the expression (5.210) is

$$\mathsf{J}_i \cdot \dot{\boldsymbol{\omega}}_i + \boldsymbol{\omega}_i \times \mathsf{J}_i \cdot \boldsymbol{\omega}_i + \sum_{k=i}^{n} m_k \mathbf{d}_{ik} \times \ddot{\mathbf{d}}_{ik} = \mathsf{K}_i \cdot \dot{\boldsymbol{\omega}}_i + \boldsymbol{\omega}_i \times \mathsf{K}_i \cdot \boldsymbol{\omega}_i . \tag{5.212}$$

In (5.208) the term involving $\ddot{\mathbf{r}}_0$ is reduced with the help of (5.95) and (5.96) to

$$\sum_{j=1}^{n} \mathbf{d}_{ij} \times m_j \ddot{\mathbf{r}}_0 = \sum_{j=1}^{n} m_j (\mathbf{b}_{i0} - \mathbf{b}_{ij}) \times \ddot{\mathbf{r}}_0 = M \mathbf{b}_{i0} \times \ddot{\mathbf{r}}_0 . \tag{5.213}$$

In the sum over $\mathbf{d}_{ij} \times \mathbf{F}_j$ in (5.208) the vectors $\mathbf{d}_{ij}$ are zero for the index combinations given in (5.94). Therefore, this sum is

$$\sum_{j=1}^{n} \mathbf{d}_{ij} \times \mathbf{F}_j = \mathbf{d}_{ii} \times \mathbf{F}_i + \sum_{j:\, \mathrm{v}_i < \mathrm{v}_j}^{n} \mathbf{d}_{ij} \times \mathbf{F}_j . \tag{5.214}$$

With this expression and with (5.213) and (5.212) the equations of motion (5.208) take the form

$$\begin{aligned} &\mathsf{K}_i \cdot \dot{\boldsymbol{\omega}}_i + \boldsymbol{\omega}_i \times \mathsf{K}_i \cdot \boldsymbol{\omega}_i + M \left[\sum_{j:\, \mathrm{v}_i < \mathrm{v}_j}^{n} \mathbf{d}_{ij} \times \ddot{\mathbf{b}}_{j0} + \mathbf{b}_{i0} \times \left(-\ddot{\mathbf{r}}_0 + \sum_{j:\, \mathrm{v}_j < \mathrm{v}_i}^{n} \ddot{\mathbf{d}}_{ji} \right) \right] \\ &\qquad + \mathbf{d}_{ii} \times \mathbf{F}_i + \sum_{j:\, \mathrm{v}_i < \mathrm{v}_j}^{n} \mathbf{d}_{ij} \times \mathbf{F}_j = \mathbf{M}_i \qquad (i = 1, \ldots, n) . \end{aligned} \tag{5.215}$$

The final form is obtained by substituting the expressions

$$\left.\begin{aligned} \ddot{\mathbf{b}}_{j0} &= \dot{\boldsymbol{\omega}}_j \times \mathbf{b}_{j0} + \boldsymbol{\omega}_j \times (\boldsymbol{\omega}_j \times \mathbf{b}_{j0}) \\ \ddot{\mathbf{d}}_{ji} &= \dot{\boldsymbol{\omega}}_j \times \mathbf{d}_{ji} + \boldsymbol{\omega}_j \times (\boldsymbol{\omega}_j \times \mathbf{d}_{ji}) \end{aligned}\right\} \qquad (i, j = 1, \ldots, n) . \tag{5.216}$$

This results in the following set of coupled differential equations for the angular velocities of the bodies:

$$\begin{aligned}
&\mathsf{K}_i \cdot \dot{\boldsymbol{\omega}}_i + M \left[\sum_{j:\,\mathrm{v}_i < \mathrm{v}_j}^{n} \mathbf{d}_{ij} \times (\dot{\boldsymbol{\omega}}_j \times \mathbf{b}_{j0}) + \mathbf{b}_{i0} \times \sum_{j:\,\mathrm{v}_j < \mathrm{v}_i}^{n} \dot{\boldsymbol{\omega}}_j \times \mathbf{d}_{ji} \right] \\
&+ \boldsymbol{\omega}_i \times \mathsf{K}_i \cdot \boldsymbol{\omega}_i \\
&+ M \left\{ \sum_{j:\,\mathrm{v}_i < \mathrm{v}_j}^{n} \mathbf{d}_{ij} \times [\boldsymbol{\omega}_j \times (\boldsymbol{\omega}_j \times \mathbf{b}_{j0})] + \mathbf{b}_{i0} \times \sum_{j:\,\mathrm{v}_j < \mathrm{v}_i}^{n} \boldsymbol{\omega}_j \times (\boldsymbol{\omega}_j \times \mathbf{d}_{ji}) \right\} \\
&= M\mathbf{b}_{i0} \times \ddot{\mathbf{r}}_0 + \mathbf{M}_i - \mathbf{d}_{ii} \times \mathbf{F}_i - \sum_{j:\,\mathrm{v}_i < \mathrm{v}_j}^{n} \mathbf{d}_{ij} \times \mathbf{F}_j \quad (i = 1, \ldots, n)\,. \quad (5.217)
\end{aligned}$$

In the first line the terms involving angular accelerations are collected. These terms can be represented in the form

$$\sum_{j=1}^{n} \mathsf{K}_{ij} \cdot \dot{\boldsymbol{\omega}}_j \qquad (i = 1, \ldots, n) \qquad (5.218)$$

with the tensors

$$\mathsf{K}_{ij} = \begin{cases} \mathsf{K}_i & (i = j) \\ M(\mathbf{b}_{j0} \cdot \mathbf{d}_{ij}\, \mathsf{I} - \mathbf{b}_{j0}\mathbf{d}_{ij}) & (\mathrm{v}_i < \mathrm{v}_j) \\ M(\mathbf{d}_{ji} \cdot \mathbf{b}_{i0}\, \mathsf{I} - \mathbf{d}_{ji}\, \mathbf{b}_{i0}) & (\mathrm{v}_j < \mathrm{v}_i) \\ 0 & (\text{else})\,. \end{cases} \qquad (5.219)$$

These tensors satisfy the relationship $\mathsf{K}_{ji} = \overline{\mathsf{K}}_{ij}$ (conjugate of K_{ij}).

The n first-order differential equations (5.217) (equivalent to $3n$ scalar equations) have to be supplemented by scalar differential equations which relate angular velocities $\boldsymbol{\omega}_i$ to the time derivatives of generalized coordinates. If Euler–Rodrigues parameters are chosen the kinematic differential equations are n sets of equations, each having the form (2.119) with an index i attached to $\omega_{1,2,3}$ and to $q_{0,1,2,3}$.

This section is closed with an interpretation of (5.215). An individual equation labeled i is rearranged in the form

$$M(-\mathbf{b}_{i0}) \times \left(\ddot{\mathbf{r}}_0 - \sum_{j:\,\mathrm{v}_j < \mathrm{v}_i}^{n} \ddot{\mathbf{d}}_{ji} \right) + \mathsf{K}_i \cdot \dot{\boldsymbol{\omega}}_i + \boldsymbol{\omega}_i \times \mathsf{K}_i \cdot \boldsymbol{\omega}_i = \mathbf{M}_i^{\mathrm{A}} \qquad (5.220)$$

with

$$\mathbf{M}_i^{\mathrm{A}} = \mathbf{M}_i - \mathbf{d}_{ii} \times \mathbf{F}_i - \sum_{j:\,\mathrm{v}_i < \mathrm{v}_j}^{n} \mathbf{d}_{ij} \times (M\ddot{\mathbf{b}}_{j0} + \mathbf{F}_j)\,. \qquad (5.221)$$

It will now be shown that this has the form of the angular momentum theorem for a single rigid body if as reference point for angular momentum and for external torques a body-fixed point other than the center of mass is chosen.

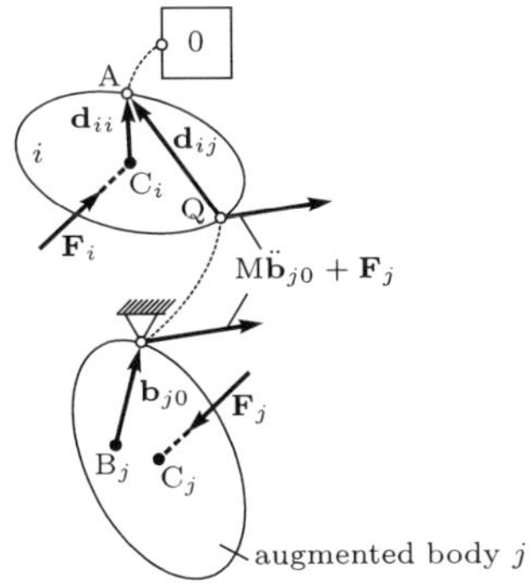

Fig. 5.29. Interpretation of $M\ddot{\mathbf{b}}_{j0} + \mathbf{F}_j$ as force applied to the suspension point of a pendulum

In (3.35) this law has been written in the form

$$m\boldsymbol{\varrho}_{\mathrm{C}} \times \ddot{\mathbf{r}}_{\mathrm{A}} + \mathsf{J}^{\mathrm{A}} \cdot \dot{\boldsymbol{\omega}} + \boldsymbol{\omega} \times \mathsf{J}^{\mathrm{A}} \cdot \boldsymbol{\omega} = \mathbf{M}^{\mathrm{A}} \,, \tag{5.222}$$

where A is a body-fixed reference point, $\ddot{\mathbf{r}}_{\mathrm{A}}$ its absolute acceleration, J^{A} the inertia tensor and $\mathbf{M}^{\mathrm{A}}$ the external torque, both with respect to A, and $\boldsymbol{\varrho}_{\mathrm{C}}$ the vector from A to the center of mass. Equation (5.220) has this form if the rigid body is understood to be the augmented body i and if, furthermore, the inboard articulation point of the body is chosen as reference point A. The mass of the body is then M and its center of mass is the barycenter B_i. The vector from the inboard articulation point to the barycenter is $-\mathbf{b}_{i0}$ (see Fig. 5.14). Equation (5.92) shows that the sum of vectors in parentheses is the absolute acceleration of the inboard articulation point A. Thus, the left-hand side of (5.220) has the desired form. Consider now the torque $\mathbf{M}_i^{\mathrm{A}}$. It contains, first, the resultant torque $\mathbf{M}_i$ about the body i center of mass C_i. By definition, the line of action of the external force $\mathbf{F}_i$ is passing through the body center of mass so that $-\mathbf{d}_{ii} \times \mathbf{F}_i$ is its torque with respect to the inboard articulation point. Also the remaining terms have the desired form in that the vectors $-\mathbf{d}_{ij}$ are pointing away from the inboard articulation point on body i. For values of j satisfying the condition vertex $i <$ vertex j the torque $-\mathbf{d}_{ij} \times (M\ddot{\mathbf{b}}_{j0} + \mathbf{F}_j)$ can be interpreted as follows. Imagine that the augmented body j is isolated from the system and suspended as a pendulum in inertial space at its own inboard articulation point. In Fig. 5.29 this pendulum is shown together with bodies i and 0 and with the paths between them. The augmented body j is subject to the external force $\mathbf{F}_j$. If now the augmented body j with mass M and with center of mass B_j is rotating with its actual angular velocity and angular acceleration then it is exerting on its suspension the force $M\ddot{\mathbf{b}}_{j0} + \mathbf{F}_j$. This force has to be shifted until its line of action is passing through the articulation point of body i which is leading toward body j (point Q in Fig. 5.29). It then produces on body i the torque $-\mathbf{d}_{ij} \times$

$(M\ddot{\mathbf{b}}_{j0} + \mathbf{F}_j)$ with respect to point A. This is the physical interpretation of the last term in $\mathbf{M}_i^{\mathrm{A}}$.

Problem 5.12. The n Eqs. (5.217) are combined in the matrix form $\underline{\mathsf{K}}\cdot\underline{\dot{\boldsymbol{\omega}}} = \underline{\mathbf{M}}+\underline{\mathbf{Q}}$. Here, $\underline{\mathsf{K}}$ is the matrix of the tensors K_{ij} defined in (5.218), and $\underline{\mathbf{Q}}$ is the abbreviation for the terms other than external torques. Imagine that each spherical joint is converted into a revolute joint by directing through the center of the spherical joint an axis fixed on the two bodies. The axis constrains the relative angular velocity of the bodies to have the direction of the axis. It causes a constraint torque normal to the axis. In joint a $(a = 1,\ldots,n)$ the unit vector along the axis is called $\mathbf{p}_a$; the angle of rotation of body $i^-(a)$ relative to body $i^+(a)$ is called q_a and the constraint torque acting on body $i^+(a)$ is called $\mathbf{Y}_a$. Starting from the equation $\underline{\mathsf{K}}\cdot\underline{\dot{\boldsymbol{\omega}}} = \underline{\mathbf{M}}+\underline{\mathbf{Q}}$ develop equations of motion of the form $\underline{A}\,\ddot{\underline{q}} = \underline{B}$. Compare with the equations of motion developed in Sect. 5.5.3 for the same system.

5.7.2 Systems Without Coupling to a Carrier Body

Subject of investigation are systems of the kind shown in Fig. 5.27 without coupling to a carrier body 0 by a spherical joint 1. Starting point for the development of equations of motion is (5.202):

$$[\mathbf{b}_{ij}] \times \underline{m}\left[\ddot{\mathbf{b}}_{ij}\right]^T \underline{1} - [\mathbf{b}_{ij}] \times \underline{\mathbf{F}} + \underline{\mathsf{J}}\cdot\underline{\dot{\boldsymbol{\omega}}} - \underline{\mathbf{M}}^* = \underline{\mathbf{0}}\,. \tag{5.223}$$

In what follows the matrix $[\mathbf{b}_{ij}] \times \underline{m}\left[\ddot{\mathbf{b}}_{ij}\right]^T$ is considered. Its elements, abbreviated $\mathbf{g}_{ij}$, are the vectors

$$\mathbf{g}_{ij} = \sum_{k=1}^{n} m_k \mathbf{b}_{ik} \times \ddot{\mathbf{b}}_{jk} \qquad (i,j = 1,\ldots,n)\,. \tag{5.224}$$

In the case $i \neq j$ this can be simplified substantially. For this purpose the directed system graph is divided into two parts by drawing a line across an arbitrary arc on the path between the vertices i and j (see Fig. 5.30). Let the set of indices of all vertices of the part containing vertex i be denoted by I and the set of indices of all vertices of the other part by II. Then, for all indices k belonging to I (abbreviated $k \in \mathrm{I}$) the identity $\mathbf{b}_{jk} = \mathbf{b}_{ji}$ holds and for all indices $k \in \mathrm{II}$ the identity $\mathbf{b}_{ik} = \mathbf{b}_{ij}$ holds. Hence, $\mathbf{g}_{ij}$ becomes

$$\mathbf{g}_{ij} = \left(\sum_{k\in\mathrm{I}} m_k \mathbf{b}_{ik}\right) \times \ddot{\mathbf{b}}_{ji} + \mathbf{b}_{ij} \times \sum_{k\in\mathrm{II}} m_k \ddot{\mathbf{b}}_{jk} \qquad (i \neq j)\,. \tag{5.225}$$

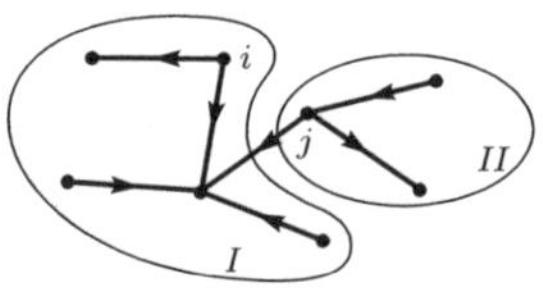

Fig. 5.30. Sets I and II of vertex indices related to two vertices i and $j \neq i$

The two sums are written in the forms

$$\left.\begin{aligned}\sum_{k\in\mathrm{I}} m_k\mathbf{b}_{ik} &= \sum_{k=1}^{n} m_k\mathbf{b}_{ik} - \sum_{k\in\mathrm{II}} m_k\mathbf{b}_{ik}\ ,\\ \sum_{k\in\mathrm{II}} m_k\ddot{\mathbf{b}}_{jk} &= \sum_{k=1}^{n} m_k\ddot{\mathbf{b}}_{jk} - \sum_{k\in\mathrm{I}} m_k\ddot{\mathbf{b}}_{jk}\ .\end{aligned}\right\} \tag{5.226}$$

According to (5.95) the first sum on the right-hand side in each equation equals zero. In the second sums the identities from above are used. The equations then read

$$\sum_{k\in\mathrm{I}} m_k\mathbf{b}_{ik} = -\mathbf{b}_{ij}\sum_{k\in\mathrm{II}} m_k\ ,\qquad \sum_{k\in\mathrm{II}} m_k\ddot{\mathbf{b}}_{jk} = -\ddot{\mathbf{b}}_{ji}\sum_{k\in\mathrm{I}} m_k\ . \tag{5.227}$$

With these expressions, $\mathbf{g}_{ij}$ takes the final form

$$\mathbf{g}_{ij} = -\left(\sum_{k\in\mathrm{I}} m_k + \sum_{k\in\mathrm{II}} m_k\right)\mathbf{b}_{ij}\times\ddot{\mathbf{b}}_{ji} = -M\mathbf{b}_{ij}\times\ddot{\mathbf{b}}_{ji} \qquad (i\neq j)\ . \tag{5.228}$$

These expressions are substituted into (5.223). At this point the matrix formulation is abandoned replacing the equation again by n individual vector equations. The quantity $\mathbf{M}_i^*$ is replaced by its original expression (5.190). The equations then read

$$\sum_{j=1}^{n}\mathbf{g}_{ij} - \sum_{j=1}^{n}\mathbf{b}_{ij}\times\mathbf{F}_j + \mathsf{J}_i\cdot\dot{\boldsymbol{\omega}}_i + \boldsymbol{\omega}_i\times\mathsf{J}_i\cdot\boldsymbol{\omega}_i = \mathbf{M}_i \tag{5.229}$$

$(i=1,\ldots,n)$ or, with the previous results for $\mathbf{g}_{ij}$,

$$\begin{aligned}&\mathsf{J}_i\cdot\dot{\boldsymbol{\omega}}_i + \boldsymbol{\omega}_i\times\mathsf{J}_i\cdot\boldsymbol{\omega}_i + \sum_{k=1}^{n} m_k\mathbf{b}_{ik}\times\ddot{\mathbf{b}}_{ik}\\ &-M\sum_{\substack{j=1\\ \neq i}}^{n}\mathbf{b}_{ij}\times\ddot{\mathbf{b}}_{ji} = \mathbf{M}_i + \sum_{j=1}^{n}\mathbf{b}_{ij}\times\mathbf{F}_j \qquad (i=1,\ldots,n)\ .\end{aligned} \tag{5.230}$$

Compare this with (5.208) for systems coupled to a carrier body 0 by a spherical joint 1. Through the arguments leading to (5.210) it has been proven that the sum of the three leading terms in (5.208) represents the time derivative of the angular momentum of the augmented body i with respect to the inboard articulation point. When $-\mathbf{d}_{ik}$ is replaced by $\mathbf{b}_{ik}$ the sum of the three leading terms in (5.230) is obtained. Whereas the vectors $-\mathbf{d}_{ik}$ locate the mass points m_k relative to the inboard articulation point the vectors $\mathbf{b}_{ik}$ locate the same mass points relative to the barycenter B_i. From this it follows that the sum of the three leading terms in (5.230) represents the time derivative of the angular momentum of the augmented body i with respect to B_i. Let $\hat{\mathsf{K}}_i$

be the inertia tensor of the augmented body i with respect to B_i. It is related to the central inertia tensor J_i of the original body i through the equation

$$\hat{\mathsf{K}}_i = \mathsf{J}_i + \sum_{k=i}^{n} m_k(\mathbf{b}_{ik}^2 \mathsf{I} - \mathbf{b}_{ik}\mathbf{b}_{ik}) \qquad (i = 1, \ldots, n)\,. \tag{5.231}$$

In terms of $\hat{\mathsf{K}}_i$ equations (5.230) have the form

$$\hat{\mathsf{K}}_i \cdot \dot{\boldsymbol{\omega}}_i + \boldsymbol{\omega}_i \times \hat{\mathsf{K}}_i \cdot \boldsymbol{\omega}_i - M \sum_{\substack{j=1\\ \neq i}}^{n} \mathbf{b}_{ij} \times \ddot{\mathbf{b}}_{ji} = \mathbf{M}_i + \sum_{j=1}^{n} \mathbf{b}_{ij} \times \mathbf{F}_j \tag{5.232}$$

$(i = 1, \ldots, n)$. For $\ddot{\mathbf{b}}_{ji}$ the expression $\dot{\boldsymbol{\omega}}_j \times \mathbf{b}_{ji} + \boldsymbol{\omega}_j \times (\boldsymbol{\omega}_j \times \mathbf{b}_{ji})$ is substituted. This results in the following set of coupled differential equations for the angular velocities of the bodies:

$$\begin{aligned} &\hat{\mathsf{K}}_i \cdot \dot{\boldsymbol{\omega}}_i - M \sum_{\substack{j=1\\ \neq i}}^{n} \mathbf{b}_{ij} \times (\dot{\boldsymbol{\omega}}_j \times \mathbf{b}_{ji}) \\ &+\boldsymbol{\omega}_i \times \hat{\mathsf{K}}_i \cdot \boldsymbol{\omega}_i - M \sum_{\substack{j=1\\ \neq i}}^{n} \mathbf{b}_{ij} \times [\boldsymbol{\omega}_j \times (\boldsymbol{\omega}_j \times \mathbf{b}_{ji})] \\ &= \mathbf{M}_i + \sum_{j=1}^{n} \mathbf{b}_{ij} \times \mathbf{F}_j \qquad (i = 1, \ldots, n)\,. \end{aligned} \tag{5.233}$$

In the first sum $\mathbf{b}_{ij} \times (\dot{\boldsymbol{\omega}}_j \times \mathbf{b}_{ji}) = (\mathbf{b}_{ji} \cdot \mathbf{b}_{ij}\ \mathsf{I} - \mathbf{b}_{ji}\mathbf{b}_{ij}) \cdot \dot{\boldsymbol{\omega}}_j$. The terms involving angular accelerations are collected in the form

$$\sum_{j=1}^{n} \mathsf{K}_{ij} \cdot \dot{\boldsymbol{\omega}}_j \qquad (i = 1, \ldots, n) \tag{5.234}$$

with tensors

$$\mathsf{K}_{ij} = \begin{cases} \hat{\mathsf{K}}_i & (i = j) \\ -M(\mathbf{b}_{ji} \cdot \mathbf{b}_{ij}\ \mathsf{I} - \mathbf{b}_{ji}\mathbf{b}_{ij}) & (i \neq j)\,. \end{cases} \tag{5.235}$$

These tensors satisfy the relationship $\mathsf{K}_{ji} = \overline{\mathsf{K}}_{ij}$ (conjugate of K_{ij}).

Equations (5.233) are complemented by the single Eq. (5.34) for the motion of the composite system center of mass. In addition, kinematic differential equations have to be formulated. They are identical with the ones added to (5.218) so that no further comment is needed.

In what follows an interpretation is given to (5.232). A single equation labeled i is written in the form

$$\hat{\mathsf{K}}_i \cdot \dot{\boldsymbol{\omega}}_i + \boldsymbol{\omega}_i \times \hat{\mathsf{K}}_i \cdot \boldsymbol{\omega}_i = \mathbf{M}_i^{\mathrm{B}} \qquad (i = 1, \ldots, n) \tag{5.236}$$

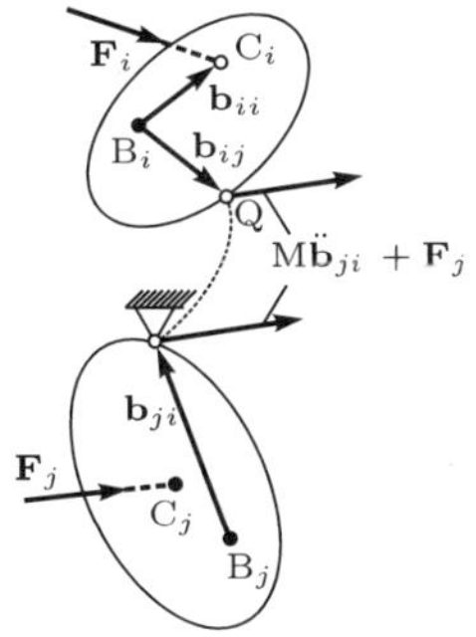

Fig. 5.31. Interpretation of $M\ddot{\mathbf{b}}_{ji} + \mathbf{F}_j$ as force applied to the suspension point of a pendulum

with

$$\mathbf{M}_i^{\mathrm{B}} = \mathbf{M}_i + \mathbf{b}_{ii} \times \mathbf{F}_i + \sum_{\substack{j=1 \\ \neq i}}^{n} \mathbf{b}_{ij} \times (M\ddot{\mathbf{b}}_{ji} + \mathbf{F}_j)\,. \qquad (5.237)$$

This has the form of the angular momentum theorem for a single rigid body in the special case when the center of mass is used as reference point for angular momentum and for external torques. The role of the rigid body is played by the augmented body i. The reference point for $\hat{\mathsf{K}}_i$ is, indeed, its center of mass, i.e. the barycenter B_i. The torque $\mathbf{M}_i^{\mathrm{B}}$ contains, first, the external torque $\mathbf{M}_i$. By definition, the line of action of the external force $\mathbf{F}_i$ is passing through the body center of mass so that $\mathbf{b}_{ii} \times \mathbf{F}_i$ is its torque with respect to B_i. Also the remaining terms have the desired form in that the vectors $\mathbf{b}_{ij}$ are pointing away from B_i. The torque $\mathbf{b}_{ij} \times (M\ddot{\mathbf{b}}_{ji} + \mathbf{F}_j)$ can be interpreted as follows. Imagine that the augmented body j is isolated from the system and suspended as a pendulum in inertial space at its articulation point which is leading toward body i. In Fig. 5.31 this pendulum is shown together with body i and with the path between the two bodies. The augmented body j is subject to the external force $\mathbf{F}_j$. If now the augmented body j with mass M and with center of mass B_j is rotating with its actual angular velocity and angular acceleration then it is exerting on its suspension the force $M\ddot{\mathbf{b}}_{ji} + \mathbf{F}_j$. This force has to be shifted until its line of action is passing through the point Q in Fig. 5.31. It then produces on body i the torque $\mathbf{b}_{ij} \times (M\ddot{\mathbf{b}}_{ji} + \mathbf{F}_j)$ with respect to B_i. This is the physical interpretation of the last term in $\mathbf{M}_i^{\mathrm{B}}$.

This section closes with an observation which establishes a simple relationship between systems with and without coupling to a carrier body (Wittenburg [93]). Imagine that to a system without such coupling a point mass $m_0 \to \infty$ is attached at some point A fixed on an arbitrarily chosen body, say body 1. This point mass has the effect that point A of body 1 is fixed in inertial space (it is assumed that m_0 has zero velocity). Furthermore, m_0 does not impose any kinematical constraint on the rotational motion of body 1.

Thus, the infinite mass is equivalent to a spherical joint 1 which connects body 1 at point A with inertial space. This means that the equations of motion of a system without spherical joint 1 but with a point mass $m_0 \to \infty$ are identical with the equations of motion of a system with a spherical joint 1 connecting body 1 to a stationary carrier body. Formally, this is shown as follows. With $m_0 \to \infty$ fixed at point A on body 1 this point becomes the center of mass C_1 of body 1. Furthermore, on each augmented body i the barycenter B_i coincides with the inboard articulation point (see Fig. 5.14). From this it follows that $\mathbf{b}_{i0} = \mathbf{0}$ $(i = 1, \ldots, n)$. Equation (5.96) yields

$$\mathbf{b}_{ij} = -\mathbf{d}_{ij} \qquad (i, j = 1, \ldots, n;\ m_0 \to \infty)\ . \tag{5.238}$$

This proves that the Eqs. (5.202) for the system with point mass m_0 are identical with (5.201) for the system with spherical joint 1 in the case $\ddot{\mathbf{r}}_0 \equiv \mathbf{0}$. In the final Eqs. (5.217) the vectors $\mathbf{d}_{ij}$ appear in combination with the vectors $\mathbf{b}_{j0}$. The vector $\mathbf{b}_{j0}$ is defined on the augmented body j for the system without point mass m_0 whereas $\mathbf{d}_{ij}$ is, according to (5.238), the vector $-\mathbf{b}_{ij}$ on the augmented body i of the system with infinite mass m_0. The matrix $\underline{\mu}$ of this system (see (5.38)) has the elements $\mu_{ij} = \delta_{ij} - \delta_{i1}$ $(i, j = 1, \ldots, n)$.

Equations (5.233) were published by Roberson and Wittenburg [66]. The equations marked the starting point for the development of the entire material presented in Chaps. 5 and 6 of this book. The simple form of the equations makes it possible to solve some multibody problems in analytical form. Sections 5.7.3 and 5.7.4 are devoted to two such problems. The first numerical simulation based on the equations was made by NASA in 1968. The system under investigation was a satellite consisting of a small central body with four very long and very flexible spaghetti-like antennas. It was feared that slow rotational motions of the satellite might cause violent bending vibrations of the antennas. Each antenna was modeled as a chain of four rigid rods interconnected with one another and with the central body by spherical joints and by springs and dampers in the joints. Numerical simulations of the resulting system of seventeen bodies with a variety of initial conditions showed that no built-up of large bending vibrations was to be expected. The simulation required 180 min of computation time per minute of real time. The subsequent performance of the satellite in orbit was successful.

Problem 5.13. n identical homogeneous rods of length ℓ, mass m and central moment of inertia J (about an axis perpendicular to the rod) are connected by spherical joints to form a chain. Give a formula for the central moment of inertia $\hat{K}_i$ (about an axis perpendicular to the rod) of the ith augmented body $(i = 1, \ldots, n)$.

Problem 5.14. Write a computer program for the calculation of the constant coordinates of the vectors $\mathbf{d}_{ij}$ and $\mathbf{b}_{ij}$ and of the tensors K_i and $\hat{\mathsf{K}}_i$ $(i, j = 1, \ldots, n)$ in the respective body-fixed bases $\underline{\mathbf{e}}^i$. Use as input data the masses and inertia components of the individual bodies and the constant coordinates of the vectors $\mathbf{c}_{ia}$ $(i, a = 1, \ldots, n)$.

5.7.3 Permanent Rotations of a Two-Body System

Subject of investigation is a system of two rigid bodies which are interconnected by a spherical joint. The system is not constrained to inertial space. It is free of external forces and torques, and there is no internal torque in the joint. Rotational motions of the system are governed by (5.233). Omitting the hat of $\hat{\mathsf{K}}_i$ they read

$$\begin{aligned} &\mathsf{K}_i \cdot \dot{\boldsymbol{\omega}}_i + M\mathbf{b}_{ij} \times (\mathbf{b}_{ji} \times \dot{\boldsymbol{\omega}}_j) \\ &= -\boldsymbol{\omega}_i \times \mathsf{K}_i \cdot \boldsymbol{\omega}_i + M\mathbf{b}_{ij} \times [\boldsymbol{\omega}_j \times (\boldsymbol{\omega}_j \times \mathbf{b}_{ji})] \quad (i,j=1,2;\ j \neq i)\,. \end{aligned} \tag{5.239}$$

It is to be investigated whether the system can be in a state of motion called permanent rotation. By this is meant that both bodies have identical and constant angular velocities $\boldsymbol{\omega}_1 \equiv \boldsymbol{\omega}_2 \equiv \omega\mathbf{e} = \text{const}$. During permanent rotation the system moves as if it were a single rigid body. The relative attitude of the two bodies is unknown, however, and so is the direction of $\mathbf{e}$ relative to the two bodies. The unknowns must be determined from (5.239) which under the said conditions reduce to

$$\mathbf{e} \times \mathsf{K}_i \cdot \mathbf{e} - M\mathbf{b}_{ij} \times [\mathbf{e} \times (\mathbf{e} \times \mathbf{b}_{ji})] = \mathbf{0} \qquad (i,j=1,2;\ j \neq i)\,. \tag{5.240}$$

In what follows only the most general case is considered in which none of the bodies is inertia-symmetric and in which the spherical joint is neither on a principal axis of inertia nor in the plane of two principal axes of any of the two bodies[7]. Such a system has altogether 13 parameters, namely three principal moments of inertia for each augmented body, three components for each of the vectors $\mathbf{b}_{12}$ and $\mathbf{b}_{21}$ and the total system mass $M = m_1 + m_2$. In Fig. 5.32 two bodies and the relevant vectors are schematically shown. On each body the spherical joint is on the dotted line connecting the body center of mass and the barycenter. The vectors $\mathbf{b}_{12}$ and $\mathbf{b}_{21}$ are pointing from the barycenters to the spherical joint.

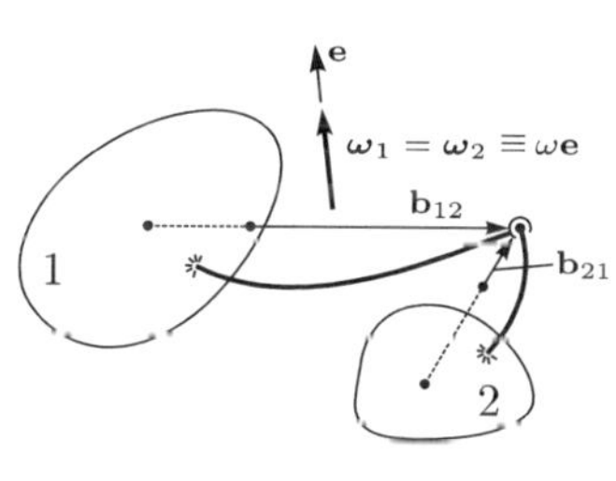

Fig. 5.32. Two-body system in permanent rotation

[7] For special cases and for more details on the general case see Wittenburg [95]. The method of solution presented here is simpler.

Scalar multiplications of (5.240) with $\mathbf{e}$ and with $\mathbf{b}_{ij}$ $(j \neq i)$ produce the three equations

$$\mathbf{e} \cdot \mathbf{b}_{12} \times \mathbf{b}_{21} = 0 \;, \tag{5.241}$$

$$\mathbf{b}_{12} \cdot \mathbf{e} \times \mathsf{K}_1 \cdot \mathbf{e} = 0 \;, \qquad \mathbf{b}_{21} \cdot \mathbf{e} \times \mathsf{K}_2 \cdot \mathbf{e} = 0 \;. \tag{5.242}$$

These equations state that the five vectors $\mathbf{b}_{12}$, $\mathbf{b}_{21}$, $\mathbf{e}$, $\mathsf{K}_1 \cdot \mathbf{e}$ and $\mathsf{K}_2 \cdot \mathbf{e}$ are coplanar. The common plane is the plane of the centers of mass of the two bodies and of the spherical joint. For coplanar vectors the identity holds $\mathbf{b}_{ij} \times [\mathbf{e} \times (\mathbf{e} \times \mathbf{b}_{ji})] = -\mathbf{e} \times \mathbf{b}_{ji}\mathbf{b}_{ij} \cdot \mathbf{e}$. For verification write $\mathbf{e} \times \mathbf{b}_{ji} = \mathbf{c}$ and use that $\mathbf{b}_{ij} \cdot \mathbf{c} = 0$. Hence, (5.240) becomes

$$\mathbf{e} \times (\mathsf{K}_i + M\mathbf{b}_{ji}\mathbf{b}_{ij}) \cdot \mathbf{e} = 0 \qquad (i, j = 1, 2;\ j \neq i) \;. \tag{5.243}$$

The sum of these two equations reads

$$\mathbf{e} \times [\mathsf{K}_1 + \mathsf{K}_2 + M(\mathbf{b}_{12}\mathbf{b}_{21} + \mathbf{b}_{21}\mathbf{b}_{12})] \cdot \mathbf{e} = 0 \;. \tag{5.244}$$

It is left to the reader to show that the symmetric tensor in brackets represents the central inertia tensor of the composite system. The equation states what is known anyway, namely that $\mathbf{e}$ has the direction of a principal axis of the quasi-rigid two-body system. The equation will not be used further. The tensors in the two Eqs. (5.243) are not symmetric.

In what follows Eqs. (5.242) are considered. Omitting indices each of the equations requires that on an augmented body three vectors $\mathbf{b}$, $\mathbf{e}$ and $\mathsf{K} \cdot \mathbf{e}$ be coplanar. In the x, y, z-system of principal axes of inertia only the vector $\mathbf{e}$ is unknown. It can be predicted that the condition is satisfied by a one-parametric manifold of vectors $\mathbf{e}$ with the following properties. The manifold is the intersection curve of a unit sphere with a (noncircular) double cone because the vector $-\mathbf{e}$ satisfies the condition if the vector $\mathbf{e}$ does. Particular vectors belonging to the manifold are the eight vectors along principal axes and along $\pm\mathbf{b}$. An explicit expression for the manifold is found by resolving the three vectors in the x, y, z-system. The vector $\mathbf{e}$ on the unit sphere surrounding the origin is specified by its geographical longitude u and latitude v (the x, y-plane is considered as equatorial plane with $u = 0$ on the x-axis). With these definitions and with principal moments of inertia K_x, K_y, K_z of the augmented body the vectors $\mathbf{e}$ and $\mathsf{K} \cdot \mathbf{e}$ have the coordinates

$$\mathbf{e}: \quad [\cos u \cos v \;,\; \sin u \cos v \;,\; \sin v] \;, \tag{5.245}$$

$$\mathsf{K} \cdot \mathbf{e}: \quad [K_x \cos u \cos v \;,\; K_y \sin u \cos v \;,\; K_z \sin v] \;. \tag{5.246}$$

A unit vector $\mathbf{e}^*$ along the given vector $\mathbf{b}$ has given coordinates (u^*, v^*). Its x, y, z-coordinates are those of $\mathbf{e}$ with the asterisk added everywhere. Coplanarity of the three vectors requires that the (3×3)-determinant of the coordinates be zero. This equation is factored in the two equations $\cos v = 0$

and

$$[(K_x - K_z)\sin u^*\cos u + (K_z - K_y)\cos u^*\sin u]\cos v^*\sin v \\ -(K_x - K_y)\sin v^*\sin u\cos u\cos v = 0\;. \quad (5.247)$$

The equation $\cos v = 0$ yields as solutions only the poles on the z-axis. In (5.247) the originally suppressed index i is given back to all parameters and variables. Each of the resulting two equations is solved for v_i as function of u_i:

$$v_i(u_i) = \tan^{-1}\left(\frac{p_i \sin u_i}{\tan u_i - q_i}\right) \qquad (i = 1, 2) \qquad (5.248)$$

with constants

$$p_i = \frac{(K_{ix} - K_{iy})\tan v_i^*}{(K_{iz} - K_{iy})\cos u_i^*}\,, \qquad q_i = \frac{K_{ix} - K_{iz}}{K_{iy} - K_{iz}}\tan u_i^* \qquad (i = 1, 2)\;. \qquad (5.249)$$

Each equation describes the predicted one-parametric manifold on the respective augmented body i. The poles on the z-axis are the points where $\tan u_i = q_i$.

For every value of u_1 (5.248) yields two vectors $\mathbf{e}$. Each vector is associated with a plane fixed on body 1 in which $\mathbf{b}_{12}$, $\mathbf{e}$ and $\mathsf{K}_1 \cdot \mathbf{e}$ are located. Likewise, for every value of u_2 (5.248) yields two vectors $\mathbf{e}$ and with each of them a plane fixed on body 2 in which $\mathbf{b}_{21}$, $\mathbf{e}$ and $\mathsf{K}_2 \cdot \mathbf{e}$ are located. Consider one of the planes on body 1 and one of the planes on body 2. Equation (5.241) requires that (i) the two planes are coplanar and that (ii) the vector $\mathbf{e}$ in one plane coincides with the vector $\mathbf{e}$ in the other plane. These requirements are satisfied in two relative configurations which differ by a 180°-rotation of one plane about $\mathbf{e}$.

Now, the Eqs. (5.243) are considered. The vectors $\mathbf{e} \times \mathsf{K}_i \cdot \mathbf{e}$ and $\mathbf{e} \times \mathbf{b}_{ji}$ ($i = 1, 2,\ j \neq i$) are collinear (perpendicular to the plane). Hence, the two scalar equations must be satisfied:

$$\left.\begin{aligned} (\mathbf{e} \times \mathsf{K}_1 \cdot \mathbf{e})^2 - (M\mathbf{e} \times \mathbf{b}_{21}\mathbf{b}_{12} \cdot \mathbf{e})^2 = 0\;, \\ (\mathbf{e} \times \mathsf{K}_2 \cdot \mathbf{e})^2 - (M\mathbf{e} \times \mathbf{b}_{12}\mathbf{b}_{21} \cdot \mathbf{e})^2 = 0\;. \end{aligned}\right\} \qquad (5.250)$$

For the vectors coordinates are calculated from (5.245) and (5.246) with v_i given by (5.248). Thus, the two equations are equations for the two unknowns u_1 and u_2. They must be solved numerically. For a simple algorithm see Wittenburg [95]. Nothing is known about the number of real solutions. In some numerical examples 36 different states of permanent rotation were found.

5.7.4 Multibody Satellite in a Circular Orbit

The problem to be investigated is explained, first, for the simple case when instead of a multibody system a single rigid body is considered. The body is moving as a satellite in a circular orbit about the Earth. The gravitational force is given by Newton's law. This means that a mass particle $\mathrm{d}m$ of the satellite at a radius vector $\mathbf{r}$ from the center of Earth is attracted by the force

$$\mathrm{d}\mathbf{F} = -\kappa \, \frac{\mathrm{d}m \, \mathbf{r}}{r^3} \tag{5.251}$$

where κ denotes the product of the universal gravitational constant and the mass of Earth. The relationship points out the physical phenomenon to be examined. Particles of identical mass but at different locations within the satellite experience different gravitational attraction forces. A typical length of the satellite is on the order of several meters. If the radius of the orbit trajectory is 6.500 km then the ratio between the two lengths is on the order of 10^{-6}. The difference between the gravitational attraction forces acting on two particles of identical mass is, therefore, exceedingly small compared with the attraction force itself. The difference can safely be neglected when the orbit trajectory is determined. In this part of the problem, therefore, for all mass particles the radius vector $\mathbf{r}$ is replaced by the radius vector $\mathbf{r}_\mathrm{C}$ of the satellite center of mass. This simplification results in a Keplerian orbit for the center of mass. In the present case, in particular, it is assumed that the orbit is circular so that the magnitude of $\mathbf{r}_\mathrm{C}$ is independent of time. The satellite is moving along its trajectory with a constant orbital angular velocity $\boldsymbol{\omega}_0$ whose magnitude depends on the orbit radius $\mathbf{r}_\mathrm{C}$. The relationship is given by Kepler's third law

$$\omega_0^2 = \frac{\kappa}{r_\mathrm{C}^3} \, . \tag{5.252}$$

The nonhomogeneity of the gravitational field over the volume occupied by the satellite must not be neglected, however, when rotational motions of the satellite are of concern. The force $\mathrm{d}\mathbf{F}$ applied to a mass particle $\mathrm{d}m$ causes a torque about the body center of mass. When this is integrated over the entire mass a resultant *gravity gradient* torque is obtained which, in general, is not zero. Although this torque is extremely small it must not be neglected for the simple reason that it is the only external torque on the body (it is assumed here that there are no other torques such as those caused by atmospheric drag, by solar pressure on the satellite surface or by interaction with the Earth's magnetic field, for instance). From the explanation given for

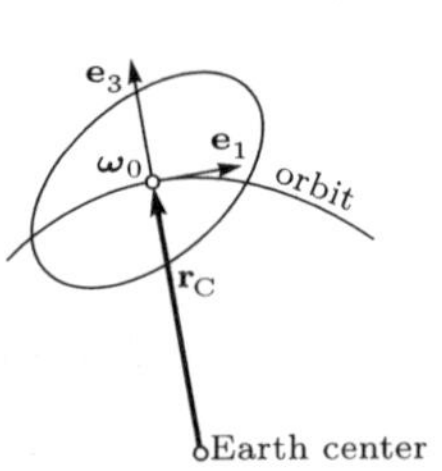

Fig. 5.33. Body in circular Earth orbit. Orbital reference frame $\underline{e}$

the resultant gravity gradient torque it follows at once that it is a function of the angular orientation of the body relative to Earth. For the description of this orientation the orbital reference frame $\underline{\mathbf{e}}$ shown in Fig. 5.33 is used. Its origin coincides at all times with the satellite center of mass, and it is rotating relative to Earth with the orbital angular velocity $\boldsymbol{\omega}_0$. The base vector $\mathbf{e}_3$ is directed along the local outward vertical and the vector $\mathbf{e}_2$ along $\boldsymbol{\omega}_0$. Let J and $\boldsymbol{\omega}$ be the central inertia tensor of the body and its absolute angular velocity, respectively. If $\mathbf{M}_{\text{grav}}$ denotes the resultant gravity gradient torque then rotational motions of the body are governed by the equation

$$\mathsf{J} \cdot \dot{\boldsymbol{\omega}} + \boldsymbol{\omega} \times \mathsf{J} \cdot \boldsymbol{\omega} = \mathbf{M}_{\text{grav}} . \tag{5.253}$$

Relative to the orbital reference frame the body is rotating with an angular velocity $\boldsymbol{\omega}_{\text{rel}}$. Hence, $\boldsymbol{\omega} = \boldsymbol{\omega}_0 + \boldsymbol{\omega}_{\text{rel}}$ and $\dot{\boldsymbol{\omega}} = \dot{\boldsymbol{\omega}}_{\text{rel}}$. The rotation relative to the orbital reference frame is, therefore, governed by the equation

$$\mathsf{J} \cdot \dot{\boldsymbol{\omega}}_{\text{rel}} + (\boldsymbol{\omega}_0 + \boldsymbol{\omega}_{\text{rel}}) \times \mathsf{J} \cdot (\boldsymbol{\omega}_0 + \boldsymbol{\omega}_{\text{rel}}) = \mathbf{M}_{\text{grav}} . \tag{5.254}$$

It is interesting to investigate whether this equation has the solution $\boldsymbol{\omega}_{\text{rel}} \equiv \mathbf{0}$, i. e. whether the satellite can remain stationary relative to the orbital reference frame. Such a state exists and is called relative equilibrium position. From (5.254) follows as condition for relative equilibrium the equation

$$\boldsymbol{\omega}_0 \times \mathsf{J} \cdot \boldsymbol{\omega}_0 = \mathbf{M}_{\text{grav}} . \tag{5.255}$$

The quantity on the right-hand side has been shown to be a function of the angular orientation of the body relative to the orbital reference frame. The same is true for the term on the left-hand side since $\boldsymbol{\omega}_0$ has constant coordinates in the orbital reference frame whereas J has constant inertia components in a body-fixed frame of reference. The equation is, therefore, determining the unknown angular orientation in the state of relative equilibrium. Relative equilibrium positions of this kind can be observed in nature. The moon is in relative equilibrium in its Earth orbit and the planet Mercury in its orbit about the sun. Relative equilibrium positions have considerable practical importance for the performance of orbiting spacecraft. In the design phase of orbiting artificial satellites for observation and signal transmission purposes the relative equilibrium positions must be known in advance. Only then can cameras and antennas be mounted in such a way that during flight they are always pointing vertically toward Earth.

After these introductory remarks the general problem to be treated can be formulated. Given is a multibody system with tree structure and with spherical joints without any internal joint torques. Each individual body is a gyrostat with rotors whose angular velocities relative to the carrier body are kept constant by control devices. The entire system is moving as a satellite in a circular orbit about the Earth. The questions to be answered are: Does the system possess relative equilibrium positions in the sense that all carriers of the system are simultaneously in a state of relative equilibrium with the rotors rotating relative to the carriers? If so, how do the relative equilibrium positions depend upon the parameters of the system, in particular upon the angular momenta of the rotors relative to the carriers? Mutual gravitational attraction forces between bodies of the system can be neglected.

The solution will be found by following the line of arguments described above for the single-body satellite. First, equations of rotational motions of the system will be formulated. For the external gravitational forces and torques explicit expressions will be developed. From the resulting equations equilibrium conditions will be obtained by introducing the identities $\boldsymbol{\omega}_i \equiv \boldsymbol{\omega}_0$ $(i = 1, \ldots, n)$ for the absolute angular velocities of all carriers of the system. The equations of rotational motions are developed from (5.233). The external forces $\mathbf{F}_i$ and $\mathbf{M}_i$ $(i = 1, \ldots, n)$ caused by the Earth's gravitational field will be examined later. The only other point requiring attention is the presence of rotors on the bodies. Equation (5.233) governs a system without rotors. It is a simple matter, however, to add terms which render the equations applicable to the present case. For this purpose it must be remembered that, except for the formulation, (5.233) and (5.230) are identical. In (5.230) the terms $\mathsf{J}_i \cdot \dot{\boldsymbol{\omega}}_i + \boldsymbol{\omega}_i \times \mathsf{J}_i \cdot \boldsymbol{\omega}_i$ represent the time derivative $\dot{\mathbf{L}}_i$ of the absolute angular momentum $\mathbf{L}_i$ of body i with respect to its center of mass. If body i is a gyrostat consisting of a carrier and of rotors with constant angular velocities relative to the carrier then the absolute angular momentum is composed of two parts (see Sect. 4.6). The first part is the angular momentum of the body (carrier plus rotors) when all rotors are "frozen". This part is the quantity called $\mathbf{L}_i$. The second part is the resultant angular momentum relative to the carrier of all rotors mounted on the carrier. It is a vector $\mathbf{h}_i$ whose coordinates in a carrier-fixed reference frame are constant. Its time derivative in inertial space is $\boldsymbol{\omega}_i \times \mathbf{h}_i$. Thus, $\dot{\mathbf{L}}_i$ has to be replaced by $\dot{\mathbf{L}}_i + \boldsymbol{\omega}_i \times \mathbf{h}_i$. The vector-cross product is the only additional term caused by the rotors. With this addition and with (5.234) the rotational equations (5.233) become

$$\sum_{j=1}^{n} \mathsf{K}_{ij} \cdot \dot{\boldsymbol{\omega}}_j + \boldsymbol{\omega}_i \times (\hat{\mathsf{K}}_i \cdot \boldsymbol{\omega}_i + \mathbf{h}_i) - M \sum_{\substack{j=1 \\ \neq i}}^{n} \mathbf{b}_{ij} \times [\boldsymbol{\omega}_j \times (\boldsymbol{\omega}_j \times \mathbf{b}_{ji})]$$
$$= \sum_{j=1}^{n} \mathbf{b}_{ij} \times \mathbf{F}_j + \mathbf{M}_i \qquad (i = 1, \ldots, n) . \tag{5.256}$$

Next, expressions are developed for $\mathbf{F}_i$ and $\mathbf{M}_i$ $(i = 1, \ldots, n)$. They are independent of the rotations of the rotors relative to the carriers since only the mass distribution of the system is relevant. Therefore, the rotors are assumed, for the moment, to be "frozen". Figure 5.34 shows the system together with the orbital reference frame $\underline{\mathbf{e}}$ whose origin is at the composite system center of mass at the radius vector $\mathbf{r}_C$ from the Earth's center. The magnitude r_C of $\mathbf{r}_C$ is constant by assumption. As before, the vector from the system center of mass to the body i center of mass is called $\mathbf{R}_i$ $(i = 1, \ldots, n)$. The location of the mass particle dm on body i is defined by the body-fixed radius vector $\boldsymbol{\varrho}$. The gravitational force acting on the mass element is

$$\mathrm{d}\mathbf{F}_i = -\kappa \frac{\mathbf{r}_\mathrm{C} + \mathbf{R}_i + \boldsymbol{\varrho}}{|\mathbf{r}_\mathrm{C} + \mathbf{R}_i + \boldsymbol{\varrho}|^3} \, \mathrm{d}m \ . \tag{5.257}$$

The denominator is developed into the Taylor series

$$\begin{aligned} |\mathbf{r}_\mathrm{C} + \mathbf{R}_i + \boldsymbol{\varrho}|^3 = [(\mathbf{r}_\mathrm{C} + \mathbf{R}_i + \boldsymbol{\varrho})^2]^{3/2} &= r_\mathrm{C}^3 \left[1 + \frac{2\mathbf{e}_3 \cdot (\mathbf{R}_i + \boldsymbol{\varrho})}{r_\mathrm{C}} + \ldots \right]^{3/2} \\ &= r_\mathrm{C}^3 \left[1 + \frac{3\mathbf{e}_3 \cdot (\mathbf{R}_i + \boldsymbol{\varrho})}{r_\mathrm{C}} + \ldots \right] . \end{aligned} \tag{5.258}$$

Dots indicate terms of second and higher order in $|\mathbf{R}_i + \boldsymbol{\varrho}|/r_\mathrm{C}$ which can be neglected (in the numerical example given earlier the second-order terms are on the order of 10^{-12}). With this expression d$\mathbf{F}_i$ takes the form

$$\mathrm{d}\mathbf{F}_i = -\frac{\kappa}{r_\mathrm{C}^3} (\mathbf{r}_\mathrm{C} + \mathbf{R}_i + \boldsymbol{\varrho}) \left[1 - \frac{3\mathbf{e}_3 \cdot (\mathbf{R}_i + \boldsymbol{\varrho})}{r_\mathrm{C}}\right] \mathrm{d}m + \ldots \ . \tag{5.259}$$

When this is multiplied out the term $(\mathbf{R}_i + \boldsymbol{\varrho})^2/r_\mathrm{C}$ can be neglected against $|\mathbf{R}_i + \boldsymbol{\varrho}|$ as a second-order term. The factor in front of the first bracket is $-\omega_0^2$ (see (5.252)). Hence,

$$\mathrm{d}\mathbf{F}_i = -\omega_0^2 [\mathbf{r}_\mathrm{C} - 3\mathbf{e}_3\mathbf{e}_3 \cdot (\mathbf{R}_i + \boldsymbol{\varrho}) + \mathbf{R}_i + \boldsymbol{\varrho}] \, \mathrm{d}m + \ldots \ . \tag{5.260}$$

Now, the integration is performed over the total mass of body i. Recognizing that $\int_{m_i} \boldsymbol{\varrho} \, \mathrm{d}m = \mathbf{0}$ we get

$$\mathbf{F}_i = -\omega_0^2 m_i (\mathbf{r}_\mathrm{C} + \mathbf{R}_i - 3\mathbf{e}_3 \mathbf{R}_i \cdot \mathbf{e}_3) \qquad (i = 1, \ldots, n) \ . \tag{5.261}$$

The force would simply be $-\omega_0^2 m_i \mathbf{r}_\mathrm{C} = -\kappa m_i \mathbf{r}_\mathrm{C}/r_\mathrm{C}^3$ if the total mass of body i were to be concentrated at the composite system center of mass. The

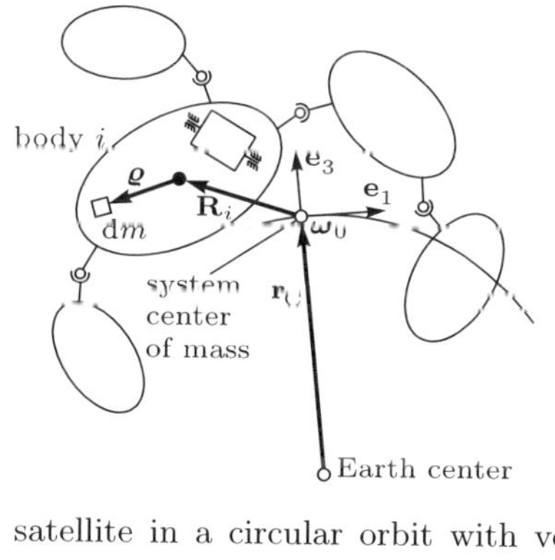

Fig. 5.34. Multi-body satellite in a circular orbit with vectors locating a mass particle on body i

remaining terms which are smaller by a factor on the order of 10^{-6} in the example given earlier are caused by the finite dimensions of the system.

Next, the torque $\mathbf{M}_i$ with respect to the body i center of mass is evaluated. It is represented by the integral

$$\mathbf{M}_i = \int_{m_i} \boldsymbol{\varrho} \times \mathrm{d}\mathbf{F}_i \qquad (i = 1, \ldots, n) \tag{5.262}$$

or with (5.260)

$$\mathbf{M}_i = -\omega_0^2 \int_{m_i} \boldsymbol{\varrho} \times [\mathbf{r}_C - 3\mathbf{e}_3\mathbf{e}_3 \cdot (\mathbf{R}_i + \boldsymbol{\varrho}) + \mathbf{R}_i + \boldsymbol{\varrho}]\, dm\,. \tag{5.263}$$

Because of $\int_{m_i} \boldsymbol{\varrho}\, dm = \mathbf{0}$ this is equal to

$$\mathbf{M}_i = 3\omega_0^2 \int_{m_i} \boldsymbol{\varrho} \times \mathbf{e}_3\mathbf{e}_3 \cdot \boldsymbol{\varrho}\, dm = -3\omega_0^2 \mathbf{e}_3 \times \int_{m_i} \boldsymbol{\varrho}\boldsymbol{\varrho}\, dm \cdot \mathbf{e}_3\,. \tag{5.264}$$

This can also be written in the form

$$\mathbf{M}_i = 3\omega_0^2 \mathbf{e}_3 \times \int_{m_i} (\boldsymbol{\varrho}^2 \mathsf{I} - \boldsymbol{\varrho}\boldsymbol{\varrho})\, dm \cdot \mathbf{e}_3 \tag{5.265}$$

since $\mathbf{e}_3 \times \mathsf{I} \cdot \mathbf{e}_3 = \mathbf{e}_3 \times \mathbf{e}_3$ is equal to zero. The integral represents the central inertia tensor J_i of body i. Thus, the final result is obtained

$$\mathbf{M}_i = 3\omega_0^2 \mathbf{e}_3 \times \mathsf{J}_i \cdot \mathbf{e}_3 \qquad (i = 1, \ldots, n)\,. \tag{5.266}$$

It confirms the statement made earlier that in a nonhomogeneous gravitational field a body of finite dimensions is, in general, subject to a very small torque which is a function of the angular orientation of the body relative to the orbital reference frame. The magnitude of the torque is mainly determined by ω_0 which for near-Earth orbits is on the order of $2\pi/(100 \text{ min}) \approx 10^{-3}/\text{s}$.

Before substituting the expressions for $\mathbf{F}_i$ and $\mathbf{M}_i$ into the equations of motion let us briefly return to the special case of a single rigid body in orbit. Its equation of motion (5.253) is now

$$\mathsf{J} \cdot \dot{\boldsymbol{\omega}} + \boldsymbol{\omega} \times \mathsf{J} \cdot \boldsymbol{\omega} = 3\omega_0^2 \mathbf{e}_3 \times \mathsf{J} \cdot \mathbf{e}_3\,. \tag{5.267}$$

From this equation it follows that two different bodies with identical ratios $J_1{:}J_2{:}J_3$ of principal moments of inertia move with identical angular velocities $\boldsymbol{\omega}(t)$ provided ω_0 and the initial conditions are also identical. The size of the bodies has no influence. With $\boldsymbol{\omega} = \omega_0 \mathbf{e}_2$ the equilibrium condition (5.255) for the single rigid body becomes

$$\mathbf{e}_2 \times \mathsf{J} \cdot \mathbf{e}_2 = 3\mathbf{e}_3 \times \mathsf{J} \cdot \mathbf{e}_3\,. \tag{5.268}$$

Either side of the equation is zero if $\mathbf{e}_2$ as well as $\mathbf{e}_3$ are parallel to principal axes of inertia of the body. Then, all three principal axes of inertia are parallel to the base vectors $\mathbf{e}_{1,2,3}$. It can be shown that these positions of relative equilibrium are the only solutions of the equilibrium condition. For this purpose the equation is decomposed into three scalar equations for coordinates in the orbital reference frame. The coordinate matrices of $\mathbf{e}_2$, $\mathbf{e}_3$ and J are

$$\underline{e}_2 = \begin{bmatrix} 0 \\ 1 \\ 0 \end{bmatrix}, \qquad \underline{e}_3 = \begin{bmatrix} 0 \\ 0 \\ 1 \end{bmatrix}, \qquad \underline{J} = \begin{bmatrix} J_{11} & -J_{12} & -J_{13} \\ -J_{12} & J_{22} & -J_{23} \\ -J_{13} & -J_{23} & J_{33} \end{bmatrix}. \tag{5.269}$$

The elements of $\underline{J}$ are, of course, still unknown because the equilibrium position is unknown. Equation (5.268) yields $\tilde{\underline{e}}_2 \underline{J}\, \underline{e}_2 = 3\tilde{\underline{e}}_3 \underline{J}\, \underline{e}_3$ or after multiplying out both sides

$$\begin{bmatrix} J_{23} \\ 0 \\ J_{12} \end{bmatrix} = 3 \begin{bmatrix} -J_{23} \\ J_{13} \\ 0 \end{bmatrix}. \tag{5.270}$$

Hence, in a position of relative equilibrium all three products of inertia J_{12}, J_{13} and J_{23} are zero. This proves that in positions of relative equilibrium all principal axes of inertia are parallel to the base vectors $\mathbf{e}_{1,2,3}$. It would now be necessary to investigate the stability of these equilibrium positions. This will not be done here. The reader is referred to Magnus [51] and Beletski [5].

We return now to the equations of motion (5.256) with $\mathbf{F}_i$ and $\mathbf{M}_i$ given by (5.261) and (5.266), respectively. First, the sum $\sum_{k=1}^n \mathbf{b}_{ik} \times \mathbf{F}_k$ is examined. For $\mathbf{R}_k$ the expression $\sum_{j=1}^n \mathbf{b}_{jk}$ is substituted (see (5.98)). This yields

$$\sum_{k=1}^n \mathbf{b}_{ik} \times \mathbf{F}_k = -\omega_0^2 \sum_{k=1}^n \mathbf{b}_{ik} \times \left(\mathbf{r}_C + \sum_{j=1}^n \mathbf{b}_{jk} - 3\mathbf{e}_3 \sum_{j=1}^n \mathbf{b}_{jk} \cdot \mathbf{e}_3 \right) m_k\,. \tag{5.271}$$

The contribution of the term involving $\mathbf{r}_C$ is zero because of the relationship $\sum_{k=1}^n \mathbf{b}_{ik} m_k = \mathbf{0}$ (see (5.95)). The remaining expression is rewritten in the form

$$\sum_{k=1}^n \mathbf{b}_{ik} \times \mathbf{F}_k = -\omega_0^2 \sum_{j=1}^n \sum_{k=1}^n m_k \mathbf{b}_{ik} \times \mathbf{b}_{jk} - 3\omega_0^2 \mathbf{e}_3 \times \sum_{j=1}^n \sum_{k=1}^n m_k \mathbf{b}_{ik} \mathbf{b}_{jk} \cdot \mathbf{e}_3\,. \tag{5.272}$$

The first term contains the sum $\sum_{k=1}^n m_k \mathbf{b}_{ik} \times \mathbf{b}_{jk}$. Except for the absence of differentiation dots it is identical with $\mathbf{g}_{ij}$ in (5.224). In the case $j \neq i$ this has been reduced to (5.228). The same line of arguments is applicable here. Hence,

$$\sum_{k=1}^n m_k \mathbf{b}_{ik} \times \mathbf{b}_{jk} = -M \mathbf{b}_{ij} \times \mathbf{b}_{ji} \qquad (i, j = 1, \ldots, n;\ i \neq j)\,. \tag{5.273}$$

In the case $j = i$ the sum is zero. The second term in (5.272) contains the sum $\sum_{k=1}^n m_k \mathbf{b}_{ik} \mathbf{b}_{jk}$, which differs from the one just considered only by the multiplication symbol. The arguments leading from (5.224) to (5.228) can, again, be used analogously. This yields

$$\sum_{k=1}^{n} m_k \mathbf{b}_{ik}\mathbf{b}_{jk} = \begin{cases} \sum_{k=1}^{n} m_k \mathbf{b}_{ik}\mathbf{b}_{ik} & (i=j) \\ -M\mathbf{b}_{ij}\mathbf{b}_{ji} & (i \neq j) \end{cases} \qquad (i,j=1,\ldots,n)\,. \quad (5.274)$$

Combining this and (5.273) with (5.272) one obtains the expression

$$\sum_{k=1}^{n} \mathbf{b}_{ik} \times \mathbf{F}_k = -3\omega_0^2 \mathbf{e}_3 \times \sum_{k=1}^{n} m_k \mathbf{b}_{ik}\mathbf{b}_{ik} \cdot \mathbf{e}_3 + \omega_0^2 M \sum_{\substack{j=1\\ \neq i}}^{n} (\mathbf{b}_{ij} \times \mathbf{b}_{ji} + 3\mathbf{e}_3 \times \mathbf{b}_{ij}\mathbf{b}_{ji} \cdot \mathbf{e}_3) \qquad (5.275)$$

$(i = 1, \ldots, n)$. The first term and the torque $\mathbf{M}_i$ from (5.266) are combined to yield

$$3\omega_0^2 \mathbf{e}_3 \times \left(\mathsf{J}_i - \sum_{k=1}^{n} m_k \mathbf{b}_{ik}\mathbf{b}_{ik} \right) \cdot \mathbf{e}_3\,. \qquad (5.276)$$

This is identical with

$$3\omega_0^2 \mathbf{e}_3 \times \left[\mathsf{J}_i - \sum_{k=1}^{n} m_k (\mathbf{b}_{ik}^2\, \mathsf{I} - \mathbf{b}_{ik}\mathbf{b}_{ik}) \right] \cdot \mathbf{e}_3 \qquad (5.277)$$

since $\mathbf{e}_3 \times \mathsf{I} \cdot \mathbf{e}_3 = \mathbf{e}_3 \times \mathbf{e}_3$ is equal to zero. Comparison with (5.231) shows that the expression in square brackets represents the central inertia tensor $\hat{\mathsf{K}}_i$ of the augmented body i. The entire expression is, therefore, simply $3\omega_0^2 \mathbf{e}_3 \times \hat{\mathsf{K}}_i \cdot \mathbf{e}_3$. Substituting this and the second term of (5.275) into the equations of motion (5.256) we get

$$\sum_{j=1}^{n} \mathsf{K}_{ij} \cdot \dot{\boldsymbol{\omega}}_j + \boldsymbol{\omega}_i \times (\hat{\mathsf{K}}_i \cdot \boldsymbol{\omega}_i + \mathbf{h}_i) - M \sum_{\substack{j=1\\ \neq i}}^{n} \mathbf{b}_{ij} \times [\boldsymbol{\omega}_j \times (\boldsymbol{\omega}_j \times \mathbf{b}_{ji})] = 3\omega_0^2 \mathbf{e}_3 \times \hat{\mathsf{K}}_i \cdot \mathbf{e}_3 + \omega_0^2 M \sum_{\substack{j=1\\ \neq i}}^{n} (\mathbf{b}_{ij} \times \mathbf{b}_{ji} + 3\mathbf{e}_3 \times \mathbf{b}_{ij}\mathbf{b}_{ji} \cdot \mathbf{e}_3) \qquad (5.278)$$

$(i = 1, \ldots, n)$. Conditions for relative equilibrium are now obtained by substituting $\dot{\boldsymbol{\omega}}_i \equiv \mathbf{0}$, $\boldsymbol{\omega}_i \equiv \omega_0 \mathbf{e}_2$ for $i = 1, \ldots, n$. This yields

$$\mathbf{e}_2 \times \left(\hat{\mathsf{K}}_i \cdot \mathbf{e}_2 + \frac{\mathbf{h}_i}{\omega_0} \right) - M \sum_{\substack{j=1\\ \neq i}}^{n} \mathbf{b}_{ij} \times [\mathbf{e}_2 \times (\mathbf{e}_2 \times \mathbf{b}_{ji})] = 3\mathbf{e}_3 \times \hat{\mathsf{K}}_i \cdot \mathbf{e}_3 + M \sum_{\substack{j=1\\ \neq i}}^{n} (\mathbf{b}_{ij} \times \mathbf{b}_{ji} + 3\mathbf{e}_3 \times \mathbf{b}_{ij}\mathbf{b}_{ji} \cdot \mathbf{e}_3) \qquad (5.279)$$

$(i = 1, \ldots, n)$. A further simplification is achieved when the triple vector-cross product on the left-hand side is rewritten in the form

$$\begin{aligned} \mathbf{b}_{ij} \times [\mathbf{e}_2 \times (\mathbf{e}_2 \times \mathbf{b}_{ji})] &= \mathbf{b}_{ij} \times (\mathbf{e}_2 \mathbf{b}_{ji} \cdot \mathbf{e}_2 - \mathbf{b}_{ji}) \\ &= -\mathbf{e}_2 \times \mathbf{b}_{ij} \mathbf{b}_{ji} \cdot \mathbf{e}_2 - \mathbf{b}_{ij} \times \mathbf{b}_{ji} \,. \end{aligned} \qquad (5.280)$$

The second term in this expression produces the sum $M \sum_{j=1 \atop \neq i}^{n} \mathbf{b}_{ij} \times \mathbf{b}_{ji}$ on the left-hand side of the equilibrium conditions (5.279). It cancels the same expression on the right-hand side. With the remaining terms the conditions read

$$\omega_0 (\mathbf{e}_2 \times \mathsf{B}_i \cdot \mathbf{e}_2 - 3\mathbf{e}_3 \times \mathsf{B}_i \cdot \mathbf{e}_3) + \mathbf{e}_2 \times \mathbf{h}_i = \mathbf{0} \qquad (i = 1, \ldots, n) \qquad (5.281)$$

where B_i is an abbreviation for the tensor

$$\mathsf{B}_i = \hat{\mathsf{K}}_i + M \sum_{j=1 \atop \neq i}^{n} \mathbf{b}_{ij} \mathbf{b}_{ji} \qquad (i = 1, \ldots, n) \,. \qquad (5.282)$$

This tensor B_i contains vectors which are fixed on different bodies. Its coordinates in a vector base fixed on body i change when the orientation of bodies changes.

The equilibrium conditions (5.281) can be considered from the following point of view. As has been mentioned earlier the relative angular momentum vector $\mathbf{h}_i$ on carrier i has constant coordinates in a vector base fixed on this carrier. It is assumed that there are at least three rotors on each carrier whose axes are not coplanar. Then, it is possible to give $\mathbf{h}_i$ any desired magnitude and direction in the carrier-fixed vector base by a proper choice of rotor angular velocities relative to the carrier. The vectors $\mathbf{h}_1, \ldots, \mathbf{h}_n$ influence the relative equilibrium positions. This suggests the following question. Is it possible to choose $\mathbf{h}_1, \ldots, \mathbf{h}_n$ in such a way that the system possesses a relative equilibrium position with certain prescribed characteristics? The practical significance of this problem is illustrated by the following example. Suppose a satellite with a camera mounted on one of the bodies is in orbit. As a result of a change in the mass distribution of the system caused by fuel consumption the original relative equilibrium position is disturbed. This causes the camera axis to leave its nominal vertical orientation. It would be desirable to change the system parameters in such a way that a new relative equilibrium position is created in which the camera axis is, again, in its nominal orientation. The only system parameters that can be changed on command from Earth are the relative rotor angular velocities. It is, therefore, desirable to deduce from the equilibrium conditions (5.281) two sets of equations. One set represents the explicit solution for $\mathbf{h}_1, \ldots, \mathbf{h}_n$. The other set should be a set of equilibrium conditions which is free of $\mathbf{h}_1, \ldots, \mathbf{h}_n$. The first goal is achieved by cross-multiplication of (5.281) with $\mathbf{e}_2$ and the other by scalar multiplication with $\mathbf{e}_2$.

First, the explicit solution for $\mathbf{h}_1, \ldots, \mathbf{h}_n$ is developed. Equations (5.281) show that equilibrium positions are unaffected by the component of $\mathbf{h}_i$ in the direction of $\mathbf{e}_2$, i.e. of the orbit angular velocity $\boldsymbol{\omega}_0$. This had to be expected. Let $\lambda_i \mathbf{e}_2$ be this component of $\mathbf{h}_i$ $(i = 1, \ldots, n)$. The component in the plane normal to $\mathbf{e}_2$ is obtained by cross-multiplication of (5.281) with $\mathbf{e}_2$. The explicit solution for $\mathbf{h}_i$ reads

$$\mathbf{h}_i = \lambda_i \mathbf{e}_2 + \omega_0 \mathbf{e}_2 \times (\mathbf{e}_2 \times \mathsf{B}_i \cdot \mathbf{e}_2 - 3\mathbf{e}_3 \times \mathsf{B}_i \cdot \mathbf{e}_3) \qquad (i = 1, \ldots, n). \quad (5.283)$$

Next, the scalar multiplication of (5.281) with $\mathbf{e}_2$ is carried out. This multiplication eliminates the first and the third term. The second term yields $\mathbf{e}_2 \cdot \mathbf{e}_3 \times \mathsf{B}_i \cdot \mathbf{e}_3 = \mathbf{e}_2 \times \mathbf{e}_3 \cdot \mathsf{B}_i \cdot \mathbf{e}_3 = \mathbf{e}_1 \cdot \mathsf{B}_i \cdot \mathbf{e}_3$. Thus, one gets the equations

$$\mathbf{e}_1 \cdot \mathsf{B}_i \cdot \mathbf{e}_3 = 0 \qquad (i = 1, \ldots, n). \quad (5.284)$$

Equations (5.283) and (5.284) together are equivalent to the original equilibrium conditions (5.281). Each solution of (5.284) determines a relative equilibrium position provided the vectors $\mathbf{h}_1, \ldots, \mathbf{h}_n$ satisfy (5.283) with free parameters $\lambda_1, \ldots, \lambda_n$. The vectors $\mathsf{B}_i \cdot \mathbf{e}_3$ in (5.284) are (see (5.282))

$$\mathsf{B}_i \cdot \mathbf{e}_3 = \hat{\mathsf{K}}_i \cdot \mathbf{e}_3 + M \sum_{\substack{j=1 \\ \neq i}}^{n} \mathbf{b}_{ij} (\mathbf{b}_{ji} \cdot \mathbf{e}_3) \qquad (i = 1, \ldots, n). \quad (5.285)$$

Suppose that on each carrier one camera is fixed with an axis of arbitrary direction and that all n camera axes are aligned parallel to the base vector $\mathbf{e}_3$ in the local vertical. This does not yet uniquely determine the position of the system in the orbital reference frame. Each carrier can still be rotated about $\mathbf{e}_3$ through an arbitrary angle. These n angles have no influence on the projections $\mathbf{b}_{ji} \cdot \mathbf{e}_3$ of the body-fixed vectors $\mathbf{b}_{ji}$ $(i, j = 1, \ldots, n)$ onto the direction of $\mathbf{e}_3$ and no influence on the coordinates of the vector $\hat{\mathsf{K}}_i \cdot \mathbf{e}_3$ $(i = 1, \ldots, n)$ in a vector base fixed on body i. From this it follows that during such rotations about $\mathbf{e}_3$ the vector $\mathsf{B}_i \cdot \mathbf{e}_3$ is fixed on carrier i. On each carrier there is one such vector. In the following, first, the general case is assumed that none of these vectors is zero or parallel to $\mathbf{e}_3$. Then, all n equations (5.284) can be satisfied by choosing the still undetermined rotation angles in such a way that on each carrier $i = 1, \ldots, n$ the vector $\mathsf{B}_i \cdot \mathbf{e}_3$ is normal to $\mathbf{e}_1$. This determines two positions for each body and, hence, 2^n different positions for the entire system. Each one of them is a relative equilibrium position provided the vectors $\mathbf{h}_1, \ldots, \mathbf{h}_n$ are chosen in accordance with (5.283). For the degenerate case in which the vector $\mathsf{B}_i \cdot \mathbf{e}_3$ is either zero or parallel to $\mathbf{e}_3$ for one or several bodies the solution is obvious. For these particular bodies (5.284) is satisfied independent of the rotation angle about $\mathbf{e}_3$. Each such body has an infinite number of relative equilibrium positions. For each of the remaining bodies two relative equilibrium positions exist as before. The results just obtained can be summarized as follows. There exist, in general, 2^n different relative equilibrium positions (an infinite number in degenerate cases) which satisfy the requirement that n arbitrarily chosen camera axes, one on each carrier, are parallel to the local vertical. For each such position the necessary vectors $\mathbf{h}_1, \ldots, \mathbf{h}_n$ are determined explicitly from (5.283). The necessary stability analysis is not presented here. For this the reader is referred to Wittenburg/Lilov [103]. Let it only be mentioned that in the stability analysis the free parameters $\lambda_1, \ldots, \lambda_n$ play an essential role.

Problem 5.15. Formulate expressions for the kinetic energy T and for the potential energy V of the satellite considered in this section.

5.8 Plane Motion

In this section arbitrary tree-structured systems with spherical joints of the general form shown in Fig. 5.27 are considered again. A system is either coupled to a carrier body 0 by a spherical joint 1 or it is free of this constraint. Systems with this constraint are governed by the equations of motion (5.217), and systems without are governed by the equations of motion (5.233). In what follows both sets of equations are adapted to the special case that the system is in plane motion. Plane motion can be achieved in various ways. The bodies can have flat surfaces which are moving in a plane fixed in inertial space. As an example one may think of a system of interconnected discs moving on an inclined plane. In contact zones constraint forces normal to the plane are acting. Plane motion can also be achieved by replacing every spherical joint by a revolute joint with all joint axes aligned parallel to one another. In each revolute joint a constraint torque is acting normal to the joint axis.

The characteristic feature of plane motion is that the angular velocities and accelerations of bodies have the forms

$$\boldsymbol{\omega}_i = \dot{\phi}_i \mathbf{e}\,, \qquad \dot{\boldsymbol{\omega}}_i = \ddot{\phi}_i \mathbf{e} \qquad (i = 0, \ldots, n)\,. \tag{5.286}$$

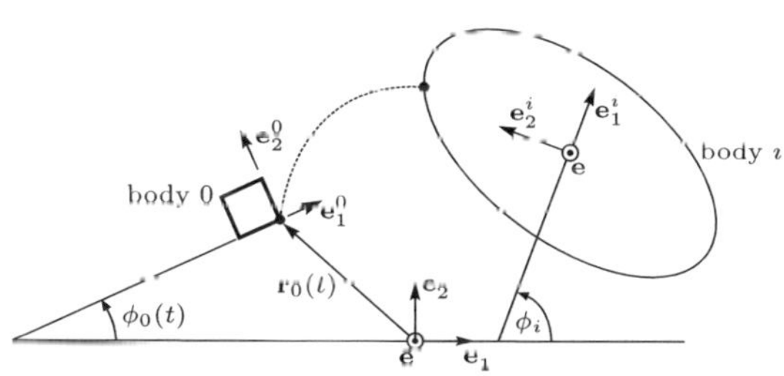

Fig. 5.35. Variables ϕ_i $(i = 0, \ldots, n)$ for a system with spherical joints in plane motion. Inertial reference base with unit vectors $\mathbf{e}_1$, $\mathbf{e}_2$ in the plane and $\mathbf{e}$ normal to the plane. Body-fixed base with unit vectors $\mathbf{e}_1^i$, $\mathbf{e}_2^i$ in the plane and $\mathbf{e}$ normal to the plane. Carrier body 0 with prescribed motion $\mathbf{r}_0(t)$ and $\phi_0(t)$

Here, $\mathbf{e}$ is the unit vector normal to the plane of motion and ϕ_i is the single angular variable of body i. It is defined as angle from a reference base vector $\mathbf{e}_1$ fixed in inertial space to the base vector $\mathbf{e}_1^i$ fixed on body i. Both unit vectors are in the plane of motion. This is schematically illustrated in Fig. 5.35. Only body 0 and one other body i are shown. The angular variable ϕ_0 of the carrier body 0 is a prescribed function of time $\phi_0(t)$. A set of scalar differential equations for the variables $\phi_1, \ldots, \phi_n$ is obtained when the expressions (5.286) are substituted into the equations of motion for general three-dimensional motions. This is done first for systems coupled to a carrier body.

5.8.1 Systems Coupled to a Carrier Body

Substitution of (5.286) into (5.217) produces the equations

$$\begin{aligned}
&\mathsf{K}_i \cdot \mathbf{e}\ddot{\phi}_i + M\left[\sum_{j:\, s_i < s_j}^{n} \mathbf{d}_{ij} \times (\mathbf{e} \times \mathbf{b}_{j0})\,\ddot{\phi}_j + \mathbf{b}_{i0} \times \sum_{j:\, s_j < s_i}^{n} \mathbf{e} \times \mathbf{d}_{ji}\,\ddot{\phi}_j\right] \\
&+\mathbf{e} \times \mathsf{K}_i \cdot \mathbf{e}\,\dot{\phi}_i^2 \\
&+M\left\{\sum_{j:\, s_i < s_j}^{n} \mathbf{d}_{ij} \times [\mathbf{e} \times (\mathbf{e} \times \mathbf{b}_{j0})]\,\dot{\phi}_j^2 + \mathbf{b}_{i0} \times \sum_{j:\, s_j < s_i}^{n} \mathbf{e} \times (\mathbf{e} \times \mathbf{d}_{ji})\,\dot{\phi}_j^2\right\} \\
&= M\mathbf{b}_{i0} \times \ddot{\mathbf{r}}_0 + \mathbf{M}_i - \sum_{j=1}^{n} \mathbf{d}_{ij} \times \mathbf{F}_j + \sum_{a=1}^{n} S_{ia}\mathbf{Y}_a \quad (i = 1, \ldots, n)\,. \qquad (5.287)
\end{aligned}$$

For the sum over $\mathbf{d}_{ij} \times \mathbf{F}_j$ see (5.94). In the newly introduced last term $\mathbf{Y}_a$ is an internal spring or damper torque acting in joint a ($+\mathbf{Y}_a$ on body $i^+(a)$ and $-\mathbf{Y}_a$ on body $i^-(a)$). The sum is the resultant of these torques acting on body i.

Each Eq. (5.287) is scalar-multiplied by $\mathbf{e}$. This results in the desired scalar differential equations for the angular coordinates $\phi_1, \ldots, \phi_n$. Multiplication of the first term produces the expression $K_{i3}\ddot{\phi}_i$ with the axial moment of inertia K_{i3}. The fourth term yields $\mathbf{e} \cdot \mathbf{e} \times \mathsf{K}_i \cdot \mathbf{e} = 0$. Consider, next, the single term $\mathbf{e} \cdot \mathbf{d}_{ij} \times [\mathbf{e} \times (\mathbf{e} \times \mathbf{b}_{j0})]$ which is contributed by the third sum on the left-hand side. The scalar-multiplication $\mathbf{e} \cdot \mathbf{d}_{ij} \times [\ldots]$ eliminates the component of $\mathbf{d}_{ij}$ along $\mathbf{e}$ and the cross-multiplication $\mathbf{e} \times \mathbf{b}_{j0}$ eliminates the component of $\mathbf{b}_{j0}$ along $\mathbf{e}$. With the same arguments it is shown that of all vectors $\mathbf{d}_{ij}$ and of all vectors $\mathbf{b}_{i0}$ $(i, j = 1, \ldots, n)$ only the projections onto the plane of motion contribute to the equation. In what follows $\mathbf{d}_{ij}$ and $\mathbf{b}_{i0}$ are understood to be these projections. The projected vector $\mathbf{d}_{ij}$ is fixed on body i. It is defined by its absolute value d_{ij} and by its constant angle α_{ij} against $\mathbf{e}_1^i$. Also the projected vector $\mathbf{b}_{i0}$ is fixed on body i. It is defined by its absolute value b_{i0} and by its constant angle β_i against $\mathbf{e}_1^i$. Figure 5.36 shows the orientation of the vectors $\mathbf{e}_1^i$, $\mathbf{d}_{ij}$ and $\mathbf{b}_{i0}$ relative to $\mathbf{e}_1$. In the

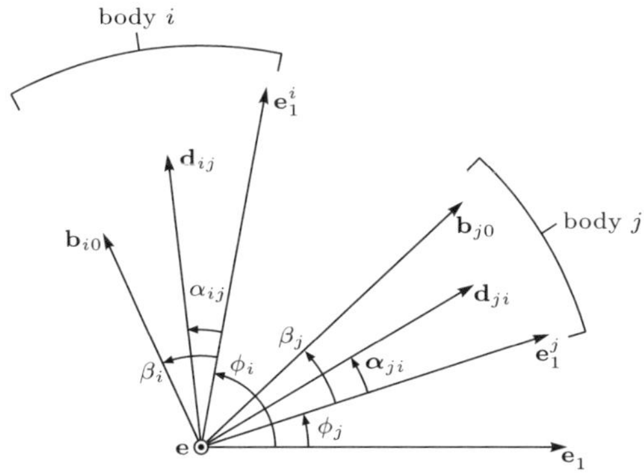

Fig. 5.36. In-plane vectors $\mathbf{e}_1^i$, $\mathbf{d}_{ij}$, $\mathbf{b}_{i0}$ on body i and $\mathbf{e}_1^j$, $\mathbf{d}_{ji}$, $\mathbf{b}_{j0}$ on body j. Constant parameters α_{ij}, β_i, α_{ji}, β_j and variables ϕ_i, ϕ_j

same figure also the vector $\mathbf{e}_1^j$ and the projected vectors $\mathbf{d}_{ji}$ and $\mathbf{b}_{j0}$ fixed on body j are shown. The angle enclosed by $\mathbf{d}_{ij}$ and $\mathbf{b}_{j0}$ is $\phi_i - \phi_j + \alpha_{ij} - \beta_j$. From this it follows that

$$\begin{aligned} \mathbf{d}_{ij} \cdot \mathbf{b}_{j0} &= d_{ij} b_{j0} \cos(\phi_i - \phi_j + \alpha_{ij} - \beta_j)\,, && (5.288)\\ \mathbf{e} \cdot \mathbf{d}_{ij} \times \mathbf{b}_{j0} &= -d_{ij} b_{j0} \sin(\phi_i - \phi_j + \alpha_{ij} - \beta_j)\,. && (5.289) \end{aligned}$$

The first sum in (5.287) contributes the product

$$\begin{aligned} \mathbf{e} \cdot \mathbf{d}_{ij} \times (\mathbf{e} \times \mathbf{b}_{j0}) &= \mathbf{e} \cdot [(\mathbf{d}_{ij} \cdot \mathbf{b}_{j0})\mathbf{e} - \underbrace{(\mathbf{d}_{ij} \cdot \mathbf{e})}_{=0}\,\mathbf{b}_{j0})] = \mathbf{d}_{ij} \cdot \mathbf{b}_{j0}\\ &= d_{ij} b_{j0} \cos(\phi_i - \phi_j + \alpha_{ij} - \beta_j) \qquad (5.290) \end{aligned}$$

and the third sum contributes the product

$$\begin{aligned} \mathbf{e} \cdot \mathbf{d}_{ij} \times [\mathbf{e} \times (\mathbf{e} \times \mathbf{b}_{j0})] &= \mathbf{e} \cdot \mathbf{d}_{ij} \times [\underbrace{(\mathbf{e} \cdot \mathbf{b}_{j0})}_{=0}\,\mathbf{e} - \mathbf{b}_{j0}] = -\mathbf{e} \cdot \mathbf{d}_{ij} \times \mathbf{b}_{j0}\\ &= d_{ij} b_{j0} \sin(\phi_i - \phi_j + \alpha_{ij} - \beta_j)\,. \qquad (5.291) \end{aligned}$$

The contributions from the second and from the fourth sum are obtained by interchanging the indices i and j.

Next, the scalar-product of $\mathbf{e}$ with the right-hand side terms in (5.287) is formulated. Of the torques $\mathbf{M}_i$ and $\mathbf{Y}_a$ components normal to $\mathbf{e}$ are eliminated. Among these are constraint torques in revolute joints which replace spherical joints. The components of $\mathbf{M}_i$ and $\mathbf{Y}_a$ along $\mathbf{e}$ are called M_i and Y_a, respectively. The contribution of a single force $\mathbf{F}_j$ is $-\mathbf{e} \cdot \mathbf{d}_{ij} \times \mathbf{F}_j$. The component along $\mathbf{e}$ is eliminated. Among the eliminated forces are constraint forces normal to the plane of motion. Let F_{j1} and F_{j2} be the in-plane components of $\mathbf{F}_j$ along the axes $\mathbf{e}_1$ and $\mathbf{e}_2$ fixed in inertial space. Then, $-\mathbf{e} \cdot \mathbf{d}_{ij} \times \mathbf{F}_j = d_{ij}[F_{j1} \sin(\phi_i + \alpha_{ij}) - F_{j2} \cos(\phi_i + \alpha_{ij})]$. Similarly, the in-plane components of $\ddot{\mathbf{r}}_0$ along $\mathbf{e}_1$ and $\mathbf{e}_2$ are called $\ddot{r}_{01}$ and $\ddot{r}_{02}$, respectively.

Then, $\mathbf{e} \cdot M\mathbf{b}_{i0} \times \ddot{\mathbf{r}}_0 = -Mb_{i0}[\ddot{r}_{01} \sin(\phi_i + \beta_i) - \ddot{r}_{02} \cos(\phi_i + \beta_i)]$. All terms are collected in the scalar differential equations

$$\sum_{j=1}^{n} \left(A_{ij}\ddot{\phi}_j + B_{ij}\dot{\phi}_j^2 \right) = Q_i + \sum_{a=1}^{n} S_{ia}Y_a \qquad (i = 1, \ldots, n) \tag{5.292}$$

with the abbreviations

$$A_{ij} = \begin{cases} K_{i3} & (i = j) \\ Md_{ij}b_{j0}\cos(\phi_i - \phi_j + \alpha_{ij} - \beta_j) & (\mathrm{v}_i < \mathrm{v}_j) \\ Md_{ji}b_{i0}\cos(\phi_i - \phi_j - \alpha_{ji} + \beta_i) & (\mathrm{v}_j < \mathrm{v}_i) \\ 0 & (\text{else}) \,, \end{cases} \quad (i, j = 1, \ldots, n) \,, \tag{5.293}$$

$$B_{ij} = \begin{cases} Md_{ij}b_{j0}\sin(\phi_i - \phi_j + \alpha_{ij} - \beta_j) & (\mathrm{v}_i < \mathrm{v}_j) \\ Md_{ji}b_{i0}\sin(\phi_i - \phi_j - \alpha_{ji} + \beta_i) & (\mathrm{v}_j < \mathrm{v}_i) \\ 0 & (\text{else}) \,, \end{cases} \quad (i, j = 1, \ldots, n) \,, \tag{5.294}$$

$$\begin{aligned} Q_i = {} & M_i - Mb_{i0}[\ddot{r}_{01}\sin(\phi_i + \beta_i) - \ddot{r}_{02}\cos(\phi_i + \beta_i)] \\ & + \sum_{j=1}^{n} d_{ij}[F_{j1}\sin(\phi_i + \alpha_{ij}) - F_{j2}\cos(\phi_i + \alpha_{ij})] \; (i = 1, \ldots, n) \,. \end{aligned} \tag{5.295}$$

Equations (5.292) can be written in the matrix form

$$\underline{A} \begin{bmatrix} \ddot{\phi}_1 \\ \vdots \\ \ddot{\phi}_n \end{bmatrix} + \underline{B} \begin{bmatrix} \dot{\phi}_1^2 \\ \vdots \\ \dot{\phi}_n^2 \end{bmatrix} = \underline{Q} + \underline{S}\,\underline{Y} \,. \tag{5.296}$$

The matrix $\underline{A}$ is symmetric and $\underline{B}$ is skew-symmetric.

Of particular interest are mechanical systems with torsional springs and dampers in the joints. Let k_a and d_a be the constant spring and damper coefficients, respectively, in joint a $(a = 1, \ldots, n)$. So far, nothing has been said about the zero position $\phi_1 = \phi_2 = \ldots = \phi_n = 0$ of the system. Now, it is defined to be the position in which all springs are unstressed. For this to be the case the vector bases $\underline{\mathbf{e}}^i$ $(i = 1, \ldots, n)$ must be fixed on the bodies $i = 1, \ldots, n$ in such a way that they are aligned parallel to the base $\underline{\mathbf{e}}^0$ fixed on the carrier body 0 when the springs are unstressed. In this case the internal torque Y_a is

$$Y_a = -k_a \left(\phi_{i^+(a)} - \phi_{i^-(a)} \right) - d_a \left(\dot{\phi}_{i^+(a)} - \dot{\phi}_{i^-(a)} \right) \quad (a = 1, \ldots, n) \,. \tag{5.297}$$

The minus sign follows from the convention that $+Y_a$ is the torque applied to body $i^+(a)$ and that in the present case it is a torque resisting the growth of the two differences shown in brackets. In what follows it is assumed that the single joint located on body 0 is joint 1. With this assumption and with the incidence matrix one can write

$$\begin{aligned} Y_a &= -k_a \sum_{i=0}^{n} S_{ia}\phi_i - d_a \sum_{i=0}^{n} S_{ia}\dot{\phi}_i \\ &= -(k_1\phi_0 + d_1\dot{\phi}_0)S_{0a} - k_a \sum_{i=1}^{n} S_{ia}\phi_i - d_a \sum_{i=1}^{n} S_{ia}\dot{\phi}_i \; (a = 1, \ldots, n) \,. \end{aligned} \tag{5.298}$$

The column matrix of all joint torques is

$$\underline{Y} = -(k_1\phi_0 + d_1\dot{\phi}_0)\underline{S}_0^T - \underline{K}\ \underline{S}^T\underline{\phi} - \underline{D}\ \underline{S}^T\dot{\underline{\phi}} \tag{5.299}$$

where $\underline{K}$ and $\underline{D}$ are diagonal $(n \times n)$-matrices of the spring and damper coefficients, respectively. Substitution into (5.296) yields

$$\underline{A}\begin{bmatrix}\ddot{\phi}_1\\ \vdots\\ \ddot{\phi}_n\end{bmatrix} + \underline{B}\begin{bmatrix}\dot{\phi}_1^2\\ \vdots\\ \dot{\phi}_n^2\end{bmatrix} + \underline{S}\ \underline{D}\ \underline{S}^T\begin{bmatrix}\dot{\phi}_1\\ \vdots\\ \dot{\phi}_n\end{bmatrix} + \underline{S}\ \underline{K}\ \underline{S}^T\begin{bmatrix}\phi_1\\ \vdots\\ \phi_n\end{bmatrix}$$
$$= \underline{Q} + (d_1\dot{\phi}_0 + k_1\phi_0)\begin{bmatrix}1\\0\\ \vdots\\0\end{bmatrix}. \tag{5.300}$$

The last column matrix $[1\ 0\ \ldots\ 0]^T$ is the product $-\underline{S}\,\underline{S}_0^T$. The matrices $\underline{A}$, $\underline{S}\ \underline{D}\ \underline{S}^T$, $\underline{S}\ \underline{K}\ \underline{S}^T$ are symmetric and $\underline{B}$ is skew-symmetric. The initial assumption that the springs and dampers have constant coefficients can now be dropped. The equations are obviously still valid if k_a and d_a are functions of $\phi_{i^+(a)} - \phi_{i^-(a)}$ and of $\dot{\phi}_{i^+(a)} - \dot{\phi}_{i^-(a)}$, respectively.

The equations of motion in the special form (5.300) and in the more general form (5.296) are applicable to many mechanical systems of interest. In Sect. 5.8.3 they are used for modeling a cantilever beam with large deformations. In Sect. 5.8.4 the stability of an upright multibody pendulum with ground excitation is investigated. Another example is the problem of gait of an anthropomorphic figure. The individual links of the human body are executing motions which can with reasonable accuracy be considered as plane motions. Equation (5.296) is valid for a phase of motion in which one foot has contact with the ground. This is the situation shown in Fig. 5.4b. Equations of motion for phases of motion with no ground contact are developed in the next section. Two feet on the ground create a closed kinematic chain with constraint equations for the variables of this chain. The incorporation of such constraint equations into the equations of motion has been explained in Sect. 5.6.1.

5.8.2 Systems Without Coupling to a Carrier Body

In this section equations of plane motions of systems are developed which are not joint-connected to a carrier body 0. The development follows the same line of arguments which started from (5.287). In the present case the corresponding equations are obtained by substituting (5.286) into (5.233) (supplemented by the last term with internal spring and damper torques $\mathbf{Y}_a$):

$$
\begin{aligned}
&\hat{\mathsf{K}}_i \cdot \mathbf{e}\,\ddot{\phi}_i - M \sum_{\substack{j=1\\ \neq i}}^{n} \mathbf{b}_{ij} \times (\mathbf{e} \times \mathbf{b}_{ji})\,\ddot{\phi}_j \\
&+ \mathbf{e} \times \hat{\mathsf{K}}_i \cdot \mathbf{e}\,\dot{\phi}_i^2 - M \sum_{\substack{j=1\\ \neq i}}^{n} \mathbf{b}_{ij} \times [\mathbf{e} \times (\mathbf{e} \times \mathbf{b}_{ji})]\,\dot{\phi}_j^2 \\
&= \mathbf{M}_i + \sum_{j=1}^{n} \mathbf{b}_{ij} \times \mathbf{F}_j + \sum_{a=1}^{n} S_{ia} \mathbf{Y}_a \qquad (i = 1, \ldots, n)\,. \qquad (5.301)
\end{aligned}
$$

In order to produce a set of differential equations for $\phi_1, \ldots, \phi_n$ each of these equations is scalar-multiplied by $\mathbf{e}$. Of the vectors $\mathbf{b}_{ij}$ only the projections onto the plane of motion contribute. The projected vector $\mathbf{b}_{ij}$ is defined by its absolute value b_{ij} and by its constant angle β_{ij} against the base vector $\mathbf{e}_1^i$ fixed on body i. Repeating the arguments leading to (5.290) and (5.291) one gets the desired equations of plane motions in the form (compare (5.296))

$$
\underline{A} \begin{bmatrix} \ddot{\phi}_1 \\ \vdots \\ \ddot{\phi}_n \end{bmatrix} + \underline{B} \begin{bmatrix} \dot{\phi}_1^2 \\ \vdots \\ \dot{\phi}_n^2 \end{bmatrix} = \underline{Q} + \underline{S}\,\underline{Y}\,. \qquad (5.302)
$$

The symmetric matrix $\underline{A}$, the skew-symmetric matrix $\underline{B}$ and the column matrix $\underline{Q}$ have the elements (compare with (5.295))

$$
\begin{aligned}
A_{ij} &= \begin{cases} \hat{K}_{i3} & (i = j) \\ -M b_{ij} b_{ji} \cos(\phi_i - \phi_j + \beta_{ij} - \beta_{ji}) & (i \neq j) \end{cases} \qquad (i, j = 1, \ldots, n)\,, \\
B_{ij} &= -M b_{ij} b_{ji} \sin(\phi_i - \phi_j + \beta_{ij} - \beta_{ji}) \qquad (i, j = 1, \ldots, n)\,, \\
Q_i &= M_i - \sum_{j=1}^{n} b_{ij} [F_{j1} \sin(\phi_i + \beta_{ij}) - F_{j2} \cos(\phi_i + \beta_{ij})] \quad (i = 1, \ldots, n)\,.
\end{aligned}
\qquad (5.303)
$$

Systems with springs and dampers in the joints are governed by (5.300) with the new matrices $\underline{A}$, $\underline{B}$ and $\underline{Q}$. The only other difference is that the spring constant k_1 and the damping constant d_1 associated with joint 1 are zero.

Numerical solutions for $\underline{\phi}(t)$ and $\underline{\dot{\phi}}(t)$ must be complemented by expressions for positions, velocities and accelerations of the body centers of mass, i.e. for the in-plane coordinates of the column matrices $\underline{\mathbf{r}}$ and $\underline{\dot{\mathbf{r}}}$. These are obtained from Newton's equation of motion for the composite system center of mass, $M\ddot{\mathbf{r}}_C = \sum_j^n \mathbf{F}_j$, and from (5.32) and (5.98):

$$
\mathbf{r}_i = \mathbf{r}_C + \sum_{j=1}^{n} \mathbf{b}_{ji}\,, \qquad \dot{\mathbf{r}}_i = \dot{\mathbf{r}}_C + \mathbf{e} \times \sum_{j=1}^{n} \dot{\phi}_j \mathbf{b}_{ji} \qquad (i = 1, \ldots, n)\,. \quad (5.304)
$$

Scalar multiplications with $\mathbf{e}_1$ and $\mathbf{e}_2$ yield the desired coordinate equations:

$$\ddot{r}_{C1} = \frac{1}{M}\sum_{j=1}^{n} F_{j1}\,, \qquad \ddot{r}_{C2} = \frac{1}{M}\sum_{j=1}^{n} F_{j2}\,, \tag{5.305}$$

$$\left.\begin{aligned} r_{i1} &= r_{C1} + \sum_{j=1}^{n} b_{ji}\cos(\phi_j + \beta_{ji})\,, & \dot{r}_{i1} &= \dot{r}_{C1} - \sum_{j=1}^{n} \dot{\phi}_j b_{ji}\sin(\phi_j + \beta_{ji})\,, \\ r_{i2} &= r_{C2} + \sum_{j=1}^{n} b_{ji}\sin(\phi_j + \beta_{ji})\,, & \dot{r}_{i2} &= \dot{r}_{C2} + \sum_{j=1}^{n} \dot{\phi}_j b_{ji}\cos(\phi_j + \beta_{ji}) \end{aligned}\right\} \quad (i = 1, \ldots, n)\,. \tag{5.306}$$

5.8.3 Cantilever Beam with Large Deformations

A simple system governed by (5.300) is shown in Fig. 5.37. It is the model of a cantilever beam. The beam is shown in the undeformed state and in a highly deformed state in which equations for beams known from elasticity theory are not valid. The system consists of n identical rigid elements of mass m and length ℓ which are coupled by revolute joints. The carrier body 0 is fixed in inertial space ($\phi_0 \equiv 0$). This is the body in which the beam is clamped. The labeling of bodies and of joints is regular, and the arcs in the graph are assumed to be directed toward body 0. Identical torsional springs with spring constant k (and no dampers) are attached to all joints. No external forces and torques are acting. The vector bases $\underline{\mathbf{e}}^i$ ($i = 0, \ldots, n$) are oriented as shown. The vector base $\underline{\mathbf{e}}^0$ on body 0 serves as reference base fixed in inertial space (in Fig. 5.35 the base $\mathbf{e}_1, \mathbf{e}_2, \mathbf{e}$). In the undeformed state all angles $\phi_1, \ldots, \phi_n$ are zero. Under these conditions the right-hand side of (5.300) is identically zero. From (5.94) and from Fig. 5.36 it follows that $\mathbf{d}_{ij} = -\ell\mathbf{e}_1^i$ for $i < j$, $\mathbf{d}_{ii} = -\frac{1}{2}\ell\mathbf{e}_1^i$ and $\mathbf{d}_{ij} = \mathbf{0}$ for $i > j$. The nonzero vectors have angles $\alpha_{ij} = \pi$. The vector $\mathbf{b}_{i0}$ on body i leads from the barycenter to the inboard articulation point. Hence, also $\beta_i = \pi$. The augmented body i carries, in addition to its own mass, the point mass $(i-1)m$ at joint i and the point mass $(n-i)m$ at joint $i+1$. From this one calculates $b_{i0} = \ell(n-i+\frac{1}{2})/n$. The total system mass is $M = nm$. With these expressions the matrix elements

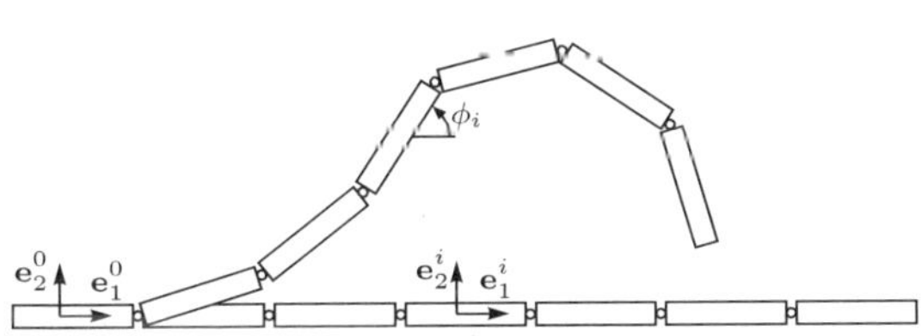

Fig. 5.37. Model of a cantilever beam with body 0 fixed in inertial space

in (5.293) and (5.294) are

$$A_{ij} = a_{ij}\cos(\phi_i - \phi_j)\,, \qquad B_{ij} = a_{ij}\sin(\phi_i - \phi_j) \qquad (i, j = 1, \ldots, n) \tag{5.307}$$

with

$$a_{ij} = a_{ji} = \begin{cases} K_{i3} & (i = j) \\ m\ell^2(n - j + \frac{1}{2}) & (i < j) \end{cases} \qquad (j = 1, \ldots, n)\,. \tag{5.308}$$

Note that the formula for B_{ij} gives the correct value $B_{ii} = 0$. The stiffness matrix is

$$\underline{S}\,\underline{K}\,\underline{S}^T = k\,\underline{S}\,\underline{S}^T = k\begin{bmatrix} 2 & -1 & 0 & . & . & . & . \\ -1 & 2 & -1 & 0 & . & . & . \\ 0 & -1 & 2 & -1 & 0 & . & . \\ . & . & . & . & . & . & . \\ . & . & . & 0 & -1 & 2 & -1 \\ . & . & . & . & 0 & -1 & 1 \end{bmatrix}\,. \tag{5.309}$$

The elements of $\underline{A}$ and $\underline{B}$ show that linearization of the equations requires that all differences $\phi_i - \phi_j$ $(i, j = 1, \ldots, n)$ be small. It is not sufficient that the differences are small for pairs of neighboring bodies. The linearized equations have the standard form

$$[a_{ij}]\ddot{\underline{\phi}} + k\,\underline{S}\,\underline{S}^T\underline{\phi} = \underline{0} \tag{5.310}$$

with a constant positive definite mass matrix $[a_{ij}]$ and with a constant positive definite stiffness matrix $k\,\underline{S}\,\underline{S}^T$. These equations can be used to determine the spring constant k. It should have a value which yields for the lowest eigenfrequency of the system the result known from other linearized beam equations. Once k has been determined the nonlinear equations of motion can be treated. In a Taylor expansion of the equations the linear terms are followed by third-order terms of the form $-a_{ij}(\phi_i - \phi_j)^2\ddot{\phi}_j/2$ and $a_{ij}(\phi_i - \phi_j)\dot{\phi}_j^2$ $(i, j = 1, \ldots, n)$.

5.8.4 Stabilized Upright Multibody Pendulum

The system shown in Fig. 5.38 is an upright n-body pendulum without springs and dampers in the joints. The bodies are identical (mass m, length ℓ). The carrier body 0 is oscillating in vertical direction. Its prescribed law of motion is given in the form $\mathbf{r}_0 = u_0\mathbf{e}_1\cos\Omega t$ with constant amplitude u_0 and constant circular frequency Ω. It is to be investigated whether the upright position of the pendulum can be stabilized by a proper choice of u_0 and Ω.
Solution: The chain of bodies is identical with the one in Fig. 5.37. The same variables are used again. This time, the upright position is the position $\phi_1 = \phi_2 = \ldots = \phi_n = 0$. Nonlinear equations of motion for $\phi_1, \ldots, \phi_n$ are obtained from (5.292). The left-hand side terms are identical with those for

the cantilever beam in Fig. 5.37. In what follows the right-hand side terms are formulated. The joint torques Y_a are zero. In the expression for Q_i in (5.295) the external torque M_i is zero. From $\mathbf{r}_0 = u_0\mathbf{e}_1 \cos \Omega t$ it follows that $\ddot{r}_{01} = -u_0\Omega^2 \cos \Omega t$ and $\ddot{r}_{02} = 0$. External forces are caused by weight only. These forces are $F_{j1} = -mg$ and $F_{j2} = 0$ independent of j. From the cantilever beam in Fig. 5.37 it is known that $\beta_i = \pi$ and $\alpha_{ij} = \pi$ independent of i and j. With these expressions Q_i becomes

$$Q_i = \left(Mb_{i0}\ddot{r}_{01} + mg \sum_{j=1}^{n} d_{ij} \right) \sin \phi_i \qquad (i = 1, \ldots, n) \,. \tag{5.311}$$

From the cantilever beam it is known, furthermore, that $Mb_{i0} = m\ell(n-i+\frac{1}{2})$ and that the sum over j equals $\ell(n-i+\frac{1}{2})$. This yields for Q_i the expression

$$Q_i = \left(1 - \frac{u_0\Omega^2}{g} \cos \Omega t\right) mg\ell(n-i+\tfrac{1}{2}) \sin \phi_i \qquad (i = 1, \ldots, n) \,. \tag{5.312}$$

For the required stability analysis the equations of motion are linearized which means that $\sin \phi_i$ is replaced by ϕ_i. Linearization of the equations of motion for the cantilever beam resulted in (5.310). In the present case the second term is missing. Thus, the linearized equations for the upright pendulum are

$$[a_{ij}]\ddot{\underline{\phi}} + \left(-1 + \frac{u_0\Omega^2}{g} \cos \Omega t\right) \underline{K}\, \underline{\phi} = \underline{0} \,. \tag{5.313}$$

For the matrix $[a_{ij}]$ see (5.308). $\underline{K}$ is the diagonal matrix with elements $K_{ii} = mg\ell(n-i+\frac{1}{2})$ $(i = 1, \ldots, n)$. Both matrices are positive definite. Let $\underline{\Phi}$ be the modal matrix associated with the simpler equation $[a_{ij}]\ddot{\underline{\phi}} + \underline{K}\,\underline{\phi} = \underline{0}$. It is determined from the eigenvalue problem $([a_{ij}] - \lambda\underline{K})\underline{z} = \underline{0}$. Because of the

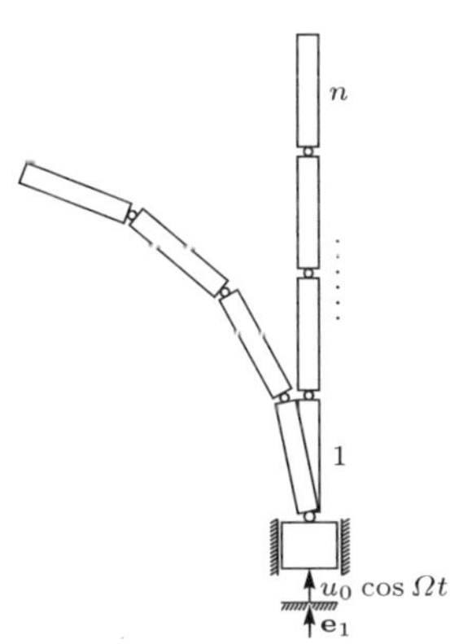

Fig. 5.38. Upright multibody pendulum on excited base. Notation as in Fig. 5.37

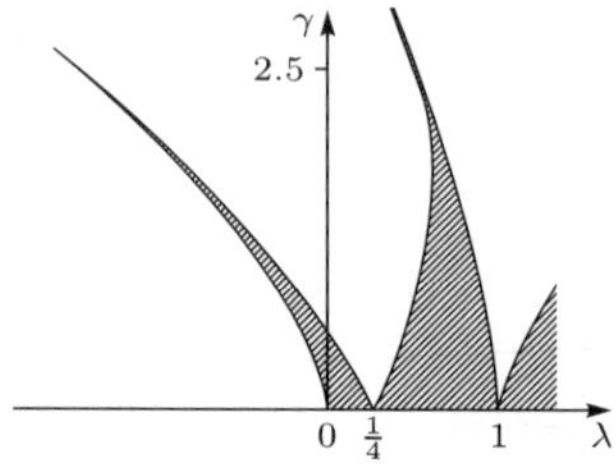

Fig. 5.39. Ince–Strutt stability diagram for the Mathieu equation

positive-definiteness of both matrices all eigenvalues are positive quantities ω_i^2 $(i = 1, \ldots, n)$. Let them be ordered such that $\omega_1 \le \omega_2 \le \ldots \le \omega_n$. The transformation $\underline{\phi} = \underline{\Phi}\,\underline{x}$ produces the set of decoupled equations $\ddot{x}_i + \omega_i^2 x_i = 0$ $(i = 1, \ldots, n)$. The same transformation applied to (5.313) produces the equations

$$\ddot{x}_i + \omega_i^2 \left(-1 + \frac{u_0 \Omega^2}{g} \cos \Omega t \right) x_i = 0 \qquad (i = 1, \ldots, n) \,. \tag{5.314}$$

Each of these equations is Mathieu's equation of a single upright pendulum. Another transformation $\tau = \Omega t$ gives the equation the standard form $x_i'' + (\lambda_i + \gamma_i \cos \tau) x_i = 0$ with

$$\lambda_i = -\omega_i^2 / \Omega^2 < 0 \,, \qquad \gamma_i = u_0 \omega_i^2 / g \qquad (i = 1, \ldots, n) \tag{5.315}$$

and with x' denoting the derivative with respect to τ. The n-body pendulum is stable if all n single-body pendulums are stable. Stability depends upon λ_i and γ_i. The Ince–Strutt diagram with axes λ and γ separates stable and unstable regions. Figure 5.39 shows the pertinent section of this diagram. Only the small shaded stability region to the left of $\lambda = 0$ is of interest. According to (5.315) the altogether n points (λ_i, γ_i) $(i = 1, \ldots, n)$ are located on a straight line in the interval between (λ_1, γ_1) and (λ_n, γ_n). The n-body pendulum is stable if this interval lies inside the shaded region. This can be achieved by a proper choice of the parameters u_0 and Ω. This result was obtained first by Otterbein [55] where a pendulum composed of n identical mass points interconnected by massless rods was investigated. For the more general case of an n-body pendulum composed of nonidentical bodies see Wittenburg [99]. Experiments with a four-body pendulum reveal that the pendulum returns to the vertical position even after severe disturbances.

5.9 Linear Vibrations of Chains of Bodies

In this section it is demonstrated that the incidence matrix and the path matrix are useful not only in nonlinear systems but in linear systems as well (Wittenburg [99], [98]). The systems to be investigated consist of bodies

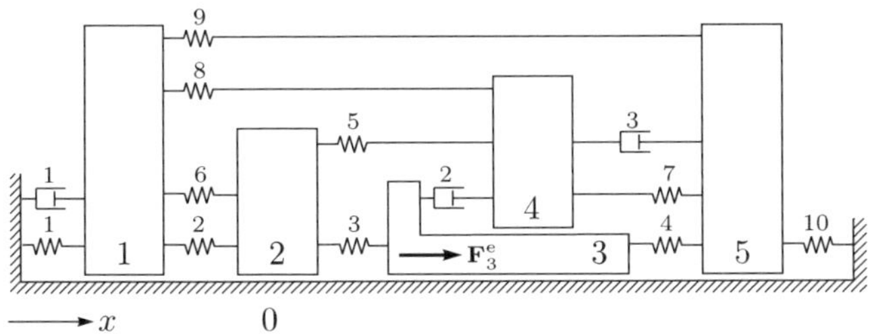

Fig. 5.40. Chain of bodies connected by springs and dampers

labeled $1, \ldots, n$ each having the single degree of freedom of translation along a common x-axis. The carrier body 0 is inertial space. The bodies $0, \ldots, n$ are interconnected by m^{S} linear springs and by m^{D} linear dampers. The numbers n, m^{S} and m^{D} are arbitrary. In Fig. 5.40 a simple example with $n = 5$, $m^{\mathrm{S}} = 10$ and $m^{\mathrm{D}} = 3$ is shown. Each body $i = 1, \ldots, n$ is subject to a given external force $F_i^{\mathrm{e}}(t)$. Only F_3^{e} is shown. The formulations to come apply also to systems with rotatory instead of translatory oscillators such as gears with the wheels being the rigid bodies and with the connecting shafts being springs and dampers.

5.9.1 Spring Graph. Damper Graph. Coordinate Graph

For the connections by springs and by dampers separate directed graphs are defined (senses of direction arbitrary). Figure 5.41a depicts the spring graph. This graph happens to be connected. The vertices 1 and 2 are connected by two arcs. Figure 5.41b depicts the damper graph. This graph is not connected. Each of the two graphs has its own incidence matrix (see (5.3) and Problem 5.3). Since the carrier body 0 is inertial space row 0 will not be used. Only the submatrix of rows $1, \ldots, n$ is of interest. This is the submatrix called $\underline{S}$ in (5.6). For the spring graph and the damper graph in

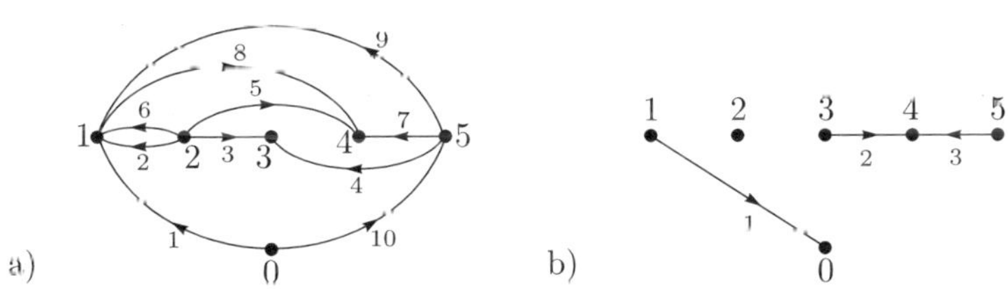

Fig. 5.41. Spring graph **(a)** and damper graph **(b)**

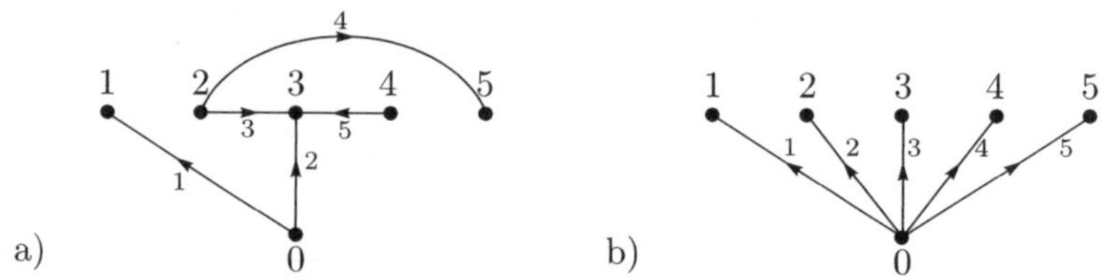

Fig. 5.42. Two coordinate graphs for the system in Fig. 5.40

Figs. 5.41a and b the two matrices are

$$\underline{S}^{\mathrm{S}} = \begin{bmatrix} -1 & -1 & 0 & 0 & 0 & -1 & 0 & 1 & -1 & 0 \\ 0 & 1 & 1 & 0 & 1 & 1 & 0 & 0 & 0 & 0 \\ 0 & 0 & -1 & -1 & 0 & 0 & 0 & 0 & 0 & 0 \\ 0 & 0 & 0 & 0 & -1 & 0 & -1 & -1 & 0 & 0 \\ 0 & 0 & 0 & 1 & 0 & 0 & 1 & 0 & 1 & -1 \end{bmatrix}, \qquad \underline{S}^{\mathrm{D}} = \begin{bmatrix} 1 & 0 & 0 \\ 0 & 0 & 0 \\ 0 & 1 & 0 \\ 0 & -1 & -1 \\ 0 & 0 & 1 \end{bmatrix}. \tag{5.316}$$

As generalized coordinates $q_1, \ldots, q_n$ relative as well as absolute displacements of bodies are accepted. The arbitrarily chosen coordinates are represented by the arcs of still another directed graph called the coordinate graph. Also this graph has vertices $i = 0, \ldots, n$. Definition:

$$\begin{aligned} q_a = {} & \text{displacement of body } i^-(a) \text{ relative} \\ & \text{to body } i^+(a) \text{ in positive } x\text{-direction} \quad (a = 1, \ldots, n)\,. \end{aligned} \tag{5.317}$$

Figure 5.42a depicts a coordinate graph in which q_1 and q_2 are absolute displacements of the bodies 1 and 3, respectively, whereas q_3, q_4 and q_5 are relative displacements. If n absolute displacements are chosen as coordinates then the coordinate graph has the form shown in Fig. 5.42b. The coordinate graph for any suitably chosen set of coordinates is connected and tree-structured. If it would consist of unconnected subgraphs then the coordinates would not be suitable for describing the location of bodies. If the graph would have a circuit then the coordinates associated with the arcs of this circuit would not be independent. Thus, for the coordinate graph there exist an incidence matrix $\underline{S}$ and a path matrix $\underline{T}$ which is the inverse of $\underline{S}$. For the coordinate graph shown in Fig. 5.42a the two matrices are

$$\underline{S} = \begin{bmatrix} -1 & 0 & 0 & 0 & 0 \\ 0 & 0 & 1 & 1 & 0 \\ 0 & -1 & -1 & 0 & -1 \\ 0 & 0 & 0 & 0 & 1 \\ 0 & 0 & 0 & -1 & 0 \end{bmatrix}, \qquad \underline{T} = \begin{bmatrix} -1 & 0 & 0 & 0 & 0 \\ 0 & -1 & -1 & -1 & -1 \\ 0 & 1 & 0 & 0 & 1 \\ 0 & 0 & 0 & 0 & -1 \\ 0 & 0 & 0 & 1 & 0 \end{bmatrix}.$$

For the coordinate graph shown in Fig. 5.42b both matrices are the unit matrix multiplied by -1.

Let x_i $(i = 1, \ldots, n)$ be the absolute displacement of body i in positive x-direction. The displacements are defined to be zero when the system is in a state of equilibrium in the absence of external forces. In this state the springs may be prestressed. We also define $x_0 \equiv 0$. In terms of the incidence matrix the definition (5.317) of generalized coordinates reads

$$q_a = x_{i^-(a)} - x_{i^+(a)} = -\sum_{i=1}^{n} S_{ia} x_i \qquad (a = 1, \ldots, n) . \tag{5.318}$$

These n equations are combined in the first equation below. The second equation is a consequence of the fact that the path matrix is the inverse of the incidence matrix.

$$\underline{q} = -\underline{S}^T \underline{x} , \qquad \underline{x} = -\underline{T}^T \underline{q} . \tag{5.319}$$

Newton's equation for body i reads

$$m_i \ddot{x}_i = F_i^{\mathrm{e}} + R_i^{\mathrm{S}} + R_i^{\mathrm{D}} \qquad (i = 1, \ldots, n) . \tag{5.320}$$

Here, R_i^{S} is the resultant of all spring forces on body i, and R_i^{D} is the resultant of all damper forces. In what follows spring forces and, therefore, the spring graph will be considered. The spring a $(a = 1, \ldots, m^{\mathrm{S}})$ with spring constant k_a connects the bodies $i^+(a)$ and $i^-(a)$. In analogy to the displacements q_a defined in (5.318) for the coordinate graph we define for the spring graph the quantities

$$\Delta\ell_a = x_{i^-(a)} - x_{i^+(a)} = -\sum_{j=1}^{n} S_{ja}^{\mathrm{S}} x_j \qquad (a = 1, \ldots, m^{\mathrm{S}}) . \tag{5.321}$$

$\Delta\ell_a$ is the change of length of spring a compared with its length in the equilibrium configuration without external forces. Independent of the sense of direction of arc a body $i^+(a)$ is subject to the force $+F_a = k_a \Delta\ell_a$ and body $i^-(a)$ is subject to the force $-F_a$ (in addition to forces already acting in the equilibrium position due to prestressing of springs). From the definition of the incidence matrix it follows that the resultant spring force R_i^{S} on body i is

$$R_i^{\mathrm{S}} = \sum_{a=1}^{m^{\mathrm{S}}} S_{ia}^{\mathrm{S}} F_a = -\sum_{a=1}^{m^{\mathrm{S}}} S_{ia}^{\mathrm{S}} k_a \sum_{j=1}^{n} S_{ja}^{\mathrm{S}} x_j \qquad (i = 1, \ldots, n) . \tag{5.322}$$

For the resultant damper force R_i^{D} on body i an equivalent expression is obtained with elements of the incidence matrix $\underline{S}^{\mathrm{D}}$ for the damper graph and with damping constants d_a $(a = 1, \ldots, m^{\mathrm{D}})$ instead of spring constants k_a.

With these expressions (5.320) takes the form

$$m_i\ddot{x}_i + \sum_{a=1}^{m^{\mathrm{D}}} S^{\mathrm{D}}_{ia} d_a \sum_{j=1}^{n} S^{\mathrm{D}}_{ja}\dot{x}_j + \sum_{a=1}^{m^{\mathrm{S}}} S^{\mathrm{S}}_{ia} k_a \sum_{j=1}^{n} S^{\mathrm{S}}_{ja} x_j = F^{\mathrm{e}}_i \tag{5.323}$$

$(i = 1, \ldots, n)$. All n equations are combined in matrix form as follows:

$$\underline{M}\,\underline{\ddot{x}} + \underline{S}^{\mathrm{D}}\underline{D}\,\underline{S}^{\mathrm{D}^T}\underline{\dot{x}} + \underline{S}^{\mathrm{S}}\underline{K}\,\underline{S}^{\mathrm{S}^T}\underline{x} = \underline{F}^{\mathrm{e}}\;. \tag{5.324}$$

Here, $\underline{M}$ is the diagonal matrix of the n masses, $\underline{D}$ is the diagonal matrix of the m^{D} damper constants, and $\underline{K}$ is the diagonal matrix of the m^{S} spring constants.

For $\underline{x}$ the expression in the second equation (5.319) is substituted. The equation is then premultiplied by $\underline{T}$. This results in the following equation of motion for the variables $\underline{q}$ with symmetric mass, damping and stiffness matrices:

$$\begin{aligned}\underline{T}\,\underline{M}\,\underline{T}^T\underline{\ddot{q}} &+ (\underline{T}\,\underline{S}^{\mathrm{D}})\underline{D}(\underline{T}\,\underline{S}^{\mathrm{D}})^T\underline{\dot{q}} \\ &+ (\underline{T}\,\underline{S}^{\mathrm{S}})\underline{K}(\underline{T}\,\underline{S}^{\mathrm{S}})^T\underline{q} = -\underline{T}\,\underline{F}^{\mathrm{e}}\;.\end{aligned} \tag{5.325}$$

In this formulation the physical parameters $\underline{M}$, $\underline{D}$ and $\underline{K}$, the parameters $\underline{S}^{\mathrm{D}}$ and $\underline{S}^{\mathrm{S}}$ describing the system topology and the parameter $\underline{T}$ describing the choice of coordinates appear separately in the equations.

5.9.2 Chains Without Coupling to Inertial Space

In a chain of bodies without spring and damper connections to body 0 all spring and damper forces are internal forces without effect on the absolute acceleration $\ddot{x}_{\mathrm{C}}$ of the composite system center of mass. This acceleration is obtained by summing all n Eqs. (5.323). With M being the total system mass this is the equation

$$\ddot{x}_{\mathrm{C}} = \frac{1}{M}\sum_{i=1}^{n} F^{\mathrm{e}}_i\;. \tag{5.326}$$

Since $\ddot{x}_{\mathrm{C}}$ is independent of the variables $\underline{q}$ it must be possible to extract from (5.323) $n-1$ independent equations of motion for as many relative displacements. In the special case of a chain with $n = 2$ bodies the procedure is known. One must multiply the first equation by m_2, the second by m_1 and then take the difference. This results in the well-known equation

$$\frac{m_1 m_2}{m_1 + m_2}\,\ddot{q} + d\dot{q} + kq = \frac{m_1 F^{\mathrm{e}}_2 - m_2 F^{\mathrm{e}}_1}{m_1 + m_2} \tag{5.327}$$

with the reduced mass $m_1m_2/(m_1+m_2)$ and with the relative displacement $q = x_2 - x_1$. In what follows the procedure will be shown for the general case $n \geq 2$. Starting point is the principle of virtual power:

$$\sum_{i=1}^{n} \delta\dot{x}_i(m_i\ddot{x}_i - F_i) = 0\,, \qquad F_i = F_i^{\mathrm{e}} + R_i^{\mathrm{S}} + R_i^{\mathrm{D}} \qquad (i = 1,\ldots,n)\,. \quad (5.328)$$

This is equivalent to Newton's equation (5.320). Auxiliary coordinates z_1, $\ldots, z_n$ are defined through the equations

$$x_i = x_{\mathrm{C}} + z_i \quad (i = 1,\ldots,n)\,, \qquad \sum_{i=1}^{n} z_i m_i = 0\,. \qquad (5.329)$$

This is the scalar form of (5.32). Therefore, (5.39) is valid:

$$\underline{z} = \underline{\mu}^T \underline{x}\,. \qquad (5.330)$$

The definition of the matrix $\underline{\mu}$ and one of its properties are (see (5.38) and (5.40))

$$\mu_{ij} = \delta_{ij} - \frac{m_i}{M} \quad (i,j = 1,\ldots,n)\,, \qquad \underline{\mu}\,\underline{M} = \underline{\mu}\,\underline{M}\,\underline{\mu}^T\,. \qquad (5.331)$$

With (5.329), (5.328) becomes

$$\delta\dot{x}_{\mathrm{C}}\left(M\ddot{x}_{\mathrm{C}} - \sum_{i=1}^{n} F_i^{\mathrm{e}}\right) + \sum_{i=1}^{n} \delta\dot{z}_i(m_i\ddot{z}_i - F_i) = 0\,. \qquad (5.332)$$

Since $\delta\dot{x}_{\mathrm{C}}$ is independent the coefficient is zero. This is (5.326). The rest of the equation is written in the form

$$\delta\dot{\underline{z}}^T(\underline{M}\,\ddot{\underline{z}} - \underline{F}) = 0\,. \qquad (5.333)$$

Into this equation the following expressions are substituted which are obtained from (5.330) and (5.324)

$$\ddot{\underline{z}} = \underline{\mu}^T\ddot{\underline{x}}\,, \qquad \delta\dot{\underline{z}}^T = \delta\dot{\underline{x}}^T\underline{\mu}\,, \qquad \underline{F} = \underline{F}^{\mathrm{e}} - \underline{S}^{\mathrm{D}}\underline{D}\,\underline{S}^{\mathrm{D}^T}\dot{\underline{x}} - \underline{S}^{\mathrm{S}}\underline{K}\,\underline{S}^{\mathrm{S}^T}\underline{x}\,. \quad (5.334)$$

This produces the equation

$$\delta\dot{\underline{x}}^T\left(\underline{\mu}\,\underline{M}\,\underline{\mu}^T\ddot{\underline{x}} + \underline{\mu}\,\underline{S}^{\mathrm{D}}\underline{D}\,\underline{S}^{\mathrm{D}^T}\dot{\underline{x}} + \underline{\mu}\,\underline{S}^{\mathrm{S}}\underline{K}\,\underline{S}^{\mathrm{S}^T}\underline{x} - \underline{\mu}\,\underline{F}^{\mathrm{e}}\right) = 0\,. \qquad (5.335)$$

Since the elements of $\delta\dot{\underline{x}}$ are independent the expression in parentheses is zero. The damping matrix and the stiffness matrix appear to be unsymmetric, but they are not. In a system without coupling to inertial space the identities hold: $\underline{\mu}\,\underline{S}^{\mathrm{D}} = \underline{S}^{\mathrm{D}}$ and $\underline{\mu}\,\underline{S}^{\mathrm{S}} = \underline{S}^{\mathrm{S}}$. Proof for $\underline{S}^{\mathrm{S}}$: When no spring is connected to body 0 then every column of $\underline{S}^{\mathrm{S}}$ contains exactly one element $+1$ and one element -1 so that $\sum_{j=1}^{n} S^{\mathrm{S}}_{ja} = 0$ $(a = 1, \ldots, n)$. Hence,

$$\left(\underline{\mu}\,\underline{S}^{\mathrm{S}}\right)_{ia} = \sum_{j=1}^{n} \left(\delta_{ij} - \frac{m_i}{M}\right) S^{\mathrm{S}}_{ja} = S^{\mathrm{S}}_{ia} \,. \tag{5.336}$$

End of proof. Hence, (5.335) yields the equation with symmetric matrices

$$\underline{\mu}\,\underline{M}\,\underline{\mu}^T \ddot{\underline{x}} + \underline{S}^{\mathrm{D}} \underline{D}\,\underline{S}^{\mathrm{D}^T} \dot{\underline{x}} + \underline{S}^{\mathrm{S}} \underline{K}\,\underline{S}^{\mathrm{S}^T} x = \underline{\mu}\,\underline{F}^{\mathrm{e}} \,. \tag{5.337}$$

This equation can be obtained in a much simpler way. Just premultiply (5.324) with $\underline{\mu}$ and make use of the identities $\underline{\mu}\,\underline{M} = \underline{\mu}\,\underline{M}\,\underline{\mu}^T$, $\underline{\mu}\,\underline{S}^{\mathrm{D}} = \underline{S}^{\mathrm{D}}$ and $\underline{\mu}\,\underline{S}^{\mathrm{S}} = \underline{S}^{\mathrm{S}}$. In other words: The insertion of $\underline{\mu}$ and $\underline{\mu}^T$ into (5.324) in the described way has the effect of eliminating the equation of motion for the composite system center of mass. This effect is known from (5.48).

For $\underline{x}$ the expression from the second Eq. (5.319) is substituted. Following this, the equation is premultiplied by $\underline{T}$. This results in the equations of motion

$$\begin{aligned}(\underline{T}\,\underline{\mu})\underline{M}\,(\underline{T}\,\underline{\mu})^T \ddot{\underline{q}} + (\underline{T}\,\underline{S}^{\mathrm{D}})\underline{D}(\underline{T}\,\underline{S}^{\mathrm{D}})^T \dot{\underline{q}}& \\ + (\underline{T}\,\underline{S}^{\mathrm{S}})\underline{K}(\underline{T}\,\underline{S}^{\mathrm{S}})^T \underline{q} = -\underline{T}\,\underline{\mu}\,\underline{F}^{\mathrm{e}} \,.&\end{aligned} \tag{5.338}$$

The damping matrix and the stiffness matrix are the same as in (5.325). Only the mass matrix and the right-hand side term are different. In what follows a physical interpretation is given to the elements of the mass matrix. This is done with the help of the numbers σ_a $(a = 1, \ldots, n)$, of the sets κ_{ab} $(a, b = 1, \ldots, n)$ of vertices and of the ordering relationship arc $a <$ arc b introduced in Sect. 5.3.2. Let m_{ab} be the total mass of all bodies which are associated with the vertices in the set κ_{ab} $(m_0 = 0)$. Examples: For the coordinate graph of Fig. 5.42a one has $\sigma_1 = \sigma_2 = -1$, $\sigma_3 = +1$, $m_{13} = m_1$, $m_{31} = m_2 + m_5$ und $m_{22} = m_2 + m_3 + m_4 + m_5$. From the definitions follow the relationships

$$\sum_{i=1}^{n} T_{ai} m_i = \sigma_a m_{aa} \,, \tag{5.339}$$

$$(\underline{T}\,\underline{\mu})_{ai} = \sum_{\ell=1}^{n} T_{a\ell} \left(\delta_{\ell i} - \frac{m_\ell}{M}\right) = T_{ai} - \frac{\sigma_a m_{aa}}{M} \,, \tag{5.340}$$

$$\sum_{i=1}^{n} T_{ai} T_{bi} m_i = \begin{cases} m_{aa} & (a = b) \\ \sigma_a \sigma_b m_{aa} & (\text{arc } a > \text{arc } b) \\ \sigma_a \sigma_b m_{bb} & (\text{arc } b > \text{arc } a) \\ 0 & (\text{else}) \,. \end{cases} \tag{5.341}$$

With this the elements of the mass matrix can be given the form

$$\begin{aligned}&\left[(\underline{T}\,\underline{\mu})\underline{M}\,(\underline{T}\,\underline{\mu})^T\right]_{ab}\\&=\sum_{i=1}^{n}(\underline{T}\,\underline{\mu})_{ai}(\underline{T}\,\underline{\mu})_{bi}m_i=\sum_{i=1}^{n}\left(T_{ai}-\frac{\sigma_a m_{aa}}{M}\right)\left(T_{bi}-\frac{\sigma_b m_{bb}}{M}\right)m_i\\&=\sum_{i=1}^{n}T_{ai}T_{bi}m_i-\sigma_a\sigma_b\,\frac{m_{aa}m_{bb}}{M}\\&=\begin{cases}m_{aa}(M-m_{aa})/M & (a=b)\\+\sigma_a\sigma_b m_{aa}(M-m_{bb})/M & (\text{arc } b<\text{arc } a)\\+\sigma_a\sigma_b m_{bb}(M-m_{aa})/M & (\text{arc } a<\text{arc } b)\\-\sigma_a\sigma_b m_{ab}m_{ba}/M & (\text{else})\,.\end{cases}\end{aligned}\tag{5.342}$$

If arc $b <$ arc a then $m_{aa} = m_{ab}$ and $M - m_{bb} = m_{ba} < M - m_{aa}$. Similarly, if arc $a <$ arc b then $m_{bb} = m_{ba} < m_{aa}$ and $M - m_{aa} = m_{ab}$. In the case denoted *else* the following is true: $m_{aa} = m_{ab}$ and $m_{bb} = m_{ba} < M - m_{aa}$. From the inequalities it follows that in every row and in every column of the mass matrix the diagonal element has the largest absolute value. The equalities give the matrix elements the symmetric form

$$\begin{aligned}&\left[(\underline{T}\,\underline{\mu})\underline{M}\,(\underline{T}\,\underline{\mu})^T\right]_{ab}\\&=\begin{cases}m_{aa}(M-m_{aa})/M & (a=b)\\+\sigma_a\sigma_b\, m_{ab}m_{ba}/M & (\text{arc } a<\text{arc } b\ \text{ or arc } b<\text{arc } a)\\-\sigma_a\sigma_b\, m_{ab}m_{ba}/M & (\text{else})\,.\end{cases}\end{aligned}\tag{5.343}$$

Every element is a reduced mass in a generalized sense. Example: The coordinate graph of Fig. 5.42a yields $\left[(\underline{T}\,\underline{\mu})\underline{M}\,(\underline{T}\,\underline{\mu})^T\right]_{13} = m_1(m_2 + m_5)/M$.

Let the coordinates $\underline{q}$ be chosen such that only q_n is an absolute displacement (of an arbitrary body). All other coordinates are relative displacements. Then, only arc n is incident with vertex 0 in the coordinate graph. Consequently, $m_{nn} = M$ and $m_{na} = 0$ $(a \neq n)$. From this it follows that all elements in the nth row and in the nth column of the mass matrix are equal to zero. Each of the remaining terms in (5.338) has the form $\underline{T}\,\underline{\mu}\,\underline{X}$ with some column matrix X (where the factor $\underline{\mu}$ is missing it can be inserted again). The nth element is the sum $\sum_{i=1}^{n}(\underline{T}\,\underline{\mu})_{ni}X_i$. With the special choice of coordinates it follows from (5.340) that $(\underline{T}\,\underline{\mu})_{ni} = 0$ $(i = 1, \ldots, n)$ since $T_{ni} = \sigma_n$ $(i = 1, \ldots, n)$ and $m_{nn} = M$. This proves that the nth equation of (5.340) is the identity $0 = 0$. Thus, one has $n - 1$ independent equations for relative displacements $q_1, \ldots, q_{n-1}$. In the special case $n = 2$ the system of $n - 1$ equations has the form (5.327).

6

Impact Problems in Multibody Systems

The previous chapter was devoted to the mathematical description of continuous motions of multi-body systems. In contrast, in the present chapter the case is studied in which a system experiences discontinuous changes of velocities and angular velocities. Such phenomena occur when a multi-body system collides with a single body or with another multi-body system or when two bodies belonging to one and the same multi-body system collide with one another. Typical examples are illustrated in Fig. 6.1. The actual physical processes during impact are highly complex. In order to render the problem amenable to mathematical treatment some simplifying assumptions must be made. Whether these are acceptable must be judged in every particular case of application of the resulting formalism. The assumptions concern the dynamic behavior of bodies under the action of impulsive forces and the phenomena in the immediate vicinity of the point of collision, in particular. They are the same which are made in the classical treatment of the collision between two rigid bodies. In addition, some assumptions are made about the nature of constraints in joints. These latter assumptions do not go beyond the ones made in the last chapter.

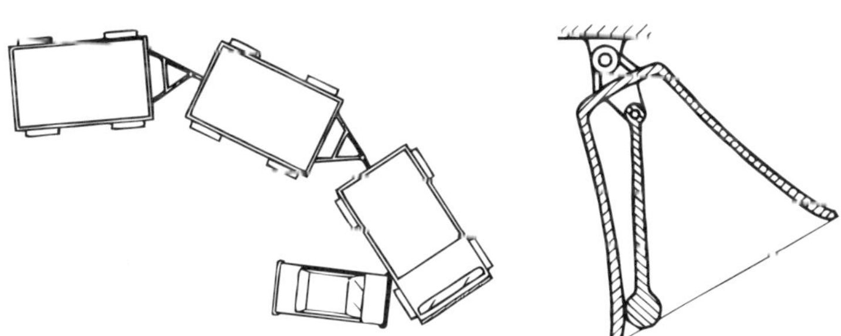

a) b)

Fig. 6.1. Collision between two systems **(a)** and collision within a system **(b)**

6.1 Basic Assumptions

As regards the behavior of bodies it is postulated that the impact is taking place in such a short time interval Δt that in the mathematical description the idealization $\Delta t \to 0$ is possible. This implies that any propagation of waves of deformations and stresses through bodies is neglected since such processes require finite periods of time. Hence, all bodies of a system are treated as rigid bodies during impact. During the infinitesimally short time interval positions and angular orientations of bodies remain unchanged since all velocities and angular velocities remain finite. Springs and dampers in a system do not play any role since they exert forces and torques of finite magnitude whose integrals over the infinitesimally short time interval are zero. Only impulsive forces can cause discontinuous changes of velocities and angular velocities. A force $\mathbf{F}(t)$ is called impulsive force if the integral $\int_{t_0}^{t_0+\Delta t} \mathbf{F}(t)\,\mathrm{d}t$ over the time interval Δt converges toward a finite quantity $\hat{\mathbf{F}}$ when Δt approaches zero. For this to be the case the magnitude of $\mathbf{F}(t)$ must tend toward infinity during Δt in the limit case $\Delta t \to 0$. The quantity $\hat{\mathbf{F}}$ is called impulse. The torque of an impulse is called impulse couple.

The assumption of rigid bodies applies, in particular, to mechanisms in joints. This has the consequence that an impulsive force acting at a point of collision causes simultaneously acting impulsive constraint forces and impulsive constraint torques in the joints of a system. As in the previous chapter on continuous motion joints are assumed to be frictionless. This together with the assumption $\Delta t \to 0$ has the consequence that impulsive constraint forces and impulsive constraint torques have constant directions during the time interval of collision. As in Sect. 5.6.5 the constraint force in joint a is called $\mathbf{X}_a$, and the torque about the articulation point a at the tip of the vector $\mathbf{c}_{ia}$ is called $\mathbf{Y}_a$. Figure 6.2 depicts a single body of a multibody system which is isolated from the rest of the system. At the point of collision located by the vector $\boldsymbol{\varrho}$ the impulsive force $\mathbf{F}$ is acting, and in a single joint a the constraint force $\mathbf{X}_a$ and the constraint torque $\mathbf{Y}_a$ are acting. Newton's law

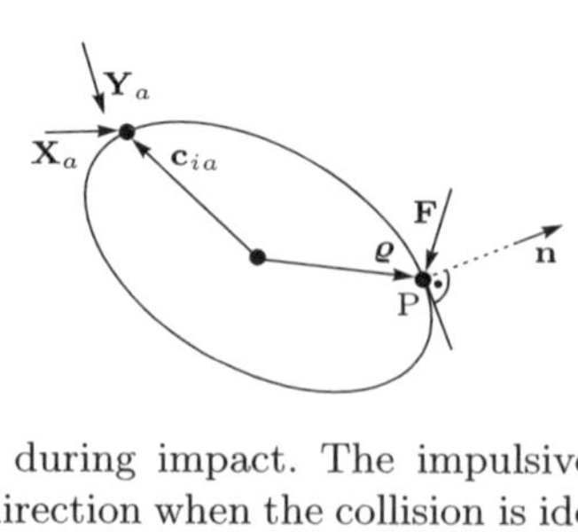

Fig. 6.2. Isolated body during impact. The impulsive force $\mathbf{F}$ at the point of collision P has unknown direction when the collision is ideally plastic, and it has the direction of the impact normal with unit vector $\mathbf{n}$ when the collision is frictionless. Constraint reactions $\mathbf{X}_a$ and $\mathbf{Y}_a$ in a single joint

and the angular momentum theorem read

$$m\ddot{\mathbf{r}} = \mathbf{F} + \mathbf{X}_a\,, \qquad \mathsf{J}\cdot\dot{\boldsymbol{\omega}} + \boldsymbol{\omega}\times\mathsf{J}\cdot\boldsymbol{\omega} = \boldsymbol{\varrho}\times\mathbf{F} + \mathbf{c}_{ia}\times\mathbf{X}_a + \mathbf{Y}_a\,. \tag{6.1}$$

Both equations are integrated over the time interval Δt. In the limit $\Delta t \to 0$ the tensor J and the vector $\boldsymbol{\varrho}$ have constant coordinates in inertial space. The finite term $\boldsymbol{\omega}\times\mathsf{J}\cdot\boldsymbol{\omega}$ does not contribute to the integral. Therefore, the resulting equations read

$$m\Delta\dot{\mathbf{r}} = \hat{\mathbf{F}} + \hat{\mathbf{X}}_a\,, \qquad \mathsf{J}\cdot\Delta\boldsymbol{\omega} = \boldsymbol{\varrho}\times\hat{\mathbf{F}} + \mathbf{c}_{ia}\times\hat{\mathbf{X}}_a + \hat{\mathbf{Y}}_a\,. \tag{6.2}$$

They are homogeneous linear equations with constant coefficients for the increments $\Delta\dot{\mathbf{r}}$ and $\Delta\boldsymbol{\omega}$ of velocity and angular velocity, respectively, and for impulses and impulse couples.

Rigidity of the bodies is assumed everywhere with the exception of the immediate vicinity of the point of collision. In a volume surrounding this point elastic and/or plastic deformations are allowed to occur. It is assumed, however, that this volume is infinitesimally small so that the previous assumptions of rigid bodies and of an infinitesimally short time interval of impact are not violated. Two cases are distinguished:

1. Ideally plastic collisions: Plastic deformations, eventually in combination with friction at the point of collision, have the effect that the colliding bodies have zero velocity relative to one another at the point of collision immediately after impact. In this case both magnitude and direction of the impulse $\hat{\mathbf{F}}$ are unknown. Let $\mathbf{v}_1$ and $\mathbf{v}_2$ be the absolute velocities immediately before impact of the two colliding body-fixed points and let $\Delta\mathbf{v}_1$ and $\Delta\mathbf{v}_2$ be the finite increments of these velocities as the result of the collision. The condition that the velocity of the two body-fixed points relative to one another is zero immediately after impact reads

$$(\mathbf{v}_1 + \Delta\mathbf{v}_1) - (\mathbf{v}_2 + \Delta\mathbf{v}_2) = \mathbf{0}\,. \tag{6.3}$$

This vector equation is required for determining the unknown vector $\hat{\mathbf{F}}$.

2. Frictionless collisions: Under this assumption the collision force $\mathbf{F}$ has the constant direction of the *impact normal*, i.e. of the normal to the common tangent plane at the point of collision. In Fig. 6.2 the impact normal is identified by the unit vector $\mathbf{n}$. Together with $\mathbf{F}$ also the impulse $\hat{\mathbf{F}}$ has this direction, i.e.

$$\hat{\mathbf{F}} = \hat{F}\mathbf{n}\,. \tag{6.4}$$

Since $\mathbf{n}$ is known, only the scalar $\hat{F}$ is unknown.

Frictionless collisions actually represent a whole family of cases because it must be distinguished whether the deformation of the bodies in the vicinity of the point of collision is a purely plastic compression or whether the phase of compression is followed by a phase of either total or partial decompression.

Let $\hat{\mathbf{F}}^{\mathrm{c}}$ (superscript ^{c} for compression) be the impulse which is exerted on one of the colliding bodies by the other colliding body during the phase of compression. A parameter e called *coefficient of restitution* is defined. It is the ratio between the impulse in the phase of decompression and $\hat{F}^{\mathrm{c}}$. Hence, the total impulse during impact is

$$\hat{\mathbf{F}} = (1+e)\hat{\mathbf{F}}^{\mathrm{c}} . \tag{6.5}$$

The parameter e is zero if the compression is fully plastic and one if it is fully elastic. For partially elastic compressions it is in the range $0 < e < 1$.

Let $\Delta\mathbf{v}_{\mathrm{C}}^{\mathrm{c}}$ be the finite increment of the absolute velocity $\mathbf{v}_{\mathrm{C}}$ of the body center of mass in the compression phase. Similarly, let $\Delta\boldsymbol{\omega}^{\mathrm{c}}$ be the increment of the absolute angular velocity in the compression phase. Also the point of collision experiences a velocity increment in the compression phase. It is denoted $\Delta\mathbf{v}^{\mathrm{c}}$. Since the velocity of deformation of the body is zero at the beginning and at the end of the compression phase the rigid-body formula is applicable:

$$\Delta\mathbf{v}^{\mathrm{c}} = \Delta\mathbf{v}_{\mathrm{C}}^{\mathrm{c}} + \Delta\boldsymbol{\omega}^{\mathrm{c}} \times \boldsymbol{\varrho} . \tag{6.6}$$

As in the classical theory of collision between two bodies the assumption is made that in both colliding bodies the compression phase ends at the same time. Under this condition the following statement is valid. At the end of the compression phase the component of the relative velocity of the two colliding body-fixed points in the direction of the normal $\mathbf{n}$ is zero. With indices 1 and 2 for the two points this is the equation

$$[(\mathbf{v}_1 + \Delta\mathbf{v}_1^{\mathrm{c}}) - (\mathbf{v}_2 + \Delta\mathbf{v}_2^{\mathrm{c}})] \cdot \mathbf{n} = 0 \tag{6.7}$$

where $\mathbf{v}_1$ and $\mathbf{v}_2$ are, as before, the absolute velocities of the two points immediately before impact. The dynamic equations (6.2) as well as the kinematic equation (6.6) are homogeneous linear equations with constant coefficients for unknown velocity increments, impulses and impulse couples. This allows the conclusion that for all these unknowns relationships in the form of (6.5) exist. In particular, $\Delta\mathbf{v}_1 = (1+e)\Delta\mathbf{v}_1^{\mathrm{c}}$ and $\Delta\mathbf{v}_2 = (1+e)\Delta\mathbf{v}_2^{\mathrm{c}}$ are the velocity increments of the two colliding body-fixed points between the moments immediately before and immediately after impact. With these expressions, (6.7) becomes

$$\left[\left(\mathbf{v}_1 + \frac{\Delta\mathbf{v}_1}{1+e}\right) - \left(\mathbf{v}_2 + \frac{\Delta\mathbf{v}_2}{1+e}\right)\right] \cdot \mathbf{n} = 0 \tag{6.8}$$

or

$$[(\mathbf{v}_1 + \Delta\mathbf{v}_1) - (\mathbf{v}_2 + \Delta\mathbf{v}_2)] \cdot \mathbf{n} = -e(\mathbf{v}_1 - \mathbf{v}_2) \cdot n . \tag{6.9}$$

The quantity on the left-hand side is the component along $\mathbf{n}$ of the velocity of the two colliding body-fixed points relative to one another immediately after impact. It is $(-e)$ times the same quantity immediately before impact. This is the same kinematic relationship that is valid in the classical theory of collision between two rigid bodies (Routh [68]). The single scalar equation is required for determining the unknown scalar $\hat{F}$ in (6.4).

6.2 Velocity Increments. Impulses

With the assumptions specified in the previous section the problem of impact is now reduced to the determination of two groups of mechanical quantities. One group is made up of instantaneous finite increments of velocities. Velocities of interest are the first time derivatives of generalized coordinates as well as angular velocities of bodies and translational velocities of body centers of mass and of other body-fixed points. The second group comprises the impulse applied to a system at the point of collision and, in addition, constraint impulses and constraint impulse couples in joints. Only joints with holonomic constraints are considered. The unknowns must be determined from a set of equations which consists of dynamics equations of the general form (6.2), of kinematics equations of the general form (6.6) and of a single equation which has either the form (6.3) or the form (6.9), depending on the type of collision. Equations (6.2) and (6.6) are homogeneous linear equations for the unknowns, whereas (6.3) and (6.9) are inhomogeneous linear equations. The initial velocities $\mathbf{v}_1$ and $\mathbf{v}_2$ as well as the position of the system which determines the constant coefficients of the equations are known. From this it follows that all unknown constraint impulses and impulse couples in joints as well as all unknown velocity and angular velocity increments are proportional to the impulse $\hat{\mathbf{F}}$ at the point of collision. The unknown impulse $\hat{\mathbf{F}}$ itself is determined either from (6.3) or from (6.9).

In what follows the proportionality between the velocity increments of the two colliding bodies at the point of collision on the one hand and the impulse $\hat{\mathbf{F}}$ on the other hand is expressed. The general situation is schematically shown in Fig. 6.3. The two colliding bodies are labeled i and j. At the point of collision P they are shown separated from one another. Arbitrarily it is assumed that the impulse $+\hat{\mathbf{F}}$ is applied to body i, and, consequently, the impulse $-\hat{\mathbf{F}}$ to body j. As is shown in Fig. 6.1 the two bodies belong either to two different systems or to one and the same system. In the first case each system is subject to a single impulse, whereas in the second case the single system is subject to $+\hat{\mathbf{F}}$ and to $-\hat{\mathbf{F}}$ in two different places. Let $\Delta\mathbf{v}_i$ and $\Delta\mathbf{v}_j$ be the velocity increments at the point P on body i and

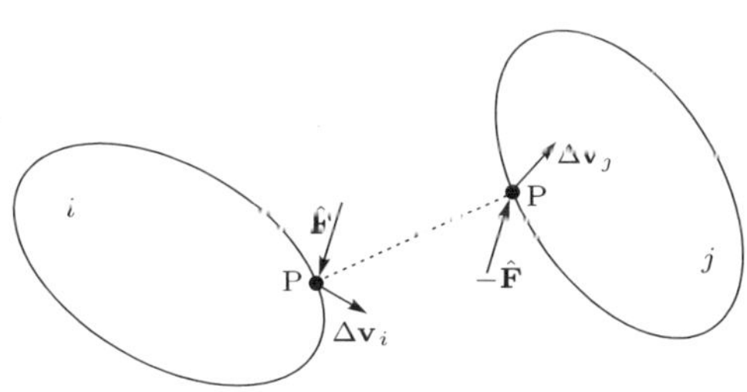

Fig. 6.3. Two colliding bodies i and j subject to impulses $+\hat{\mathbf{F}}$ and $-\hat{\mathbf{F}}$ at the point of collision P. Vectors $\boldsymbol{\varrho}_i$, $\boldsymbol{\varrho}_j$ locating P and velocity increments $\Delta\mathbf{v}_i$, $\Delta\mathbf{v}_j$ at P

on body j, respectively. The unknown proportionality factors can only be tensors because the velocity increments do not have, in general, the direction of $\hat{\mathbf{F}}$. The following notation is adopted. For two points with arbitrary labels k and ℓ the tensor relating $\Delta\mathbf{v}_k$ at point k to the impulse $\hat{\mathbf{F}}$ at point ℓ is denoted $\mathsf{U}_{k\ell}$. With this notation the two velocity increments in Fig. 6.3 have the following forms.

Collision between two multibody systems:

$$\left.\begin{aligned}\Delta\mathbf{v}_i &= \quad \mathsf{U}_{ii}\cdot\hat{\mathbf{F}}\,,\\ \Delta\mathbf{v}_j &= -\mathsf{U}_{jj}\cdot\hat{\mathbf{F}}\,.\end{aligned}\right\} \tag{6.10}$$

Collision within one multibody system:

$$\left.\begin{aligned}\Delta\mathbf{v}_i &= (\mathsf{U}_{ii}-\mathsf{U}_{ij})\cdot\hat{\mathbf{F}}\,,\\ \Delta\mathbf{v}_j &= (\mathsf{U}_{ji}-\mathsf{U}_{jj})\cdot\hat{\mathbf{F}}\,.\end{aligned}\right\} \tag{6.11}$$

In Sect. 6.3 explicit expressions for the tensors in these equations are developed. In what follows (6.3) and (6.9) are formulated in terms of these tensors. In both equations the indices 1 and 2 are replaced by i and j, respectively. Equation (6.3) for the ideally plastic collision between two multibody systems reads $(\mathbf{v}_i+\mathsf{U}_{ii}\cdot\hat{\mathbf{F}})-(\mathbf{v}_j-\mathsf{U}_{jj}\cdot\hat{\mathbf{F}})=\mathbf{0}$ or

$$(\mathsf{U}_{ii}+\mathsf{U}_{jj})\cdot\hat{\mathbf{F}}=-(\mathbf{v}_i-\mathbf{v}_j)\,. \tag{6.12}$$

In order to solve this equation for $\hat{\mathbf{F}}$ all vectors and tensors are decomposed into scalar coordinates in some common reference base. This yields the equation

$$(\underline{U}_{ii}+\underline{U}_{jj})\underline{\hat{F}}=-(\underline{v}_i-\underline{v}_j)\,. \tag{6.13}$$

The solution is

$$\text{plastic coll. betw. two systems:}\quad \underline{\hat{F}}=-(\underline{U}_{ii}+\underline{U}_{jj})^{-1}(\underline{v}_i-\underline{v}_j)\,. \tag{6.14}$$

In the same way (6.4) combined with (6.11) leads to the equation

$$\text{plastic coll. within one system:}\quad \underline{\hat{F}}=-(\underline{U}_{ii}-\underline{U}_{ij}-\underline{U}_{ji}+\underline{U}_{jj})^{-1}(\underline{v}_i-\underline{v}_j)\,. \tag{6.15}$$

Next, (6.9) for frictionless collisions is formulated. With (6.10) this is the equation $(\mathbf{v}_i-\mathbf{v}_j)\cdot\mathbf{n}+\hat{F}\mathbf{n}\cdot(\mathsf{U}_{ii}+\mathsf{U}_{jj})\cdot\mathbf{n}=-e(\mathbf{v}_i-\mathbf{v}_j)\cdot\mathbf{n}$, whence follows

$$\text{frictionless coll. betw. two systems:}\quad \hat{F}=-(1+e)\,\frac{(\mathbf{v}_i-\mathbf{v}_j)\cdot\mathbf{n}}{\mathbf{n}\cdot(\mathsf{U}_{ii}+\mathsf{U}_{jj})\cdot\mathbf{n}}\,. \tag{6.16}$$

In the same way (6.9) combined with (6.11) leads to the equation

$$\text{frictionless coll. within one system:}\quad \hat{F}=-(1+e)\,\frac{(\mathbf{v}_i-\mathbf{v}_j)\cdot\mathbf{n}}{\mathbf{n}\cdot(\mathsf{U}_{ii}-\mathsf{U}_{ij}-\mathsf{U}_{ji}+\mathsf{U}_{jj})\cdot\mathbf{n}}\,. \tag{6.17}$$

6.3 Analogy to the Law of Maxwell and Betti

In this section explicit expressions are developed for the tensors U_{ij} $(i, j = 1, \ldots, n)$ for arbitrary multibody systems. Starting point are equations of continuous motion. Tree-structured systems are governed by (5.84) and (5.83):

$$\underline{A}\,\underline{\ddot{q}} = \underline{B}\,, \tag{6.18}$$

$$\left.\begin{aligned} \underline{A} &= \underline{\mathbf{a}}_1^T \cdot \underline{m}\,\underline{\mathbf{a}}_1 + \underline{\mathbf{a}}_2^T \cdot \underline{\mathsf{J}} \cdot \underline{\mathbf{a}}_2\,, \\ \underline{B} &= \underline{\mathbf{a}}_1^T \cdot (\underline{\mathbf{F}} - \underline{m}\,\underline{\mathbf{b}}_1) + \underline{\mathbf{a}}_2^T \cdot (\underline{\mathbf{M}}^* - \underline{\mathsf{J}} \cdot \underline{\mathbf{b}}_2)\,. \end{aligned}\right\} \tag{6.19}$$

When the variables $\underline{q}$ are subject to holonomic constraints caused by the closure of kinematic chains then only a subset of variables $\underline{q}^*$ is independent. The two sets are related through (5.117):

$$\underline{\dot{q}} = \underline{G}\,\underline{\dot{q}}^* + \underline{Q}\,, \qquad \underline{\ddot{q}} = \underline{G}\,\underline{\ddot{q}}^* + \underline{H}\,. \tag{6.20}$$

The system is governed by (5.120) and (5.119):

$$\underline{A}^*\underline{\ddot{q}}^* = \underline{B}^*\,, \tag{6.21}$$

$$\underline{A}^* = \underline{G}^T\underline{A}\,\underline{G}\,, \qquad \underline{B}^* = \underline{G}^T(\underline{B} - \underline{A}\,\underline{H})\,. \tag{6.22}$$

In the absence of constraint equations the matrices are $\underline{G} = \underline{I}$ and $\underline{H} = \underline{0}$. In this case (6.21) is identical with (6.18). In what follows the more general equations (6.21) are used.

In order to find expressions for all tensors U_{ij} $(i, j = 1, \ldots, n)$ it is necessary to apply to each body i of the system one impulsive force $\mathbf{F}_i$. Forces and torques enter the equations via the term $\underline{\mathbf{a}}_1^T \cdot \underline{\mathbf{F}} + \underline{\mathbf{a}}_2^T \cdot \underline{\mathbf{M}}$ in the matrix $\underline{B}$. In the derivation of the equations of motion $\mathbf{F}_i$ was the resultant external force applied to body i at the body i center of mass, and $\mathbf{M}_i$ was the resultant external torque applied to body i. Now, the force $\mathbf{F}_i$ is applied to an arbitrary point of body i which is defined by the vector $\boldsymbol{\varrho}_i$. This has the consequence that the force produces the torque $\boldsymbol{\varrho}_i \times \mathbf{F}_i$. For reasons which will become clear later it is assumed that in addition to the impulsive force $\mathbf{F}_i$ also an impulsive torque $\mathbf{M}_i$ is acting on body i. Thus, the resultant impulsive torque on body i is $\mathbf{M}_i + \boldsymbol{\varrho}_i \times \mathbf{F}_i$. The column matrix $[\boldsymbol{\varrho}_1 \times \mathbf{F}_1 \ \ \ldots \ \ \boldsymbol{\varrho}_n \times \mathbf{F}_n]^T$ is written in the form $\underline{\boldsymbol{\varrho}} \times \underline{\mathbf{F}}$ where $\underline{\boldsymbol{\varrho}}$ is the diagonal matrix of the vectors $\boldsymbol{\varrho}_1, \ldots, \boldsymbol{\varrho}_n$. The matrix $\underline{\mathbf{M}}$ is then replaced by the expression $\underline{\mathbf{M}} + \underline{\boldsymbol{\varrho}} \times \underline{\mathbf{F}}$, and the entire expression $\underline{\mathbf{a}}_1^T \cdot \underline{\mathbf{F}} + \underline{\mathbf{a}}_2^T \cdot \underline{\mathbf{M}}$ is replaced by $(\underline{\mathbf{a}}_1^T + \underline{\mathbf{a}}_2^T \times \underline{\boldsymbol{\varrho}}) \cdot \underline{\mathbf{F}} + \underline{\mathbf{a}}_2^T \cdot \underline{\mathbf{M}}$ (the multiplication symbols $\times$ and $\cdot$ can be interchanged). Since $\underline{\boldsymbol{\varrho}}$ is a diagonal matrix this is written in the final form $(\underline{\mathbf{a}}_1 - \underline{\boldsymbol{\varrho}} \times \underline{\mathbf{a}}_2)^T \cdot \underline{\mathbf{F}} + \underline{\mathbf{a}}_2^T \cdot \underline{\mathbf{M}}$.

Equations (6.21) are integrated over the infinitesimally small time interval $\Delta t \to 0$. During this time interval the matrices $\underline{A}^*$, $\underline{\mathbf{a}}_1$ and $\underline{\mathbf{a}}_2$ are constant because they depend on $q_1, \ldots, q_n$ only. All terms containing neither $\underline{\mathbf{F}}$ nor

$\underline{\mathbf{M}}$ are finite. Therefore, the integration results in the equations

$$\underline{A}^* \Delta \underline{\dot{q}}^* = \underline{G}^T \left[(\underline{\mathbf{a}}_1 - \underline{\boldsymbol{\varrho}} \times \underline{\mathbf{a}}_2)^T \cdot \hat{\underline{\mathbf{F}}} + \underline{\mathbf{a}}_2^T \cdot \hat{\underline{\mathbf{M}}} \right] , \tag{6.23}$$

$$\Delta \underline{\dot{q}} = \underline{G}\, \Delta \underline{\dot{q}}^* = \underline{G}\, (\underline{A}^*)^{-1} \underline{G}^T \left[(\underline{\mathbf{a}}_1 - \underline{\boldsymbol{\varrho}} \times \underline{\mathbf{a}}_2)^T \cdot \hat{\underline{\mathbf{F}}} + \underline{\mathbf{a}}_2^T \cdot \hat{\underline{\mathbf{M}}} \right] . \tag{6.24}$$

The points of application of the impulses experience the finite velocity increments

$$\Delta \mathbf{v}_i = \Delta \dot{\mathbf{r}}_i - \boldsymbol{\varrho}_i \times \Delta \boldsymbol{\omega}_i \qquad (i = 1, \ldots, n) . \tag{6.25}$$

The column matrix of these velocity increments is

$$\Delta \underline{\mathbf{v}} = \Delta \underline{\dot{\mathbf{r}}} - \underline{\boldsymbol{\varrho}} \times \Delta \underline{\boldsymbol{\omega}} \tag{6.26}$$

where $\underline{\boldsymbol{\varrho}}$ is the diagonal matrix from (6.24). From the basic equations (5.41) and (5.43) it follows that

$$\Delta \underline{\dot{\mathbf{r}}} = \underline{\mathbf{a}}_1 \Delta \underline{\dot{q}} , \qquad \Delta \underline{\boldsymbol{\omega}} = \underline{\mathbf{a}}_2 \Delta \underline{\dot{q}} . \tag{6.27}$$

In combination with (6.26) this yields the equation

$$\Delta \underline{\mathbf{v}} = (\underline{\mathbf{a}}_1 - \underline{\boldsymbol{\varrho}} \times \underline{\mathbf{a}}_2) \Delta \underline{\dot{q}} . \tag{6.28}$$

In this equation and in the equation for $\Delta \underline{\boldsymbol{\omega}}$ (6.24) is substituted. Both equations are then combined in matrix form as follows:

$$\begin{bmatrix} \Delta \underline{\mathbf{v}} \\ \Delta \underline{\boldsymbol{\omega}} \end{bmatrix} = \begin{bmatrix} (\underline{\mathbf{a}}_1 - \underline{\boldsymbol{\varrho}} \times \underline{\mathbf{a}}_2) \underline{G}\, (\underline{A}^*)^{-1} \underline{G}^T (\underline{\mathbf{a}}_1 - \underline{\boldsymbol{\varrho}} \times \underline{\mathbf{a}}_2)^T & (\underline{\mathbf{a}}_1 - \underline{\boldsymbol{\varrho}} \times \underline{\mathbf{a}}_2) \underline{G}\, (\underline{A}^*)^{-1} \underline{G}^T \underline{\mathbf{a}}_2^T \\ \underline{\mathbf{a}}_2\, \underline{G}\, (\underline{A}^*)^{-1} \underline{G}^T (\underline{\mathbf{a}}_1 - \underline{\boldsymbol{\varrho}} \times \underline{\mathbf{a}}_2)^T & \underline{\mathbf{a}}_2\, \underline{G}\, (\underline{A}^*)^{-1} \underline{G}^T \underline{\mathbf{a}}_2^T \end{bmatrix} \cdot \begin{bmatrix} \hat{\underline{\mathbf{F}}} \\ \hat{\underline{\mathbf{M}}} \end{bmatrix} . \tag{6.29}$$

Every element of the $(2n \times 2n)$ coefficient matrix is a tensor. The tensors U_{ii}, U_{ij}, U_{ji} and U_{jj} in (6.14) and (6.17) relate velocity increments to impulses. These tensors are the elements (i,i), (i,j), (j,i) and (j,j), respectively, of the matrix. The tensors U_{ii} and U_{jj} in (6.16) are related to two different systems engaged in the collision. Each of these systems has its own Eq. (6.29).

The coefficient matrix in (6.29) is equal to its conjugate, which means that $\mathsf{U}_{ji} = \overline{\mathsf{U}}_{ij}$. When all vectors in (6.29) are decomposed in a common reference base the resulting scalar $(6n \times 6n)$ coefficient matrix is symmetric. The symmetry establishes an analogy between rigid body dynamics and elastostatics (Wittenburg [94]). In elastostatics the following problem is considered. A linearly elastic structure which is in a state of equilibrium is subject to forces and to couples at a number of points $\mathrm{P}_1, \ldots, \mathrm{P}_n$ (a single force $\mathbf{F}_i$ and a single couple $\mathbf{M}_i$ at each point P_i). There exists a linear relationship of the form

$$\begin{bmatrix} \underline{\mathbf{u}} \\ \underline{\boldsymbol{\phi}} \end{bmatrix} = \begin{bmatrix} \underline{\mathsf{A}}_{11} & \underline{\mathsf{A}}_{12} \\ \underline{\mathsf{A}}_{21} & \underline{\mathsf{A}}_{22} \end{bmatrix} \cdot \begin{bmatrix} \underline{\mathbf{F}} \\ \underline{\mathbf{M}} \end{bmatrix} . \tag{6.30}$$

The column matrices $\underline{\mathbf{F}}$, $\underline{\mathbf{M}}$, $\underline{\mathbf{u}}$ and $\underline{\phi}$ are composed of n vectors each. The elements of $\underline{\mathbf{u}}$ are the displacements of the points $\mathrm{P}_1, \ldots, \mathrm{P}_n$ from their locations in the unloaded state, and the elements of $\underline{\phi}$ are the rotation angles of the system at the points $\mathrm{P}_1, \ldots, \mathrm{P}_n$ (small rotation angles can be treated as vectors). Decomposition of all vectors and tensors in a common reference base results in an equation with a $(6n \times 6n)$ coefficient matrix. The law of Maxwell and Betti states that this matrix is symmetric. This means that the tensorial matrix in (6.30) is conjugate symmetric. Thus, the equation is analogous to (6.29). Note, however, the following difference between the two problems. The law of Maxwell and Betti results from energy considerations. These do not provide a closed-form expression for the matrix in (6.30). In fact, a closed-form expression is unknown for all but the simplest types of elastic structures (trusses, for instance). In contrast, the matrix in (6.29) is explicitly known. It is a highly complex expression. This is seen from (6.22) and (6.19) for $\underline{A}^*$ and $\underline{A}$ and from (5.79) and (5.82) for $\underline{\mathbf{a}}_1$ and $\underline{\mathbf{a}}_2$.

In the derivation of (6.29) it was assumed that each body of the system is subject to a single impulsive force. Motivated by the fact that in every joint of a body an impulsive constraint force is acting the generalization to an arbitrary number of impulsive forces is made. This requires modifications in two places. First, if the resultant force on body i is still denoted $\mathbf{F}_i$ then $\mathbf{F}_i = \mathbf{F}_{i1} + \ldots \mathbf{F}_{i\nu_i}$ where ν_i is the number of forces on body i. Define $\underline{\mathbf{F}}^* = [\mathbf{F}_{11} \ldots \mathbf{F}_{1\nu_1} \quad \ldots \quad \mathbf{F}_{n1} \ldots \mathbf{F}_{n\nu_n}]^T$. Then, the equation $\underline{\mathbf{F}} = \underline{D}^T \underline{\mathbf{F}}^*$ defines a block-diagonal matrix $\underline{D}$ composed of elements 0 and 1 (a block of ν_i elements 1 in column i). The resultant torque on body i is no longer $\mathbf{M}_i + \boldsymbol{\varrho}_i \times \mathbf{F}_i$, but $\mathbf{M}_i + \boldsymbol{\varrho}_{i1} \times \mathbf{F}_{i1} + \ldots \boldsymbol{\varrho}_{i\nu_i} \times \mathbf{F}_{i\nu_i}$. The column matrix of the n resultant torques is written in the form $\underline{\mathbf{M}} + \underline{\boldsymbol{\varrho}}^{*T} \times \underline{\mathbf{F}}^*$ where also $\underline{\boldsymbol{\varrho}}^*$ is a block-diagonal matrix (the block $[\boldsymbol{\varrho}_{i1} \ldots \boldsymbol{\varrho}_{i\nu_i}]^T$ in column i). With these modifications the original expression $\underline{\mathbf{a}}_1^T \cdot \underline{\mathbf{F}} + \underline{\mathbf{a}}_2^T \cdot \underline{\mathbf{M}}$ is replaced by $(\underline{\mathbf{a}}_1^T \underline{D}^T + \underline{\mathbf{a}}_2^T \times \underline{\boldsymbol{\varrho}}^{*T}) \cdot \underline{\mathbf{F}}^* + \underline{\mathbf{a}}_2^T \cdot \underline{\mathbf{M}} = (\underline{D}\,\underline{\mathbf{a}}_1 - \underline{\boldsymbol{\varrho}}^* \times \underline{\mathbf{a}}_2)^T \cdot \underline{\mathbf{F}}^* + \underline{\mathbf{a}}_2^T \cdot \underline{\mathbf{M}}$.

The second expression requiring modification is $\Delta\underline{\mathbf{v}}$ in (6.28). The point of application of $\hat{\mathbf{F}}_{ij}$ experiences a velocity increment $\Delta\underline{\mathbf{v}}_{ij}$. Define the column matrix $\Delta\underline{\mathbf{v}}^* = [\Delta\mathbf{v}_{11} \ldots \Delta\mathbf{v}_{1\nu_1} \quad \ldots \quad \Delta\mathbf{v}_{n1} \ldots \Delta\mathbf{v}_{n\nu_n}]^T$. It is left to the reader to verify that

$$\Delta\underline{\mathbf{v}}^* = (\underline{D}\,\underline{\mathbf{a}}_1 - \underline{\boldsymbol{\varrho}}^* \times \underline{\mathbf{a}}_2)\Delta\underline{\dot{q}}\,. \tag{6.31}$$

Equation (6.29) remains valid if the matrices $\Delta\underline{\mathbf{v}}$, $\underline{\hat{\mathbf{F}}}$ and $(\underline{\mathbf{a}}_1 - \underline{\boldsymbol{\varrho}} \times \underline{\mathbf{a}}_2)$ are replaced by $\Delta\underline{\mathbf{v}}^*$, $\underline{\hat{\mathbf{F}}}^*$ and $(\underline{D}\,\underline{\mathbf{a}}_1 - \underline{\boldsymbol{\varrho}}^* \times \underline{\mathbf{a}}_2)$, respectively.

Illustrative Example: This is an example for the action of two impulses on a single body. The point P_2 of the body shown in Fig. 6.4 is constrained to move along a straight guide. At the point P_1 the body is subject to a given impulse $\hat{\mathbf{F}}_1$. The points P_1 and P_2 are located by the body-fixed vectors $\boldsymbol{\varrho}_1$ and $\boldsymbol{\varrho}_2$, respectively, measured from the body center of mass C. To be determined are the reaction impulse $\hat{\mathbf{F}}_2$ exerted on the body by the guide

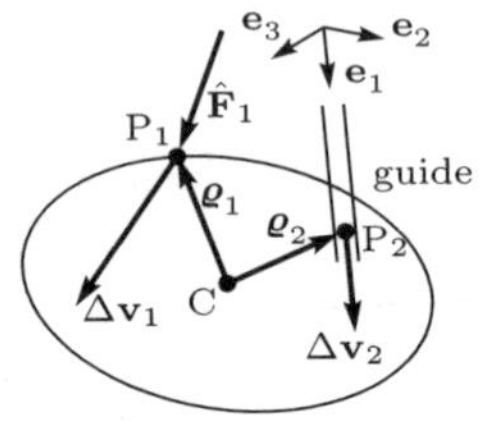

Fig. 6.4. Single body subject to impulse $\hat{\mathbf{F}}_1$ at P_1 and to a constraint at P_2

at P_2, the velocity increments $\Delta\mathbf{v}_1$ and $\Delta\mathbf{v}_2$ of P_1 and P_2, respectively, as well as the angular velocity increment $\Delta\boldsymbol{\omega}$ of the body.
Solution: Instead of adapting (6.29) to the present simple case we develop the desired relationship from basic principles of rigid body dynamics. Integration of Newton's equation and of the angular momentum theorem over the infinitesimal time interval of collision yields the equations

$$m\Delta\mathbf{v}_{\mathrm{C}} = \hat{\mathbf{F}}_1 + \hat{\mathbf{F}}_2\ , \qquad \mathsf{J}\cdot\Delta\boldsymbol{\omega} = \hat{\mathbf{M}} + \boldsymbol{\varrho}_1\times\hat{\mathbf{F}}_1 + \boldsymbol{\varrho}_2\times\hat{\mathbf{F}}_2\ . \tag{6.32}$$

The impulse couple $\hat{\mathbf{M}}$ is deliberately included although it is zero. The points of application of the impulses experience the finite velocity increments $\Delta\mathbf{v}_i = \Delta\mathbf{v}_{\mathrm{C}} - \boldsymbol{\varrho}_i\times\Delta\boldsymbol{\omega}$ $(i=1,2)$. Decomposition of all four equations in some common frame of reference yields for the coordinate matrices of $\Delta\boldsymbol{\omega}$, $\Delta\mathbf{v}_1$ and $\Delta\mathbf{v}_2$ the expressions

$$\left.\begin{aligned} \Delta\underline{\omega} &= \underline{J}^{-1}\left(\underline{\hat{M}} + \underline{\tilde{\varrho}}_1\underline{\hat{F}}_1 + \underline{\tilde{\varrho}}_2\underline{\hat{F}}_2\right)\ , \\ \Delta\underline{v}_i &= \Delta\underline{v}_{\mathrm{C}} - \underline{\tilde{\varrho}}_i\Delta\underline{\omega} \\ &= \frac{1}{m}\left(\underline{\hat{F}}_1 + \underline{\hat{F}}_2\right) - \underline{\tilde{\varrho}}_i\underline{J}^{-1}\left(\underline{\hat{M}} + \underline{\tilde{\varrho}}_1\underline{\hat{F}}_1 + \underline{\tilde{\varrho}}_2\underline{\hat{F}}_2\right) \qquad (i=1,2)\ . \end{aligned}\right\} \tag{6.33}$$

They are combined in the matrix form

$$\begin{bmatrix} \Delta\underline{v}_1 \\ \Delta\underline{v}_2 \\ \Delta\underline{\omega} \end{bmatrix} = \begin{bmatrix} \frac{1}{m}\underline{I} - \underline{\tilde{\varrho}}_1\underline{J}^{-1}\underline{\tilde{\varrho}}_1 & \frac{1}{m}\underline{I} - \underline{\tilde{\varrho}}_1\underline{J}^{-1}\underline{\tilde{\varrho}}_2 & -\underline{\tilde{\varrho}}_1\underline{J}^{-1} \\ \frac{1}{m}\underline{I} - \underline{\tilde{\varrho}}_2\underline{J}^{-1}\underline{\tilde{\varrho}}_1 & \frac{1}{m}\underline{I} - \underline{\tilde{\varrho}}_2\underline{J}^{-1}\underline{\tilde{\varrho}}_2 & -\underline{\tilde{\varrho}}_2\underline{J}^{-1} \\ \underline{J}^{-1}\underline{\tilde{\varrho}}_1 & \underline{J}^{-1}\underline{\tilde{\varrho}}_2 & \underline{J}^{-1} \end{bmatrix} \begin{bmatrix} \underline{\hat{F}}_1 \\ \underline{\hat{F}}_2 \\ \underline{\hat{M}} \end{bmatrix} . \tag{6.34}$$

This represents the scalar form of (6.29). The coefficient matrix is symmetric. If for decomposition of all vectors and tensors the base shown in Fig. 6.4 with the base vector $\mathbf{e}_1$ along the guide is used then the two column matrices are

$$[\underline{\Delta v_{11}\quad \Delta v_{12}\quad \Delta v_{13}\quad \Delta v_2}\quad 0\quad 0\quad \underline{\Delta\omega_1\quad \Delta\omega_2\quad \Delta\omega_3}]^T\ ,$$

$$[\hat{F}_{11}\quad \hat{F}_{12}\quad \hat{F}_{13}\quad 0\quad \underline{\hat{F}_{22}\quad \hat{F}_{23}}\quad 0\quad 0\quad 0\ ]^T\ .$$

The underlined quantities are the unknowns. They are determined by the matrix equation. End of example.

6.4 Constraint Impulses and Impulse Couples in Joints

In the context of (6.1) it has been said that the constraint force $\mathbf{X}_a$ and the constraint torque $\mathbf{Y}_a$ in joint a are defined as in Sect. 5.5.6. Hence, all equations formulated in Sect. 5.5.6 are valid. Equation (5.109) gave for the constraint forces and torques of a tree-structured system the explicit expressions

$$\underline{\mathbf{X}} = \underline{T}(\underline{m}\,\ddot{\underline{\mathbf{r}}} - \underline{\mathbf{F}}) \,, \qquad \underline{\mathbf{Y}} = \underline{T}(\underline{\mathsf{J}} \cdot \dot{\underline{\boldsymbol{\omega}}} - \underline{\mathbf{M}}^* - \underline{\mathbf{C}} \times \underline{\mathbf{X}}) \,. \tag{6.35}$$

By definition, $\mathbf{M}_i^* = \mathbf{M}_i - \boldsymbol{\omega}_i \times \mathsf{J}_i \cdot \boldsymbol{\omega}_i$. The second term drops out when the equation is integrated over the infinitesimally short time interval Δt. In the derivation of (6.29) it was explained that the matrix $\underline{\mathbf{M}}$ has to be replaced by $\underline{\mathbf{M}} + \underline{\boldsymbol{\varrho}} \times \underline{\mathbf{F}}$, because the force $\mathbf{F}_i$ is applied not at the body i center of mass but at a point located by the vector $\boldsymbol{\varrho}_i$ (see Fig. 6.3). Among the forces listed in $\underline{\mathbf{F}}$ only impulsive forces at the point of collision are of interest. This is either a single force $\mathbf{F}$ acting on one body or two forces $\mathbf{F}$ and $-\mathbf{F}$ acting on two bodies. All external torques in $\underline{\mathbf{M}}$ are of finite magnitude. Hence, integration of the equations over the infinitesimally short time interval Δt yields for the constraint impulses and impulse couples in joints the explicit expressions

$$\hat{\underline{\mathbf{X}}} = \underline{T}(\underline{m}\,\Delta\dot{\underline{\mathbf{r}}} - \hat{\underline{\mathbf{F}}}) \,, \qquad \hat{\underline{\mathbf{Y}}} = \underline{T}(\underline{\mathsf{J}} \cdot \Delta\underline{\boldsymbol{\omega}} - \underline{\boldsymbol{\varrho}} \times \hat{\underline{\mathbf{F}}} - \underline{\mathbf{C}} \times \hat{\underline{\mathbf{X}}}) \,. \tag{6.36}$$

This is complemented by (6.27),

$$\Delta\dot{\underline{\mathbf{r}}} = \underline{\mathbf{a}}_1 \Delta\dot{\underline{q}} \,, \qquad \Delta\underline{\boldsymbol{\omega}} = \underline{\mathbf{a}}_2 \Delta\dot{\underline{q}} \,. \tag{6.37}$$

These equations determine $\hat{\underline{\mathbf{X}}}$ and $\hat{\underline{\mathbf{Y}}}$ once all velocity increments and the impulse at the point of collision are known. The constraint impulses and impulse couples satisfy the orthogonality conditions (see (5.111))

$$\hat{\mathbf{X}}_a \cdot \mathbf{k}_{a\ell} = 0 \,, \qquad \hat{\mathbf{Y}}_a \cdot \mathbf{p}_{a\ell} = 0 \qquad (\ell = 1, \ldots, f_a) \tag{6.38}$$

(the latter ones only under the conditions stated there). These equations are used for systems with closed kinematic chains.

6.5 Chain Colliding with a Point Mass

A chain consisting of eleven identical links with revolute joints is resting on a horizontal table. The links are homogeneous rods of length ℓ, mass m and central moment of inertia $m\ell^2/12$ about the vertical axis. The angle between neighboring bodies is $9\pi/10$ for all pairs of neighboring bodies so that the chain is enveloping a semicircle with the center at 0 (Fig. 6.5a). The fourth link is hit at its own center by a point mass of mass m which is moving with

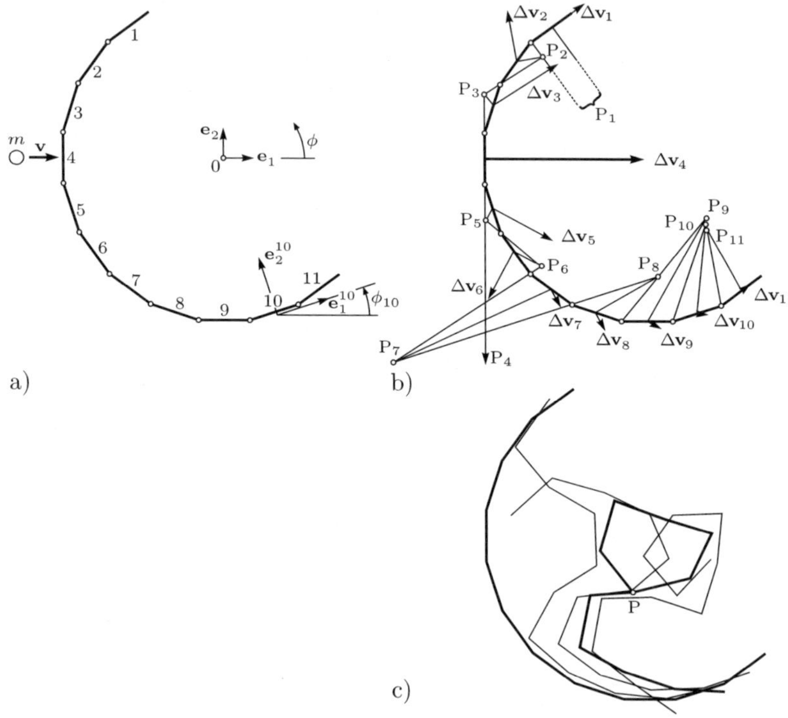

Fig. 6.5. (**a**) Chain hit by point mass m. (**b**) Velocity increments $\Delta\mathbf{v}_i$ of body centers of mass and instantaneous centers of rotation P_i $(i = 1, \ldots, 11)$. (**c**) Five positions of the chain at equal time intervals. Collision of bodies 1 and 7 at P causes new initial conditions for the subsequent motion

velocity $\mathbf{v}$ normal to the direction of the link. The collision is ideally elastic. To be determined are the velocities of the centers of mass of all links and the angular velocities of all links immediately after impact. The subsequent motion of the chain is to be computed by numerical integration of equations of motion. It is assumed that no friction occurs between the table and the chain and that no internal torques are acting in the joints.

Solution: The system is of the kind described as system with tree structure and with spherical joints not joint-connected to a carrier body 0. Its equations of motion consist of the single equation $\ddot{\mathbf{r}}_C = (1/M)\sum_{j=1}^{11} \mathbf{F}_j$ for the composite system center of mass and of equations describing rotational motions. For the present case of plane motions these latter ones are (5.302), (5.303). The angle ϕ_i $(i = 1, \ldots, n)$ is measured in the plane between a base vector $\mathbf{e}_1^i$ fixed on body i and the base vector $\mathbf{e}_1$ of an inertial frame of reference. The orientation of these base vectors is chosen as shown in Fig. 6.5a. The origin of $\underline{\mathbf{e}}$ is the point 0, $\mathbf{e}_1$ has the direction of $\mathbf{v}$ and $\mathbf{e}_1^i$ has the direction of the longitudinal axis of body i. Under these conditions all vectors $\mathbf{b}_{ij}$

($j = 1, \ldots, n$) on body i are parallel to $\mathbf{e}_1^i$ so that $\beta_{ij} = 0$ for $i, j = 1, \ldots, n$. Neither internal torques in joints nor external torques on bodies are acting. During the time interval Δt of collision body 4 is the only body subject to an external force and this force has the form $\mathbf{F}_4 = F\mathbf{e}_1$. In this special case, the dynamics equations for the composite system center of mass and (5.302), (5.303) for the angular variables have the forms (written without the hat of $\hat{K}_{i3}$)

$$\ddot{r}_{\mathrm{C}1} = \frac{1}{M} F , \qquad \underline{A} \begin{bmatrix} \ddot{\phi}_1 \\ \vdots \\ \ddot{\phi}_n \end{bmatrix} + \underline{B} \begin{bmatrix} \dot{\phi}_1^2 \\ \vdots \\ \dot{\phi}_n^2 \end{bmatrix} = F \underline{Q} , \tag{6.39}$$

$$A_{ij} = \begin{cases} K_{i3} & (i = j) \\ -Mb_{ij}b_{ji}\cos(\phi_i - \phi_j) & (i \neq j) \end{cases} \quad (i, j = 1, \ldots, n) , \tag{6.40}$$

$$B_{ij} = -Mb_{ij}b_{ji}\sin(\phi_i - \phi_j) \qquad (i, j = 1, \ldots, n) , \tag{6.41}$$

$$Q_i = -b_{i4}\sin\phi_i \quad (i = 1, \ldots, n) . \tag{6.42}$$

In addition, the kinematics equations (5.306) for the body centers of mass are required:

$$\dot{r}_{i1} = \dot{r}_{\mathrm{C}1} - \sum_{j=1}^{n} \dot{\phi}_j b_{ji} \sin\phi_j , \quad \dot{r}_{i2} = \sum_{j=1}^{n} \dot{\phi}_j b_{ji} \cos\phi_j \qquad (i = 1, \ldots, 11) . \tag{6.43}$$

Since the chain is at rest prior to the collision the velocity increments caused by the collision represent the initial velocities for the subsequent motion. They are determined through integration of the dynamics equations over the infinitesimal time interval of collision:

$$\dot{r}_{\mathrm{C}1}(0) = \frac{1}{M} \hat{F} , \qquad \underline{\dot{\phi}}(0) = \hat{F}\,[\underline{A}(0)]^{-1}\underline{Q}(0) . \tag{6.44}$$

The impulse $\hat{F}$ is determined from (6.16):

$$\hat{F} = -(1 + e)\,\frac{(\mathbf{v}_i - \mathbf{v}_j) \cdot \mathbf{n}}{\mathbf{n} \cdot (\mathsf{U}_{ii} + \mathsf{U}_{jj}) \cdot \mathbf{n}} . \tag{6.45}$$

Let the index i in this equation refer to the chain and the index j to the point mass. The tensor U_{jj} of the point mass is abbreviated U. The tensor U_{ii} of the chain is called U_{44} because it relates the velocity increment of the body 4 center of mass to the impulse applied at the same point. The velocities prior to the collision are $\mathbf{v}_i = \mathbf{0}$ (body 4 of the chain) and $\mathbf{v}_j = v\mathbf{e}_1$. Furthermore, $\mathbf{n} = \mathbf{e}_1$ and $e = 1$ (ideally elastic collision). Equation (6.45) then reads

$$\hat{F} = \frac{2v}{\mathbf{e}_1 \cdot (\mathsf{U}_{44} + \mathsf{U}) \cdot \mathbf{e}_1} . \tag{6.46}$$

From Newton's equation for a point mass it follows that $\mathsf{U} = (1/m)\,\mathsf{I}$ and, hence, $\mathbf{e}_1 \cdot \mathsf{U} \cdot \mathbf{e}_1 = 1/m$.

An expression for U_{44} is developed from Eqs. (6.39)–(6.44). By definition, U_{44} is the tensor in the equation $\Delta\dot{\mathbf{r}}_4 = \mathsf{U}_{44} \cdot \hat{F}\mathbf{e}_1$. Hence, $\mathbf{e}_1 \cdot \mathsf{U}_{44} \cdot \mathbf{e}_1 = \Delta\dot{r}_{41}/\hat{F}$. The first Eq. (6.43) yields the expression $\Delta\dot{r}_{41} = \Delta\dot{r}_{C1} - \sum_{j=1}^{n} \Delta\dot{\phi}_j b_{j4} \sin\phi_j$. Comparison with (6.42) shows that this is $\Delta\dot{r}_{41} = \Delta\dot{r}_{C1} + \sum_{j=1}^{n} Q_j \Delta\dot{\phi}_j = \Delta\dot{r}_{C1} + \underline{Q}^T \Delta\underline{\dot{\phi}}$ with the matrix $\underline{Q}$ from (6.39). All velocity and angular velocity increments in this equation represent the respective initial velocities immediately after impact. For the initial values on the right-hand side the expressions (6.44) are substituted. This yields

$$\dot{r}_{41}(0) = \hat{F}\left(\frac{1}{M} + \underline{Q}^T(0)[\underline{A}(0)]^{-1}\underline{Q}(0)\right) . \tag{6.47}$$

The expression in parentheses is the desired product $\mathbf{e}_1 \cdot \mathsf{U}_{44} \cdot \mathbf{e}_1$. With this expression and with $\mathbf{e}_1 \cdot \mathsf{U} \cdot \mathbf{e}_1 = 1/m$ (6.46) yields for the impulse $\hat{F}$ the final result

$$\hat{F} = \frac{2v}{1/m + 1/M + \underline{Q}^T(0)[\underline{A}(0)]^{-1}\underline{Q}(0)} . \tag{6.48}$$

The expression on the right-hand side is determined by the given conditions prior to the collision. With $\hat{F}$ the initial velocity $\dot{r}_{C1}(0)$ of the composite system center of mass and the initial angular velocities of all bodies are calculated from (6.44). Next, the initial velocities of all body centers of mass are calculated from (6.43). Finally, the constraint impulses in the joints are calculated from the first Eq. (6.36).

Numerical results based on the given parameters are listed in the equation below and in Table 6.1.

$$\hat{F} = 1.323\,mv\,, \qquad \dot{r}_{C1}(0) = 0.1189\,v\,. \tag{6.49}$$

Table 6.1. Initial angular velocities and initial velocities of body centers of mass

i	$\dot{\phi}_i(0)\,\ell/v$	$\dot{r}_{i1}(0)/v$	$\dot{r}_{i2}(0)/v$
1	−0.0070	0.1132	0.0837
2	−0.3716	−0.0391	0.1958
3	0.9185	0.2474	0.1630
4	−0.0133	0.6774	0.0211
5	−0.9460	0.2209	−0.1250
6	0.3384	−0.0920	−0.1717
7	−0.0269	0.0370	−0.0831
8	0.0561	0.0377	−0.0674
9	0.0239	0.0464	−0.0287
10	0.0249	0.0425	−0.0049
11	0.0257	0.0311	0.0173

Table 6.2. Apparent system mass M^* for collisions of the mass m with different bodies i

i	1	2	3	4	5	6
M^*/m	1.173	1.761	1.892	1.952	1.979	1.987

The vectors $\Delta\mathbf{v}_i$ $(i = 1, \ldots, 11)$ in Fig. 6.5b show magnitude and direction of the initial velocities of the body centers of mass $(\Delta\mathbf{v}_i = \dot{\mathbf{r}}_i(0))$. With the data in Table 6.1 the locations of the instantaneous centers of rotation P_1, ..., P_n of the bodies are calculated. The results satisfy the condition that the line connecting the centers of rotation of any two neighboring bodies passes through the joint connecting these bodies.

If the total mass of the chain were to be concentrated in a single point the interaction impulse would be $\hat{\mathbf{F}} = 2\mathbf{v}/(1/m + 1/M)$. The actual interaction impulse $\hat{\mathbf{F}}$ can be represented in the form $\hat{\mathbf{F}} = 2\mathbf{v}/(1/m + 1/M^*)$. This equation defines an apparent system mass M^*. Comparison with (6.48) yields $1/M^* = 1/M + \underline{Q}^T(0)[\underline{A}(0)]^{-1}\underline{Q}(0)$. The matrix $\underline{A}(0)$ is positive definite so that $M^* < M$. In the present case M^* equals $1.952\,m$. If the calculation is repeated for the case when the point mass does not strike the fourth body but the i th body $(i = 1, \ldots, 6)$ – always at the center of mass and normal to the body – then the results listed in Table 6.2 are obtained. The apparent system mass is always much smaller than the actual mass $M = 11\,m$, and except for $i = 1$, it depends little on the location of the point of collision.

With the given initial conditions the continuous motion of the chain following the collision is found by integrating equations of motion. These are Eqs. (6.39) with zero right-hand sides. In Fig. 6.5c the chain is shown in five positions at equal intervals of time. In the fourth position the end-point of body 1 collides with body 7 at the point P. This collision causes instantaneous changes of velocities and of angular velocities (not of $\dot{\mathbf{r}}_{\mathrm{C}}$) which are calculated from equations similar to the ones used before. The details are left to the reader. With new initial conditions thus obtained the numerical integration is continued until the next collision occurs.

The results displayed in Fig. 6.5c were tested experimentally ten years after making the calculations. The chain was supported by teflon feet on a polished glass plate. The point mass was replaced by a piston which was propelled by a prestressed spring in a fixed cylinder. The observed motion was precisely as calculated including the collision between bodies 1 and 7 and the subsequent motion.

Problem 6.1. The solar panels on the spacecraft in Fig. 6.6 are deployed by means of torsional springs in the revolute joints. The individual bodies are (unrealistically) considered as rigid. When neighboring bodies reach their final relative orientation their motion relative to one another is suddenly stopped (ideally plastic collision).

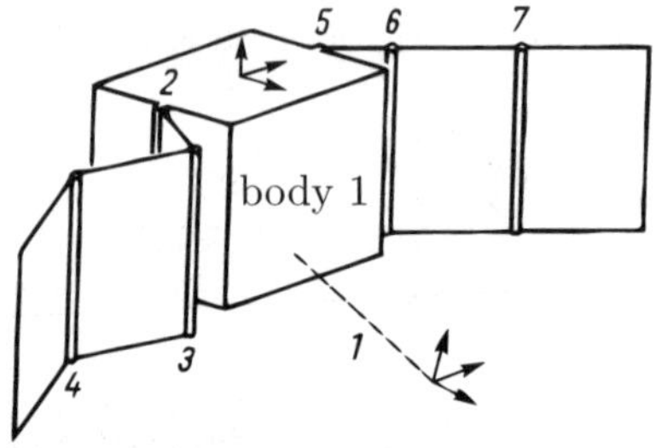

Fig. 6.6. Spacecraft unfolding solar panels at the moment of collision in joint 6

It is assumed that these collisions occur one at a time. Determine the finite angular velocity increment of the central body 1 caused by the collision in joint 6. Joints 2 to 5 are still unfolding while joint 7 has reached its final state already.

Solutions to Problems

Problem 1.1: $\underline{I}$; 3 ; $\begin{bmatrix} \mathbf{0} & \mathbf{e}_3^1 & -\mathbf{e}_2^1 \\ -\mathbf{e}_3^1 & \mathbf{0} & \mathbf{e}_1^1 \\ \mathbf{e}_2^1 & -\mathbf{e}_1^1 & \mathbf{0} \end{bmatrix}$; $\mathbf{0}$; $\underline{A}^{12}$; $\underline{A}^{21}$; $\mathrm{tr}\,\underline{A}^{21}$

Problem 1.2: $(\underline{\mathbf{a}}\,\underline{c}\cdot\underline{\mathbf{b}})_{ij} = \sum_k \sum_\ell \mathbf{a}_{i\ell} c_{\ell k}\cdot\mathbf{b}_{kj} = \sum_\ell \sum_k \mathbf{a}_{i\ell}\cdot c_{\ell k}\mathbf{b}_{kj} = (\underline{\mathbf{a}}\cdot\underline{c}\,\underline{\mathbf{b}})_{ij}$

Problem 1.3: 1. $\underline{a}^{1^T}\underline{\tilde{b}}^1\underline{A}^{12}\underline{c}^2$, 2. $-\underline{b}^{1^T}\underline{\tilde{a}}^1\underline{\tilde{b}}^1\underline{A}^{12}\underline{c}^2$, 3. $\underline{c}^{2^T}\underline{D}^2\underline{A}^{21}\underline{a}^1$,
4. $\underline{c}^{2^T}\underline{A}^{21}\underline{\tilde{b}}^1\underline{A}^{12}\underline{D}^2\underline{c}^2$, 5. $\underline{\tilde{a}}^1\underline{b}^1$, 6. $\underline{\tilde{a}}^1\underline{A}^{12}\underline{c}^2$, 7. $\underline{\tilde{a}}^1\underline{A}^{12}\underline{\tilde{c}}^2\underline{A}^{21}\underline{b}^1$,
8. $\underline{A}^{12}\underline{\tilde{c}}^2\underline{D}^2\underline{A}^{21}\underline{a}^1$, 9. $-\underline{\tilde{a}}^1\underline{A}^{12}\underline{\tilde{c}}^2\underline{D}^2\underline{A}^{21}\underline{b}^1$

Problem 1.4: If $\mathbf{p}$ and $\mathbf{q}$ are any vectors for which $\mathbf{p}\times\mathbf{q}=\mathbf{d}$ then
$\mathsf{D}_{11} = \mathbf{b}\cdot\mathbf{b}\mathsf{I} - \mathbf{b}\mathbf{b} = \bar{\mathsf{D}}_{11}$, $\mathsf{D}_{12} = (\mathbf{c}\cdot\mathbf{b}\mathsf{I} - \mathbf{c}\mathbf{b}) + (\mathbf{q}\mathbf{p} - \mathbf{p}\mathbf{q}) = \bar{\mathsf{D}}_{21}$,
$\mathsf{D}_{22} = \mathbf{c}\cdot\mathbf{c}\mathsf{I} - \mathbf{c}\mathbf{c} = \bar{\mathsf{D}}_{22}$,
$\underline{D}_{11} = \underline{b}^T\underline{b}\,\underline{I} - \underline{b}\,\underline{b}^T = \underline{D}_{11}^T$, $\underline{D}_{12} = \underline{c}^T\underline{b}\,\underline{I} - \underline{c}\,\underline{b}^T + \underline{\tilde{d}} = \underline{D}_{21}^T$,
$\underline{D}_{22} = \underline{c}^T\underline{c}\,\underline{I} - \underline{c}\,\underline{c}^T = \underline{D}_{22}^T$; the (6×6) matrix is symmetric

Problem 2.1: Eigenvalues $\lambda_1 = 1$, $\lambda_{2,3} = \cos\varphi \pm \mathrm{i}\sin\varphi$.
Eigenvectors $\underline{n}_1 = [1\ \ 0\ \ 0]$, $\underline{n}_{2,3} = [0\ \ \mp\mathrm{i}\ \ 1]$

Problem 2.2: Matrix $\underline{A}_1$: $\psi = 225°$, $\theta = \cos^{-1}\frac{1}{3}$ (principal value), $\phi = 45°$,
$q_0 = \sqrt{2/3}$, $q_1 = q_2 = -\sqrt{1/6}$, $q_3 = 0$.
Matrix $\underline{A}_2$: $\psi = \cos^{-1}\frac{7}{10}\sqrt{2}$ (principal v.), $\theta = \cos^{-1}\frac{1}{3}$ (principal v.),
$\phi = 135°$, $q_0 = \sqrt{8/15}$, $q_1 = \sqrt{3/10}$, $q_2 = -\sqrt{1/30}$, $q_3 = \sqrt{2/15}$.
Matrix $\underline{A}_3$: $\varphi = 180°$ (because $\underline{A}^{21}$ is symmetric), $q_0 = 0$, $\mathbf{q} = \mathbf{n}$,
$(\underline{A}^{21} - \underline{I})\underline{n} = \underline{0}$ yields $n_1 = 1/3$, $n_2 = 2/3$, $n_3 = 2/3$, $c_\theta = -1/9$, $s_\theta = (4/9)\sqrt{5}$,
$c_\psi = -(2/5)\sqrt{5}$, $s_\psi = (1/5)\sqrt{5}$, $c_\phi = (2/5)\sqrt{5}$, $s_\phi = (1/5)\sqrt{5}$

Problem 2.3: $\cos\phi_2 = 0$ if in (2.9) $a_{31}^{21} = \sigma$ ($+1$ or -1). This is the condition (see (2.24)) $n_1 n_3\cos\varphi - n_2\sin\varphi = n_1 n_3 - \sigma$. It has the form $A\cos\varphi + B\sin\varphi = C$. Hence, $\cos\varphi = (AC \pm BW)/N$, $\sin\varphi = (BC \mp AW)/N$ with $N = A^2 + B^2$ and $W = \sqrt{A^2+B^2-C^2} = \sqrt{n_1^2 n_3^2 + n_2^2 - (n_1 n_3 - \sigma)^2} = \sqrt{-(n_3 - \sigma n_1)^2}$. Real roots (a double root) exist only if $n_3 = \sigma n_1$, whence follows $n_2^2 = 1 - 2n_1^2$,
$\cos\varphi = -n_1^2/(1-n_1^2)$, $\sin\varphi = \sigma n_2/(1-n_1^2)$ with $n_1^2 \le 1/2$ (arbitr.)

Problem 2.4: Equations (2.61) and (2.62) formulated with $\varphi_1 = -\pi$ (equivalent to $\varphi_1 = \pi$) and with $\varphi_2 = \pi$ yield $\cos\frac{\varphi_{\text{res}}}{2} = \mathbf{n}_1 \cdot \mathbf{n}_2 = \cos\alpha$, $\mathbf{n}_{\text{res}} \sin\frac{\varphi_{\text{res}}}{2} = \mathbf{n}_1 \times \mathbf{n}_2$, whence follows $\varphi_{\text{res}} = 2\alpha$ and $\mathbf{n}_{\text{res}} = \mathbf{n}_1 \times \mathbf{n}_2 / \sin\alpha$

Problem 2.5: A body-fixed vector with the coordinate matrix $\underline{r}^2$ in $\underline{\mathbf{e}}^2$ has in $\underline{\mathbf{e}}^1$ after the first rotation the coordinate matrix $\underline{r}^1 = \underline{A}_1^{12}\underline{r}^2$, after the second $\underline{r}^1 = \underline{A}_2^{12}\underline{A}_1^{12}\underline{r}^2$ and after the third $\underline{r}^1 = \underline{A}_3^{12}\underline{A}_2^{12}\underline{A}_1^{12}\underline{r}^2$, where (with $c_i = \cos\phi_i$, $s_i = \sin\phi_i$)

$$\underline{A}_3^{12} = \begin{bmatrix} c_3 & s_3 & 0 \\ -s_3 & c_3 & 0 \\ 0 & 0 & 1 \end{bmatrix}, \quad \underline{A}_2^{12} = \begin{bmatrix} c_2 & 0 & -s_2 \\ 0 & 1 & 0 \\ s_2 & 0 & c_2 \end{bmatrix}, \quad \underline{A}_1^{12} = \begin{bmatrix} 1 & 0 & 0 \\ 0 & c_1 & s_1 \\ 0 & -s_1 & c_1 \end{bmatrix}.$$

After the third rotation $\underline{A}^{21} = (\underline{A}_3^{12}\underline{A}_2^{12}\underline{A}_1^{12})^T$ and $D_{\text{res}} = D_3 D_2 D_1$.

Special cases: $\underline{A}^{21} = \begin{bmatrix} 0 & 0 & -1 \\ 0 & 1 & 0 \\ -1 & 0 & 0 \end{bmatrix}$ and $D_{\text{res}} = D_2$ for the first two sets of angles;

$\underline{A}^{21} = \underline{I}$ and $D_{\text{res}} = (1, \mathbf{0})$ for the third set

Problem 2.6: $\mathbf{j} = \mathbf{j}_{\text{A}} + (\boldsymbol{\gamma} - \omega^2\boldsymbol{\omega}) \times \boldsymbol{\varrho} - 3(\dot{\boldsymbol{\omega}} \cdot \boldsymbol{\omega})\boldsymbol{\varrho} + 2(\boldsymbol{\omega} \cdot \boldsymbol{\varrho})\dot{\boldsymbol{\omega}} + (\dot{\boldsymbol{\omega}} \cdot \boldsymbol{\varrho})\boldsymbol{\omega}$ where $\boldsymbol{\gamma}$ is the time derivative of $\dot{\boldsymbol{\omega}}$ in the body-fixed base $\underline{\mathbf{e}}^2$

Problem 2.7:
$(\mathbf{v}_1 - \mathbf{v}_3) \times (\mathbf{v}_2 - \mathbf{v}_3) = [\boldsymbol{\omega} \times (\mathbf{r}_1 - \mathbf{r}_3)] \times (\mathbf{v}_2 - \mathbf{v}_3) = -\boldsymbol{\omega}\,(\mathbf{r}_1 - \mathbf{r}_3) \cdot (\mathbf{v}_2 - \mathbf{v}_3)$ $= -\boldsymbol{\omega}\,[(\mathbf{r}_1 - \mathbf{r}_3) \cdot \boldsymbol{\omega} \times (\mathbf{r}_2 - \mathbf{r}_3)]$. From this it follows: $\boldsymbol{\omega}$ lies in the plane of the three points if $(\mathbf{v}_1 - \mathbf{v}_3) \times (\mathbf{v}_2 - \mathbf{v}_3) = \mathbf{0}$. Assume, first, that this is not the case. Then

$$\boldsymbol{\omega} = \frac{(\mathbf{v}_1 - \mathbf{v}_3) \times (\mathbf{v}_2 - \mathbf{v}_3)}{(\mathbf{r}_3 - \mathbf{r}_1) \cdot (\mathbf{v}_2 - \mathbf{v}_3)} = \frac{\mathbf{v}_1 \times \mathbf{v}_2 + \mathbf{v}_2 \times \mathbf{v}_3 + \mathbf{v}_3 \times \mathbf{v}_1}{(\mathbf{r}_3 - \mathbf{r}_1) \cdot (\mathbf{v}_2 - \mathbf{v}_3)}.$$

Differentiation of $(\mathbf{r}_3 - \mathbf{r}_1) \cdot (\mathbf{r}_2 - \mathbf{r}_3) = \text{const}$ produces for the denominator (abbreviated D) the expression $D = -(\mathbf{v}_3 - \mathbf{v}_1) \cdot (\mathbf{r}_2 - \mathbf{r}_3)$ or, by adding D to both sides, $2D = [(\mathbf{r}_3 - \mathbf{r}_1) \cdot (\mathbf{v}_2 - \mathbf{v}_3) - (\mathbf{v}_3 - \mathbf{v}_1) \cdot (\mathbf{r}_2 - \mathbf{r}_3)]$. Hence,

$$\boldsymbol{\omega} = 2\,\frac{\mathbf{v}_1 \times \mathbf{v}_2 + \mathbf{v}_2 \times \mathbf{v}_3 + \mathbf{v}_3 \times \mathbf{v}_1}{\mathbf{v}_1 \cdot (\mathbf{r}_2 - \mathbf{r}_3) + \mathbf{v}_2 \cdot (\mathbf{r}_3 - \mathbf{r}_1) + \mathbf{v}_3 \cdot (\mathbf{r}_1 - \mathbf{r}_2)}.$$

Special case: $\boldsymbol{\omega}$ lies in the plane of the three points. Ansatz: $\boldsymbol{\omega} = \lambda(\mathbf{r}_1 - \mathbf{r}_2) + \mu(\mathbf{r}_2 - \mathbf{r}_3)$ with unknowns λ and μ. Hence, $\mathbf{v}_1 - \mathbf{v}_2 = \boldsymbol{\omega} \times (\mathbf{r}_1 - \mathbf{r}_2) = \mu\mathbf{n}$ and $\mathbf{v}_2 - \mathbf{v}_3 = \boldsymbol{\omega} \times (\mathbf{r}_2 - \mathbf{r}_3) = -\lambda\mathbf{n}$ with $\mathbf{n} = -(\mathbf{r}_1 - \mathbf{r}_2) \times (\mathbf{r}_2 - \mathbf{r}_3) = -(\mathbf{r}_1 \times \mathbf{r}_2 + \mathbf{r}_2 \times \mathbf{r}_3 + \mathbf{r}_3 \times \mathbf{r}_1)$. Hence, $\lambda = -(\mathbf{v}_2 - \mathbf{v}_3) \cdot \mathbf{n}/\mathbf{n}^2$, $\mu = (\mathbf{v}_1 - \mathbf{v}_2) \cdot \mathbf{n}/\mathbf{n}^2$, whence follows
$\boldsymbol{\omega} = (\mathbf{n}/\mathbf{n}^2) \cdot [\mathbf{v}_1(\mathbf{r}_2 - \mathbf{r}_3) + \mathbf{v}_2(\mathbf{r}_3 - \mathbf{r}_1) + \mathbf{v}_3(\mathbf{r}_1 - \mathbf{r}_2)]$

Problem 2.8: $\underline{A}^{21} = \underline{A}_3\underline{B}_2\underline{A}_2\underline{B}_1\underline{A}_1$ with $\underline{A}_3$, $\underline{A}_2$, $\underline{A}_1$ from (2.8) and with

$$\underline{B}_2 = \begin{bmatrix} 1 & 0 & 0 \\ 0 & c_\beta & -s_\beta \\ 0 & s_\beta & c_\beta \end{bmatrix}, \quad \underline{B}_1 = \begin{bmatrix} c_\alpha & s_\alpha & 0 \\ -s_\alpha & c_\alpha & 0 \\ 0 & 0 & 1 \end{bmatrix}.$$

$\underline{\omega} = \underline{A}_3\underline{B}_2\underline{A}_2\underline{B}_1[\dot\phi_1\ \ 0\ \ 0]^T + \underline{A}_3\underline{B}_2[0\ \ \dot\phi_2\ \ 0]^T + [0\ \ 0\ \ \dot\phi_3]^T$ and $[\dot\phi_1\ \ \dot\phi_2\ \ \dot\phi_3]^T =$

$$= \frac{1}{c_2 c_\alpha c_\beta}\begin{bmatrix} c_3 c_\beta & -s_3 c_\beta & 0 \\ c_2 s_3 c_\alpha + c_3(s_\alpha c_\beta + s_2 c_\alpha s_\beta) & c_2 c_3 c_\alpha - s_3(s_\alpha c_\beta + s_2 c_\alpha s_\beta) & 0 \\ -c_\alpha(s_2 c_3 + c_2 s_3 s_\beta) & c_\alpha(s_2 s_3 - c_2 c_3 s_\beta) & c_2 c_\alpha c_\beta \end{bmatrix}\begin{bmatrix} \omega_1 \\ \omega_2 \\ \omega_3 \end{bmatrix}$$

($c_i = \cos\phi_i$, $s_i = \sin\phi_i$ $(i = 1, 2, 3)$, $c_\alpha, c_\beta, s_\alpha, s_\beta = \cos\alpha, \cos\beta, \sin\alpha, \sin\beta$, respectively)

Problem 3.1: Conservation of kinetic energy requires that $J_1\dot{q}_1^2 + J_2\dot{q}_2^2 \equiv 2T =$ const. This yields the desired equation: $\dot{q}_1(q_1) = \sqrt{2T/[J_1 + J_2/i^2(q_1)]}$. With the dimensionless parameter $\lambda^2 = J_2/J_1$ and with the constant angular velocity ω_0 defined through the equation $2T = (J_1 + J_2)\omega_0^2$ it takes the form

$$\dot{q}_1(q_1) = \omega_0\sqrt{\frac{1+\lambda^2}{1+\lambda^2/i^2(q_1)}}$$

Problem 3.2:
Equation (3.27): The entire body mass must be located in the plane $\varrho_k = 0$.
Equation (3.28): The entire body mass must be located in one of the two planes which contain the axis $\mathbf{e}_i$ and which are inclined under 45° against the axes $\mathbf{e}_j$ and $\mathbf{e}_k$, so that for every mass element $\varrho_j^2 + \varrho_k^2 = 2\varrho_j\varrho_k$.
Equation (3.29): The entire body mass must be located either on the axis $\mathbf{e}_i$ or on the axis $\mathbf{e}_j$. Then $J_{ij} = 0$ and either $J_{ii} = 0$ or $J_{jj} = 0$

Problem 3.4: Center of mass at $\ell\,[a/4 \quad 1/2 \quad -a/4]$ with $a = \tan\gamma$;
mass $m = \varrho\ell^3 a^2/6$;

$$\underline{J}^{\mathrm{C}} = \frac{\varrho\ell^5 a^2}{40}\begin{bmatrix} 1/3 + a^2/4 & a/6 & -a^2/12 \\ a/6 & a^2/2 & a/6 \\ -a^2/12 & a/6 & 1/3 + a^2/4 \end{bmatrix}$$

Problem 3.5: If A is the contact point then $\boldsymbol{\varrho}_{\mathrm{C}} = -b\sin\phi\mathbf{e}_1 - (R - b\cos\phi)\mathbf{e}_2$,
$\ddot{r}_{\mathrm{A}} = -R\dot{\phi}^2\mathbf{e}_2$, $J^{\mathrm{A}}\cdot\boldsymbol{\omega} = [J^{\mathrm{C}} + m(R^2 + b^2 - 2Rb\cos\phi)]\dot{\phi}\mathbf{e}_3$, $\dot{\omega} = \ddot{\phi}\mathbf{e}_3$,
$\mathbf{M}^{\mathrm{A}} = -mgb\sin\phi\mathbf{e}_3$. Equation of motion:
$[J^{\mathrm{C}} + m(R^2 + b^2 - 2Rb\cos\phi)]\ddot{\phi} + mbR\dot{\phi}^2\sin\phi + mg\,b\sin\phi = 0$.
For using the other reference points introduce reaction forces at the point of contact and eliminate these forces from the law of angular momentum by formulating also Newton's law

Problem 3.6: $\mathbf{r}_{\mathrm{A}} = \overrightarrow{0\mathrm{P}_1} = c\sin\phi\,\mathbf{e}_1$ with $c = a/\sin\alpha$, $\delta\dot{\mathbf{r}}_{\mathrm{A}} = c\cos\phi\,\delta\dot{\phi}\,\mathbf{e}_1$,
$\delta\boldsymbol{\omega} = \delta\dot{\phi}\,\mathbf{e}_3$, $\ddot{\mathbf{r}}_{\mathrm{A}} = c(\ddot{\phi}\cos\phi - \dot{\phi}^2\sin\phi)\mathbf{e}_1$, $\ddot{\mathbf{r}}_{\mathrm{C}} = \ddot{\mathbf{r}}_{\mathrm{A}} + \ddot{\phi}\mathbf{e}_3\times\boldsymbol{\varrho}_{\mathrm{C}} - \dot{\phi}^2\boldsymbol{\varrho}_{\mathrm{C}}$, $\mathbf{M}^{\mathrm{A}} = \mathbf{a}\times\mathbf{F}_2$.
Equation of motion (with $J^{\mathrm{A}} = J^{\mathrm{C}} + m\varrho_{\mathrm{C}}^2$):
$\ddot{\phi}\{J^{\mathrm{A}} + m[c^2\cos^2\phi - 2c\,\varrho_{\mathrm{C}}\,\cos\phi\,\sin(\phi + \alpha - \beta)]\}$
$-\dot{\phi}^2 mc[c\sin\phi\cos\phi + \varrho_{\mathrm{C}}\cos(2\phi + \alpha - \beta)] = c[F_1\cos\phi + F_2\cos(\phi + \alpha)]$.
Special case $\alpha = \pi/2$, $\beta = 0$, $\varrho_{\mathrm{C}} = a/2$: $J^{\mathrm{A}}\ddot{\phi} = a(F_1\cos\phi - F_2\sin\phi)$

Problem 4.1: $J_2\dot{\omega}_2 - (J_3 - J_1)\omega_3\omega_1 < 0 \quad$ for $\quad \omega_3\omega_1 > 0$

Problem 4.2:

$$\begin{aligned} J_1\dot{\omega}_1 - (J_1 - J_3^*)\omega_2\omega_3 &= -\omega_2 L^{\mathrm{r}}(t), \\ J_1\dot{\omega}_2 - (J_3^* - J_1)\omega_3\omega_1 &= +\omega_1 L^{\mathrm{r}}(t), \\ J_3^*\dot{\omega}_3 &= -\;M^{\mathrm{r}}(t) \end{aligned}$$

with $J_3^* = J_3 - J^{\mathrm{r}}$ and $L^{\mathrm{r}}(t) = \int_{t_0}^t M^{\mathrm{r}}(\tau)\,\mathrm{d}\tau + L^{\mathrm{r}}(t_0)$. Integrals of motion:
$\omega_1^2 + \omega_2^2 = \Omega^2 =$ const and $J_3^*\omega_3 + L^{\mathrm{r}}(t) = J_3\omega_3 + h = L =$ const (axial coordinate of total absolute angular momentum).
Solution: $\omega_3 = \omega_3(t_0) - [L^{\mathrm{r}}(t) - L^{\mathrm{r}}(t_0)]/J_3^*$, $\omega_1 = \Omega\sin\alpha(t)$,
$\omega_2 = \Omega\cos\alpha(t)$ with $\alpha(t) = \alpha(t_0) + \int_{t_0}^t f(\tau)\,\mathrm{d}\tau$ and $f(t) = L/J_1 - \omega_3(t)$.
Interaction torque $M^{\mathrm{r}}(t) = ah$: Equation (4.129) is replaced by (4.114) and (4.117)

with the additional term $-ah$. This yields

$$\begin{aligned} J_1\dot\omega_1 - (J_1 - J_3)\omega_2\omega_3 \quad &= -\omega_2 h,\\ J_1\dot\omega_2 - (J_3 - J_1)\omega_3\omega_1 \quad &= \omega_1 h,\\ J_3\dot\omega_3 \qquad\qquad + \dot h &= 0,\\ J_3^{\mathrm{r}}\dot\omega_3 \qquad\qquad + \dot h &= M^{\mathrm{r}}(t) - ah\,. \end{aligned}$$

Integrals of motion $\omega_1^2 + \omega_2^2 = \Omega^2 = \text{const}$ and $J_3\omega_3 + h = L = \text{const}$ as before.
Solution:

$$h(t) = \phi(t) + [h(t_0) - \phi(t_0)]\exp(bt),$$

$$\omega_3 = \omega_3(t_0) + \frac{1}{J_3}\{\phi(t) + [h(t_0) - \phi(t_0)]\exp(-bt) - h(t_0)\}, \qquad b = a\,\frac{J_3}{J_3^*}.$$

$\phi(t)$ is the particular solution of $\dot h + bh = M^{\mathrm{r}}(t)J_3/J_3^*$; $\omega_1 = \Omega\sin\alpha(t)$, $\omega_2 = \Omega\cos\alpha(t)$ with $\alpha(t)$ as before.

Problem 5.1: Replace in (5.93) $\mathbf{c}_{ia}$ by one and adapt the subsequent arguments

Problem 5.2: $i^+(a) = a,\ i^-(a) = 0\ \ (a = 1,\dots,n),\ \underline{S}_{0\mathrm{t}} = -\underline{1}^T,\ \underline{S}_{\mathrm{t}} = \underline{T} = \underline{I}$

Problem 5.3: The transpose of the (8×3) incidence matrix is

$$\begin{bmatrix} +1 & 0 & 0 & 0 & 0 & 0 & -1 & 0\\ 0 & 0 & 0 & +1 & 0 & -1 & 0 & 0\\ 0 & 0 & 0 & -1 & 0 & +1 & 0 & 0 \end{bmatrix}$$

Problem 5.4: κ_{52} contains only vertex 3. κ_{22} is the set of vertices 3, 6 and 7

Problem 5.5: $k = 3$: 1. vertices 1, 7 ; 2. no vertex ;
$k = 5$: 1. vertex 1 ; 2. vertex 2

Problem 5.6: $\boldsymbol{\Omega}_a = \boldsymbol{\omega}_{i^-(a)} - \boldsymbol{\omega}_{i^+(a)} = -\sum_{i=1}^n S_{ia}\boldsymbol{\omega}_i\ \ (a = 1,\dots,m)$,
$\mathbf{F}_{i_{\mathrm{res}}} = \sum_{a=1}^m S_{ia}\mathbf{F}_a\ \ (i = 1,\dots,n)$,
tree structure: $\underline{\boldsymbol{\Omega}} = -\underline{S}^T\underline{\boldsymbol{\omega}} \ \leftrightarrow\ \underline{\boldsymbol{\omega}} = -\underline{T}^T\underline{\boldsymbol{\Omega}}$, $\quad\underline{\mathbf{F}}_{\mathrm{res}} = \underline{S}\,\underline{\mathbf{F}} \ \leftrightarrow\ \underline{\mathbf{F}} = \underline{T}\,\underline{\mathbf{F}}_{\mathrm{res}}$,
$\underline{\mathbf{M}}_{\mathrm{res}} = \underline{\mathbf{C}}\times\underline{\mathbf{F}}$

Problem 5.7: $\underline{A} = (\underline{\mathbf{k}}\,\underline{T})\cdot\underline{m}(\underline{\mathbf{k}}\,\underline{T})^T = \text{const}$, $\qquad\underline{B} = (\underline{\mathbf{k}}\,\underline{T})\cdot[\underline{m}(\ddot{\mathbf{r}}_0\underline{1} - \underline{T}^T\underline{\mathbf{s}}) - \underline{\mathbf{F}}]$

Problem 5.8: $\underline{A}$ is not singular

Problem 5.10: Equation (5.100) is replaced by
$-\underline{\mathbf{p}}\,\underline{T}\cdot\underline{\mathbf{M}} = -\underline{\mathbf{p}}\,\underline{T}\cdot\underline{S}\,\underline{\mathbf{p}}^T\,\underline{K}(\underline{q} - \underline{q}_0) = -\underline{\mathbf{p}}\cdot\underline{\mathbf{p}}^T\,\underline{K}(\underline{q} - \underline{q}_0)$. Diagonal matrix $\underline{K}$.
A Hooke's joint labeled a is associated with the matrix (see (5.57)) $\underline{\mathbf{p}}_a^T = [\mathbf{p}_{a1}\ \mathbf{p}_{a2}]$.
The matrix $\underline{\mathbf{p}}\cdot\underline{\mathbf{p}}^T$ has submatrices $\underline{\mathbf{p}}_a\cdot\underline{\mathbf{p}}_a^T = \begin{bmatrix} 1 & \mathbf{p}_{a1}\cdot\mathbf{p}_{a2}\\ \mathbf{p}_{a1}\cdot\mathbf{p}_{a2} & 1\end{bmatrix}$ along the diagonal. It is the unit matrix if in all Hooke's joints the joint axes are orthogonal.

Problem 5.11: $\mathbf{z}_a = -\sum_{i=0}^n S_{ia}\mathbf{c}_{ia} - \sum_{i=0}^n S_{ia}\mathbf{r}_i = -\sum_{i=0}^n \mathbf{C}_{ia} - \sum_{i=0}^n S_{ia}\mathbf{r}_i$,
$\underline{\mathbf{z}} = -\underline{\mathbf{C}}_{0\mathrm{c}}^T - \underline{\mathbf{C}}_{\mathrm{c}}^T\underline{1} - \mathbf{r}_0\underline{S}_{0\mathrm{c}}^T - \underline{S}_{\mathrm{c}}^T\underline{\mathbf{r}}$ and with (5.74)
$\underline{\mathbf{z}} = -\mathbf{r}_0(\underline{S}_{0\mathrm{c}}^T + \underline{S}_{\mathrm{c}}^T\underline{1}) - \underline{\mathbf{C}}_{0\mathrm{c}}^T - \underline{\mathbf{C}}_{\mathrm{c}}^T\underline{1} - \underline{S}_{\mathrm{c}}^T[-(\underline{\mathbf{C}}_0\underline{T})^T - (\underline{\mathbf{C}}\,\underline{T})^T\underline{1}]$
$= -(\underline{\mathbf{C}}_{0\mathrm{c}} - \underline{\mathbf{C}}_0\underline{T}\,\underline{S}_{\mathrm{c}})^T - (\underline{\mathbf{C}}_{\mathrm{c}} - \underline{\mathbf{C}}\,\underline{T}\,\underline{S}_{\mathrm{c}})^T\underline{1}$. The factor of $\mathbf{r}_0$ is zero because of (5.9).
According to (5.21) and (5.25) $\underline{T}\,\underline{S}_{\mathrm{c}} = -\underline{U}_{\mathrm{t}}^T$. Hence

$$\underline{\mathbf{z}} = -(\underline{\mathbf{C}}_{0c} + \underline{\mathbf{C}}_0 \underline{U}_t^T)^T - (\underline{\mathbf{C}}_c + \underline{\mathbf{C}}\,\underline{U}_t^T)^T \underline{1} = -\underline{U}_t \underline{\mathbf{C}}_0^T - \underline{\mathbf{C}}_{0c}^T - \underline{U}_t \underline{\mathbf{C}}^T \underline{1} - \underline{\mathbf{C}}_c^T \underline{1}$$

$$= -[\underline{U}_t \quad \underline{I}\,] \begin{bmatrix} \underline{\mathbf{C}}_0^T + \underline{\mathbf{C}}^T \underline{1} \\ \underline{\mathbf{C}}_{0c}^T + \underline{\mathbf{C}}_c^T \underline{1} \end{bmatrix} = -\underline{U}\,\underline{\mathbf{D}}$$

Problem 5.12: On body i the resultant constraint torque $\sum_{a=1}^{n} S_{ia} \mathbf{Y}_a$ is acting in addition to the external torque. Thus, $\underline{\mathsf{K}} \cdot \underline{\dot{\omega}} = \underline{\mathbf{M}} + \underline{\mathbf{Q}} + \underline{S}\,\underline{\mathbf{Y}}$. Multiplication from the left by $\underline{T}$ yields $\underline{T}\,\underline{\mathsf{K}} \cdot \underline{\dot{\omega}} = \underline{T}(\underline{\mathbf{M}} + \underline{\mathbf{Q}}) + \underline{\mathbf{Y}}$. Scalar multiplication by $\underline{\mathbf{p}}$ eliminates the constraint torques: $\underline{\mathbf{p}}\,\underline{T} \cdot \underline{\mathsf{K}} \cdot \underline{\dot{\omega}} = \underline{\mathbf{p}}\,\underline{T} \cdot (\underline{\mathbf{M}} + \underline{\mathbf{Q}})$. According to (5.65) and (5.66) $\underline{\dot{\omega}} = \underline{\mathbf{a}}_2 \underline{\ddot{q}} + \underline{\mathbf{b}}_2$ with $\underline{\mathbf{a}}_2 = -(\underline{\mathbf{p}}\,\underline{T})^T$, $\underline{\mathbf{b}}_2 = \dot{\omega}_0 \underline{1} - \underline{T}^T \underline{\mathbf{f}}$ ($\underline{\mathbf{w}} = \underline{\mathbf{0}}$ in systems with revolute joints). Substitution yields the desired equations of motion: $\underline{\mathbf{a}}_2^T \cdot \underline{\mathsf{K}} \cdot \underline{\mathbf{a}}_2\, \underline{\ddot{q}} = \underline{\mathbf{a}}_2^T \cdot (\underline{\mathbf{M}} + \underline{\mathbf{Q}} - \underline{\mathsf{K}} \cdot \underline{\mathbf{b}}_2)$. Comparison with (5.83) and (5.86) shows that $\underline{\mathbf{a}}_2^T \cdot \underline{\mathsf{K}} \cdot \underline{\mathbf{a}}_2 = \underline{\mathbf{a}}_1^T \cdot \underline{m}\,\underline{\mathbf{a}}_1 + \underline{\mathbf{a}}_2^T \cdot \underline{\mathsf{J}} \cdot \underline{\mathbf{a}}_2$ with $\underline{\mathbf{a}}_1 = (\underline{\mathbf{C}}\,\underline{T})^T \times \underline{\mathbf{a}}_2$, $\underline{\mathbf{a}}_2 = -(\underline{\mathbf{p}}\,\underline{T})^T$.

Problem 5.13: $\hat{K}_i = J_i + \frac{m\ell^2}{4n}[4(n-i)(i-1) + n - 1]$

Problem 5.15:

$$2T = M\dot{\mathbf{r}}_C^2 + \sum_{i=1}^{n} \left[\boldsymbol{\omega}_i \cdot \hat{\mathsf{K}} \cdot \boldsymbol{\omega}_i + M \sum_{\substack{j=1 \\ \neq i}}^{n} (\boldsymbol{\omega}_i \times \mathbf{b}_{ij}) \cdot (\mathbf{b}_{ji} \times \boldsymbol{\omega}_j) + 2\boldsymbol{\omega}_i \cdot \mathbf{h}_i + \sum_{k=1}^{s_i} J_{ik}\, \omega_{ik_{\mathrm{rel}}}^2 \right],$$

$$V = \frac{1}{2}\omega_0^2 \sum_{i=1}^{n} \left(3\mathbf{e}_3 \cdot \mathsf{B}_i \cdot \mathbf{e}_3 - M \sum_{\substack{j=1 \\ \neq i}}^{n} \mathbf{b}_{ij} \cdot \mathbf{b}_{ji} \right)$$

absolute velocity $\dot{\mathbf{r}}_C$ of the satellite center of mass, axial moment of inertia J_{ik} of the kth rotor on body i, angular velocity $\omega_{ik_{\mathrm{rel}}}$ of this rotor relative to its carrier, number s_i of rotors on body i, tensor B_i as in (5.282). See Wittenburg/Lilov [103]

Problem 6.1:

In what follows the joint axes are not required to be parallel to one another. At the moment immediately prior to the collision the system consists of six bodies (the bodies connected by the locked joint 7 represent a single body). The system has eleven generalized coordinates: $\underline{q} = [q_{11}\ q_{12}\ q_{13}\ q_{14}\ q_{15}\ q_{16}\ q_2\ q_3\ q_4\ q_5\ q_6]^T$ (six for body 1 relative to inertial space and one in each of the joints 2, 3, 4, 5, 6). The state of motion immediately prior to the collision is known, in particular $\dot{q}_6$ is known. After the collision $\dot{q}_6 = 0$. Thus $\Delta\dot{q}_6 = \ \dot{q}_6$ is known. The locking in joint 6 is caused by the component along the joint axis of the constraint torque $\hat{\mathbf{Y}}_6$. Joint 1 between body 1 and inertial space is free of joint reactions. Equations (6.36) and (6.37) read $\underline{\hat{\mathbf{X}}} = \underline{T}\underline{m}\,\Delta\underline{\dot{\mathbf{r}}}$, $\underline{\hat{\mathbf{Y}}} = \underline{T}(\underline{\mathsf{J}} \cdot \Delta\underline{\omega} - \underline{\mathbf{C}} \times \underline{\hat{\mathbf{X}}})$, $\Delta\underline{\mathbf{r}} = \underline{\mathbf{a}}_1 \Delta\underline{\dot{q}}$, $\Delta\underline{\omega} = \underline{\mathbf{a}}_2 \Delta\underline{\dot{q}}$ with matrices $\underline{\hat{\mathbf{X}}} = [\mathbf{0}\ \hat{\mathbf{X}}_2\ \hat{\mathbf{X}}_3\ \hat{\mathbf{X}}_4\ \hat{\mathbf{X}}_5\ \hat{\mathbf{X}}_6]^T$ and $\underline{\hat{\mathbf{Y}}} = [\mathbf{0}\ \hat{\mathbf{Y}}_2\ \hat{\mathbf{Y}}_3\ \hat{\mathbf{Y}}_4\ \hat{\mathbf{Y}}_5\ \hat{\mathbf{Y}}_6]^T$. Decomposition in $\underline{\mathbf{e}}^1$ of the equations for $\underline{\hat{\mathbf{X}}}$ and $\underline{\hat{\mathbf{Y}}}$ yields 36 scalar equations. In $\underline{\hat{\mathbf{X}}}$ 15 scalar coordinates are unknown. In $\underline{\hat{\mathbf{Y}}}$ eleven scalar coordinates are unknown, namely eight coordinates of $\hat{\mathbf{Y}}_2, \ldots, \hat{\mathbf{Y}}_5$ normal to joint axes and three coordinates of $\hat{\mathbf{Y}}_6$. The axial coordinates of $\hat{\mathbf{Y}}_2, \ldots, \hat{\mathbf{Y}}_5$ are zero. In the matrix $\Delta\underline{\dot{q}} = [\Delta\dot{q}_{11}\ \ldots\ \Delta\dot{q}_{16}\ \Delta\dot{q}_2\ \Delta\dot{q}_3\ \Delta\dot{q}_4\ \Delta\dot{q}_5\ \Delta\dot{q}_6]^T$ ten quantities (all except $\Delta\dot{q}_6$) are unknown. The total number of unknowns is 36. This equals the number of scalar equations.

References

1. Amirouche F (2006) Fundamentals of multibody dynamics. Theory and applications. Birkhäuser, Boston
2. Angeles J (1997) Fundamentals of robotic mechanical systems. Theory, methods and algorithms. Springer, Berlin, Heidelberg, New York
3. Arnold RN, Maunder L (1961) Gyrodynamics and its engineering applications. Academic Press, New York London
4. Baumgarte J (1972) Stabilization of constraints and integrals of motion in dynamical systems. Comp. Meth. in Appl. Mech. and Eng. 1:1–16
5. Beletski VV (1966) Motion of an artificial satellite about its center of mass. Jerusalem; also NASA Transl. (in Russ. 1965)
6. Boland P, Samin JC, Willems PY (1974) On the stability of interconnected deformable bodies in a topological tree. AIAA J. 12:1025–1030
7. Boland P, Samin JC, Willems PY (1975) Stability analysis of interconnected deformable bodies with closed configuration. AIAA J. 13:864–867
8. Branin FH (1962) The relation between Kron's method and the classical methods of network analysis, The Matrix and Tensor Quarterly 12:69–105
9. Bremer E, Pfeiffer F (1993) Elastische Mehrkörpersysteme. Teubner, Stuttgart
10. Bricard R (1927) Leçons de cinématique, t. II: Cinématique Appliquée. Gauthier-Villars, Paris
11. Bührle P (1996) Zur dynamischen Simulation motorgetriebener Mehrkörpersysteme. Diss. Univ. Karlsruhe, VDI-Verl. Reihe 20, 231
12. Busacker RG, Saaty TL (1965) Finite graphs and networks. An introduction with applications. Mc Graw-Hill, New York
13. Charlamov PV (1969) On the velocity distribution in a rigid body (russ.). mech. tverdovo tela 1:77–81
14. Eich-Söllner E, Führer C (1998) Numerical methods in multibody dynamics. Teubner, Stuttgart
15. Euler L (1758) Du mouvement de rotation des corps solides autour d'un axe variable. In: Füssli O (ed) (1964) Leonhard Euler. Opera Omnia: Commentationes Mechanicae ad Theoriam Corporum Rigidorum Pertinentes, volumen prius:200–236
16. Featherstone R (1987) Robot dynamics algorithms. Kluwer, Dordrecht

17. Fischer O (1906) Einführung in die Mechanik lebender Mechanismen. Teubner, Leipzig
18. Frisch HP (1975) A vector-dyadic development of the equations of motion of n-coupled flexible bodies and point masses. NASA Techn. Note TN D–8047
19. Gantmacher FR (1959) Theory of matrices. v. 1, 2. Chelsea, New York
20. Garcia de Jalón J, Bayo E (1994) Kinematic and dynamic simulation of multibody systems: The real time challenge. Springer, Berlin Heidelberg New York
21. Géradin M, Cardona A (2001) Flexible multibody dynamics: A finite element approach. Wiley, Chichester
22. Grammel R (1950) Der Kreisel. Seine Theorie und seine Anwendungen. v.1. Springer, Berlin Göttingen Heidelberg
23. Haug EJ (ed) (1984) Computer-aided analysis and optimization of mechanical system dynamics. NATO ASI ser. F, Springer, Berlin Heidelberg New York
24. Haug EJ, Deyo RC (1991) Real-time integration methods for mechanical system simulation. Springer Series F: Computer and Systems Sciences, Berlin Heidelberg New York
25. Haug EJ (1989) Computeraided kinematics and dynamics of mechanical systems. Allyn and Bacon
26. Hong Jiazhen (1999) Computational dynamics of multibody systems. High Education Press, Beijing
27. Hooker WW, Margoulis G (1965) The dynamical attitude equations for an n-body satellite. J. Astronaut. Sci. 12:123–128
28. Hooker WW (1970) A set of r dynamical attitude equations for an arbitrary n-body satellite having r rotational degrees of freedom. AIAA J. 8:1205–1207
29. Hopf H (1940) Systeme symmetrischer Bilinearformen und Euklidische Modelle der projektiven Räume. Naturf. Ges., Zürich:165–177
30. Huston RL, Liu Youwu (1987) Dynamics of multibody systems. Tianjin Univ. Press
31. Huston RL (1990) Multibody dynamics. Butterworth-Heinemann, Boston
32. Jerkovsky W (1976) The transformation operator approach to multisystem dynamics. Matrix Tensor Quat. 27:48–59
33. Jonker JB (1988) A finite element dynamic analysis of flexible spatial mechanisms and manipulators. PhD Thes. Delft Univ. of Techn.
34. Juskevic AA, Winter E (1965) Leonhard Euler und Christian Goldbach: Briefwechsel 1729–1764. Akademie-Verl., Berlin
35. Keppler R (2007) Zur Modellierung und Simulation von Mehrkörpersystemen unter Brücksichtigung von Greifkontakt bei Robotern. Diss. Univ. Karlsruhe, Univ.-Verl. Karlsruhe
36. Kane TR, Levinson DA (1985) Dynamics. Theory and applications. McGraw-Hill, New York
37. Kecskeméthy A (1993) Objektorientierte Modellierung der Dynamik von Mehrkörpersystemen mit Hilfe von Übertragungselementen. Diss. Univ. Duisburg, VDI-Verl. Reihe 20, 88
38. Kim SS, Haug EJ (1988) A recursive formulation for flexible multibody dynamics. Part I: Open-loop systems. Comp. Meth. in Appl. Mech. and Eng. 71:293–314
39. Kim SS, Haug EJ (1989) A recursive formulation for flexible multibody dynamics. Part II: Closed-loop systems. Comp. Meth. in Appl. Mech. and Eng. 74:251–269

40. Kreuzer E (1979) Symbolische Berechnung der Bewegungsgleichungen von Mehrkörpersystemen. Diss. Univ. Stuttgart, VDI-Verl. Reihe 11, 32
41. Lagally M, Franz W (1964) Vorlesungen über Vektorrechnung. Akad. Verl. Ges., Leipzig
42. Li HY (1990) Ein Verfahren zur vollständigen Lösung der Rückwärtstransformation für Industrieroboter mit allgemeiner Geometrie. Diss. Univ. Duisburg
43. Likins PW (1974) Dynamic analysis of a system of hinge-connected rigid bodies with nonrigid appendages. JPL Techn. Rep. 32–1576, Pasadena
44. Likins PW, Fleischer GE (1972) Large-deformation modal coordinates for nonrigid vehicle dynamics. JPL Techn. Rep. 32–1565, Pasadena
45. Lilov L (1993) Modelirovanie sistem svjasannych tel (Modelling of systems of interconnected bodies). Nauka, Moscow
46. Lilov L, Chirikov V (1981) On the dynamics equations of systems of interconnected bodies. PMM 45:383–390
47. Lilov L, Wittenburg J (1977) Bewegungsgleichungen für Systeme starrer Körper mit Gelenken beliebiger Eigenschaften. ZAMM 57:137–152
48. Liu Yanzhu, Hong Jiazhen, Yang Haixing (1989) Duogangti xitong donglixue (Dynamics of Systems of Multi-Rigid Bodies). High Education Press, Beijing
49. Lu Youfang (1996) Dynamics of flexible multibody systems. High Education Press, Beijing
50. Magnus K (1965) Der Kreisel. Eine Einführung in die Lehre vom Kreisel mit Anleitung zur Durchführung von Versuchen. Industrie-Druck Verl., Göttingen
51. Magnus K (1971) Kreisel. Theorie und Anwendungen. Springer, Berlin Heidelberg New York
52. Moiseev NN, Rumiancev VV (1968) Dynamic stability of bodies containing fluid. Springer, Berlin Heidelberg New York
53. Müller A, Maisser P (2003) A Lie-Group formulation of kinematics and dynamics of constrained MBS and its application to analytical mechanics. J. Multibody System Dynamics, 9:311–352
54. Nikravesh PE (1988) Computer-aided analysis of mechanical systems. Prentice Hall
55. Otterbein S (1982) Stabilisierung des n-Pendels und der indische Seiltrick. Arch. Rational Mech. Anal. 78:381–393
56. Pars LA (1979) A treatise on analytical dynamics. Ox Bow Press, Woodbridge
57. Paul PR (1981) Robot manipulators: Mathematics, programming and control. MIT Press, Cambridge Mass.
58. Pereira MS, Ambrósio J (eds) (1993) Computer-aided analysis of rigid and flexible mechanical systems. Kluwer, Dordrecht Boston London
59. Pfeiffer F, Glocker C (1996) Multibody dynamics with unilateral contacts. Wiley, Berlin
60. Pfister F (1995) Contributions à la Mécanique Analytique des Systèmes Multicorps. Ph.D. thesis INSA Lyon, Lyon
61. Popov EP et al. (1974) Synthèse de la commande des robots utilisant des modèles dynamiques des systèmes mécaniques. 6th IFAC Symp. Control in Space X:31–51
62. Reif T (1998) Zur objektorientierten Modellierung aktiver Mehrkörpersysteme. Diss. Univ. Karlsruhe, VDI-Verl. Reihe 20, 284
63. Renaud M (1975) Contribution à l'étude de la modélisation et de la commande des systèmes mécaniques articulés. Diss. INSA Toulouse

64. Roberson RE (1972) A form of the translational dynamical equations for relative motions in systems of many non-rigid bodies. Acta Mech. 14:297–308
65. Roberson RE, Schwertassek R (1988) Dynamics of multibody systems. Springer, Berlin Heidelberg New York
66. Roberson RE, Wittenburg J (1966) A dynamical formalism for an arbitrary number of interconnected rigid bodies. With reference to the problem of satellite attitude control, 3rd IFAC Congr. London, (1968) Proc. 46D.2–46D.9
67. Rodrigues O (1840) Des lois géometriques qui régissent les déplacements d'un système solide dans l'espace, et de la variation des coordonnées provenant de ces déplacement considerés indépendamment des causes qui peuvent les produire. J. Math. Pures et Appl., s. 1:380–440
68. Routh EJ (1905) Dynamics of a system of rigid bodies (Elementary part). Macmillan, London. German transl. (1898) Die Dynamik der Systeme starrer Körper. Die Elemente. Teubner, Leipzig
69. Rui Xiaoting (1995) Launch dynamics of multibody systems. National Defense Industry Press, Beijing
70. Saidov PI (1965) Theory of gyroscopes (Russ.). v.1, Vysshaya Shkola, Moscow
71. Salecker M (1991) Zur Dynamik hybrider Mehrkörpersysteme – Theorie und symbolische Programmierung. Diss. Univ. Karlsruhe
72. Samin J-C, Fisette P (2003) Symbolic modeling of multibody systems (Solid Mechanics and its Applications). Kluwer, Dordrecht
73. Schiehlen W (ed) (1990) Multibody systems handbook. Springer
74. Schiehlen W (ed) (1993) Advanced multibody system dynamics: Simulation and Software tools. Kluwer, Dordrecht
75. Schiehlen W, Ambrósio J A C (eds) Journal Multibody System Dynamics. Springer, Dordrecht
76. Schwertassek R, Wallrapp O (1999) Dynamik flexibler Mehrkörpersysteme. Vieweg, Braunschweig
77. Shabana AA (2005) Dynamics of multibody systems. Cambridge Univ. Press
78. Stuelpnagel J (1964) On the parametrization of the three-dimensional rotation group. Siam Rev. 6, No 4:422–430
79. Tan Zheng (2000) Compressive method for multibody systems. Science Press, Beijing
80. Tölke F (1967) Praktische Funtionenlehre. v. 3, 4. Springer Berlin Heidelberg New York
81. Truesdell C (1964) Die Entwicklung des Drallsatzes. ZAMM 44:149–158
82. Uicker JJ (1968) Dynamic behavior of spatial linkages. Trans. of the ASME 68, Mech.:1–15
83. Velman JR (1967) Simulation results for a dual-spin spacecraft. Aerosp. Corp. Rep. TR–0158 (3307–01)–16, El Segundo
84. Volterra V (1898) Sur la théorie des variations des latitudes. Acta Math 22:201–357
85. Vukobratović M, Kircanski N (1985) Real-time dynamics of manipulation robots. Series: Scientific fundamentals of robotis 4, Springer, Berlin
86. Vukobratović M, Potkonjak V (1982) Dynamics of manipulation robots. Springer, Berlin
87. Vukobratović M, Stepanenko J (1973) Mathematical models of general anthropomorphic systems. Math Biosci. 17:191–242
88. Wangerin A (1889) Über die Bewegung miteinander verbundener Körper. Univ.-Schrift Halle

89. Weber B (1993) Symbolische Programmierung in der Mehrkörperdynamik, Diss. Karlsruhe University, Preprint Nr. 93/3, Inst. f. Wiss. Rechnen u. Math. Modellbildung, Univ. Karlsruhe
90. Weber B, Wittenburg J (1993) Symbolical programming in system dynamics. In [74]:153–172
91. Wei H (1991) Dynamische Simulation hybrider Mehrkörpersysteme mit beliebiger Systemstruktur. Diss. Univ. Karlsruhe, VDI-Verl. Reihe 18, 108
92. Wittbrodt E, Adamiec-Wójcik I, Wojciech S (2006) Dynamics of flexible multibody systems: Rigid finite element method (Foundations of Eng. Mechanics). Springer, Berlin Heidelberg New York
93. Wittenburg J (1971) The dynamics of systems of coupled rigid bodies. A new general formalism with applications. In: Grioli G (ed) Stereodynamics. Edizione Cremonese, Roma
94. Wittenburg J (1972) Stossvorgänge in räumlichen Mechanismen. Eine Analogie zwischen Kreiseldynamik und Elastostatik. Acta Mech. 14:309–330
95. Wittenburg J (1974) Permanente Drehungen zweier durch ein Kugelgelenk gekoppelter, starrer Körper, Acta Mech. 19:215–226
96. Wittenburg J (1975) Beiträge zur Dynamik von Gyrostaten. Acc. Naz. dei Lincei, Quad. 217:217–254
97. Wittenburg J (1977) Dynamics of systems of rigid bodies. Teubner, Stuttgart. translated into Russian (1980), Chinese (1986), Vietnamese (2000)
98. Wittenburg J (1994) Topological description of articulated systems. In: [58]
99. Wittenburg J (1996) Schwingungslehre. Lineare Schwingungen. Theorie und Anwendungen. Springer, Berlin Heidelberg New York
100. Wittenburg J (2003) Closure conditions for spatial mechanisms: A non-recursive formulation. Arch. Appl. Mech. 72:933–948
101. Wittenburg J (2006) Striction point and distribution parameter on the raccording axodes of a rigid body. Proc. Wissenschaftlich-technische Jubiläumskonferenz "15 Jahre FDIBA", Sofia
102. Wittenburg J Kinematics. to appear
103. Wittenburg J, Lilov L (1975) Relative equilibrium positions and their stability for a multi-body satellite in a circular orbit. Ing-Arch. 44:269–279
104. Wittenburg J, Wolz U, Schmidt A (1990) MESA VERDE – a general-purpose program package for symbolical dynamics simulations of multibody systems. In: [73]
105. Woernle C (1988) Ein systematisches Verfahren zur Aufstellung der geometrischen Schliessbedingungen in kinematischen Schleifen mit Anwendung bei der Rückwärtstransformation für Industrieroboter. Diss. Univ. Duisburg, VDI-Verl. Reihe 18,59
106. Wolz U (1985) Dynamik von Mehrkörpersystemen – Theorie und symbolische Programmierung. Diss. Univ. Karlsruhe, VDI-Verl. Reihe 11, 75
107. Wüst K, Schmidt E (2001) Assessment of the potential of independent suspension in a light truck. ATZ 103:11–13 and 1142–1150

Index